History and Future of Plants, Planet and People: Towards a New Ecologically Sustainable Age in People's Relationships with Plants

History and Future of Plants, Planet and People: Towards a New Ecologically Sustainable Age in People's Relationships with Plants

Alan Hamilton

Fellow of the Linnean Society of London
Honorary Professor, Kunming Institute of Botany, Chinese
Academy of Sciences

and

Pei Shengji

Fellow of the Linnean Society of London
Professor, Kunming Institute of Botany, Chinese Academy of
Sciences

CABI

Disclaimer

The boundaries and names of countries on the maps in this book do not imply the expression of any opinion whatsoever on the part of the authors or publisher concerning the delimitation of their frontiers or boundaries.

CABI is a trading name of CAB International

CABI
Nosworthy Way
Wallingford
Oxfordshire OX10 8DE
UK

CABI
200 Portland Street
Boston
MA 02114
USA

Tel: +44 (0)1491 832111
E-mail: info@cabi.org
Website: www.cabi.org

Tel: +1 (617)682-9015
E-mail: cabi-nao@cabi.org

A catalogue record for this book is available from the British Library, London, UK.

ISBN-13: 9781836993285 (paperback)
         9781789248920 (hardback)
         9781789248937 (ePDF)
         9781789248944 (ePub)

DOI: 10.1079/9781789248944.0000

Commissioning Editor: Ward Cooper
Editorial Assistant: Helen Elliott
Production Editor: James Bishop

Typeset by Straive, Pondicherry, India
Printed in the USA

Contents

*Written in the first person by Alan Hamilton

Maps

Figures

Tables

Preface

This book contains suggestions about how to make plant conservation more effective (Chapter 9). These suggestions have been influenced by our own experiences in botany and conservation, some of which are described in Chapters 10–21. They include two multi-country conservation programmes, the People and Plants Initiative and the Medicinal Plants Conservation Initiative (Chapters 10 and 11). We hope that other people interested in making plant conservation more effective, coming with their own experiences, will find our suggestions helpful and be able to suggest improvements. To provide context, we briefly describe our life stories below, as they relate to the subject of this book. The stories in Chapters 10–21 are as we remember them. Others who were involved at the time may have different recollections.

Experiences in Botany and Conservation – Pei Shengji

Pei Shengji was born into an ordinary Chinese farming family in 1938 in the village of Long Kou ('Dragon-Mouth'). His hometown of Zitong County in the north of Sichuan Province is famous for Traditional Chinese Medicine (TCM), both for the production of medicinal plants (such as *Aconitum carmichaelii*, *Ophiopogon japonicus* and *Platycodon grandiflorus*) and for being the home of some well-known TCM families. As Pei was growing up, he learnt a lot about TCM by visiting the drug stores and backyard drug processing workshops belonging to the families of his friends and joining in their work. There was no Western medicine in Sichuan at the time.

In his teenage years, Pei often visited the nearby Qiqu Mountain Temple in the north of the county town in Mianyang City Prefecture, which had been founded by Zhang Yazi during the Jin dynasty (266–420 CE). Zhang Yazi was a TCM doctor who, later in life, developed his moral character and became a sage. He was one of the founders of a school of Daoist philosophy known as Wenchang Culture, which reveres scholarship and learning. This well-known saying illustrates the extent of influence of Wenchang Culture in Southern China, which is where Sichuan is situated: 'In the North there is Confucius and in the South Wenchang' (Wang Xia, 2020). Pei's contact with Qiqu Mountain Temple early in life gave him an appreciation of the worldviews of traditional China (see Section 4.3.1).

At the age of 15, Pei's father started working in a bakery in a nearby town, living in the baker's house and acquiring the art of baking through the traditional Chinese master/pupil relationship. His father passed away at the age of 60, but his mother lived until she was 84 and was an important influence on Pei, raising him in the traditional Chinese way, which is by example and through storytelling. Her farming included sericulture (silkworm rearing) and animal husbandry. One of Pei's tasks as a boy was to collect the oil-rich seeds of the tallow and tung oil trees (*Triadica sebifera* and *Vernicia fordii*) from hillslopes and the verges of roads in farmland, take them to local workshops to have their oil extracted and bring the oil home to burn in lamps (see Section 6.6). Kerosene was unavailable in Sichuan when Pei was young. Pei studied at night under the light of a lamp fuelled by the family's own oil. From his mother and the Qiqu temple, Pei learnt to respect the maxims of Daoism, such as 'loyalty and filial piety', 'benevolence', 'righteousness', 'help others', 'know how to conduct oneself in society' and 'respect scholars'.

Pei went to school at the age of five and then attended Sichuan Provincial School of Agricultural Technology (1952–1955). His family was too poor for him to go to university and, instead, he was instructed to join the Kunming Institute of Botany, Chinese Academy of Sciences, to study field botany and plant taxonomy. He was a junior member of a field team assigned to survey the vegetation of Yunnan Province, one of the richest places botanically on Earth. The survey was being undertaken under the directorship of Professor Tsai Xitao (Hse Tao Tsai) (1911–1981), a leading figure in the history of scientific botany in China (see Section 7.5.1). The purpose of the survey was to find out whether there was anywhere in China suitable for establishing plantations of rubber and other tropical crops (see Chapter 12). A domestic supply of rubber was urgently needed because the United States had placed an embargo on its import into China when the Communist Party of China took power in 1949. During the fieldwork, Pei often stayed in the homes of the ethnic minorities who live in Yunnan, which is how he came to know how much they knew about plants and to appreciate their worldviews.

Pei's first scientific paper was written under the guidance of Prof. Tsai Xitao on the integrated utilisation of wild plant resources in Yunnan (Tsai Xitao and Pei Shengji, 1959). His first ethnobotanical papers were studies of the ethnobotany of Xishuangbanna Prefecture in the south of Yunnan. Published first in Chinese and later in English, they contain the first scientific descriptions of the Nong Holy Hills of the Dai, a Buddhist people who live in the tropical rainforest zone of Yunnan (see Chapter 14) (Pei Shengji, 1982, 1985, 1988). Later, during the 1980s, he extended his work on ethnobotany to include the

Naxi and the Tibetans, as well as other ethnic minorities (see Chapter 21). Pei was a pioneer in the study of traditional conservation and Sacred Natural Sites (SNSs), aspects of conservation that are now attracting more attention (see Sections 8.1 and 8.3).

One of the traditional land use practices that Pei has investigated in Yunnan is the swidden (or cut-and-burn) agriculture practised by the Hani people who live in its hilly tropical forest zone (see Chapter 13). In the 1980s, the Yunnan Government came to regard this practice as backward and began to take steps to have it banned. However, from his work with the Hani, Pei had come to realise that rotational agriculture is sustainable in forested areas, provided that the human population density remains below about 15 people per km^2. He argued that it is a thousand-year-old practice that results in an alternation of two types of vegetation, one with a high diversity of crops and the other with a high diversity of useful wild plants. This experience triggered the start of a new research line for Pei, which was to investigate whether modifications to the Hani's rotational agriculture could both safeguard the environment and provide the people with new sources of income.

One of Tsai Xitao's achievements was to persuade the Chinese Government in 1959 to establish Xishuangbanna Tropical Botanical Garden (XTBG). The purposes assigned to the garden included supporting the development of rubber and other tropical crops in China and continuing with research on ethnobotany (see Chapter 12). During the Cultural Revolution (1966–1976), Tsai Xitao was made to live in a cowshed charged with being an intellectual who was opposed to the academic authorities. Pei was put under pressure to denounce him politically.

After the Cultural Revolution had ended, Pei became director of XTBG, a post which he held from 1978 to 1986. One of his tasks was to edit eight tropical plant families for the *Flora of China* – *Dipterocarpaceae, Guttiferae, Musaceae* (banana family), *Myristicaceae, Palmae* (*Arecaceae* – palms), *Piperaceae, Tetramelaceae* and *Zingiberaceae* (Pei Shengji et al., 1994–present). During this time, Pei used every opportunity he could find to argue the case for the conservation of tropical forest in China, for the benefits of both China and the rest of the world. In 1980–1981, the Chinese Premier, Zhao Ziyang, and the General Secretary of the Communist Party, Hu Yaobang, visited XTBG and Pei was able to put his case personally to the top Chinese leadership. He argued successfully for the scale of rubber planting to be cut back and for the establishment of more protected areas.

Tsai Xitao, Pei's mentor, was a believer in international scientific collaboration and, during the time when he was still influential in China, arranged for Pei to spend more than two years in Hawaii, working in the Department of Botany of the University of Hawaii and at the East-West Centre. The East-West Centre is an educational and research organisation established by the US Congress in 1960 to strengthen relations and understanding between the peoples and nations of Asia, the Pacific and the United States. Pei's time in Hawaii gave Pei the opportunity to meet a number of leading Americans involved in the early scientific development of ethnobotany, among them Dr Isabella Abbott and Richard Schultes (see Section 5.3.5 and Chapter 10). Facilitated by the East-West Centre, Pei served as leader of the Biodiversity Task Force of the Southeast-Asian University Agroecosystem Network (SUAN), which brought together biological and social scientists from China, Malaysia, the Philippines, Singapore, Thailand and Vietnam to study interactions between people and the environment.

Another formative influence on Pei's work was Dr S.K. Jain (1926–2021), an Indian botanist who is regarded in India as the father of ethnobotany. In 1988, Jain visited the Kunming Institute of Botany (KIB) and urged Pei, who had recently been appointed the head of its new Department of Ethnobotany, to push ahead with the development of ethnobotany in China. Jain was an advocate of cross-disciplinary science. He encouraged students of botany to seek linkages with the social sciences and students of phytochemistry to seek linkages with indigenous knowledge.

In 1990, Pei started a 9-year period of service with the International Centre for Integrated Mountain Development (ICIMOD), an international institution founded in 1983 with its headquarters in Kathmandu, Nepal. ICIMOD's purpose is to promote ecologically sound development in the Hindu Kush-Himalayas. Pei, whose job title was Head of ICIMOD's Mountain Natural Resources Division and Biodiversity Specialist, was charged specifically with training students and established scientists from the countries of the Hindu Kush-Himalayas in ethnobotany. In 1992, through the good offices of Malcolm Hadley, who was working for the Man and Biosphere (MAB) Programme in UNESCO's headquarters in Paris, a grant was received by ICIMOD from the Danish aid agency DANIDA that allowed a practical programme of capacity-building in ethnobotany to go ahead. This involved training workshops, grants to support individual field projects and the founding of national ethnobotanical associations, such as the Chinese Association of Ethnobotany (see Chapter 1). Later, both Pei and Hadley became involved in the People and Plants Initiative, a partnership programme of WWF, UNESCO and the Royal Botanic Gardens, Kew, designed to increase international capacity in applied ethnobotany (see Chapter 10).

Despite being 'retired' for many years, Pei is still active in plant conservation and sustainable development in China and regionally. He continues to be involved in the biennial national symposia of the Chinese Association of Ethnobotany (an organisation that he founded – see Chapter 1) and, in September 2022, helped run a conference on conservation and sustainable development in the six countries which have land in the catchment area of the Mekong River. He continues to follow up his interest in the Nong Holy Hills of Xishuangbanna and was delighted to learn in 2023 that the Xishuangbanna Bureau of Forestry and Pasture has declared 22 May of each year to be a Nong Holy Hill Forest Protection Day, when all the Dai villages will be encouraged to restore their Holy Forests and plant trees (see Chapter 14). In the same year, Professor Pei was awarded the honorary title of 'Scientific Advisor of Xishuangbanna Dai Autonomous Prefecture'.

Experiences in Botany and Conservation – Alan Hamilton

Written in the first person
My interest in natural history began early in life through my hobbies of tree-climbing, collecting fossils and teaching myself the rudiments of the geology of the British Isles. The geological period that especially interested me was the Quaternary – the period, extending up to today, that covers the last series of global ice ages and the emergence of the human species. In 1963, I was accepted by the University of Cambridge to study Natural Sciences (Botany, Zoology and Geology) and specialised in Botany in my final year. During the summer vacations, some of us undergraduates interested in natural history organised our own expeditions to go to remote places to do research. Two of these expeditions took me to islands off the coast of the British Isles, where I studied the vegetation and flora. The third, as a member of a four-person team, was to study the montane forest on the eastern slopes of the Peruvian Andes. The locality where we worked, which now lies within Rio Abiseo National Park, has an outstandingly rich flora – much richer, as I later saw for myself, than the equivalent tropical montane forests in East Africa. One of my extra-curricular activities at Cambridge was to help Oliver Rackham, an expert on the history of the British countryside, to coppice trees in a local woodland (see Section 5.5). Another was to serve as the secretary of the Cambridge Society for the Study of Religion.

In 1966, I was offered the chance to go to Uganda on a Nuffield Scholarship to undertake research on the environmental history of tropical Africa (see Chapter 15). This was perfect for me, even though the only things that I then knew about Uganda were that it was somewhere in the middle of Africa, had tropical rainforest and had recently been part of the British Empire. I was given space for my work in a laboratory in the Department of Botany of Makerere University, where, later on, I met my future wife, Naomi Masembe, who was an undergraduate in the same department. After completing my doctoral thesis, I worked during 1971 in the Department of Forestry at Makerere to write *A Field Guide to Ugandan Forest Trees*. Naomi and I married at the end of 1971, after which we left Uganda and eventually travelled to Britain. In 1972, I was successful in an application for a job as a Lecturer in Biogeography at the New University of Ulster (NUU) in Northern Ireland. Later, I transferred to a new Department of Environmental Science when it was started by Professor Palmer Newbould.

Idi Amin's period of presidency of Uganda (1971–1980) was marked by a degradation in the standards of public administration and a deterioration in the state of Makerere University. There was a brief period of hope immediately after Amin's ousting, during which an advertisement was placed in the British press requesting suitably qualified people to take up short-term contracts to teach at, and help rehabilitate, the university. Naomi and I thought that we should respond and, after NUU had granted me leave of absence to do so, we departed for Uganda on a one-year contract for 1981–1982 to work in the Department of Botany as Lecturer in Ecology and Demonstrator, respectively.

On arrival in Uganda, we found that no other British lecturer had responded to the call to help rehabilitate Makerere. Hopes for a peaceful new beginning for Uganda had evaporated. Guerilla forces were trying to oust Milton Obote, who had been installed as the president of the country for a second term (his first was in 1966–1971, before Amin). Every night, we could hear shooting in Kampala from our flat on the university campus and, every school day, I had to pass through several army roadblocks as I took and collected our two children from their primary school. On a visit to Makerere Printery, I discovered 200 copies of the *Field Guide* that I had written ten years earlier piled up on the floor, yet to be distributed. When Naomi and I had left Uganda ten years earlier, the metal printing plates for the book had already been etched, but Amin's coup had halted further progress. The staff of the printery told me that, it was only after the Tanzanian army had captured Kampala from Amin in 1979, that they had been able to find some paper to do the job. The books had been printed on yellow paper, because, I was told, there was no white paper available at the time.

I did manage to carry out some research during this year at Makerere. I journeyed twice to Southwest Uganda to collect samples of sediment to follow up my interest in exploring Uganda's environmental past and remeasured trees in a small permanent research plot in Mpanga Forest that I had established in the late 1960s (see Section 15.5.2). At the request of the government's Forest Department, which had lost all its transport and was experiencing difficulty with monitoring its forests, I carried out a study of the state of the country's forests (Hamilton, 1984).

In 1986, I was offered another opportunity to work in Africa on a short-term contract, this time on a conservation project for the International Union for Conservation and Natural Resources. IUCN had approached me to ask whether I was available for a year to lead a team of researchers to provide information for the preparation of a new management plan for the forests of the East Usambara Mountains in Tanzania (see Chapter 16). A logging project, assisted by a European development aid agency, was rapidly degrading the forests, which I knew from my own biogeographical research were top priorities for the conservation of biodiversity. On request, NUU did eventually agree to give me leave to go, but only on condition that I returned for a three-month period to squeeze in all my essential teaching.

Later, in 1988, I carried out a similar, but only one-month, assignment in Mexico (see Section 16.6). These experiences in practical conservation contributed to my decision in 1989 to apply for the job of Plants Conservation Officer in the international branch of the World Wide Fund for Nature (otherwise known as the World Wildlife Fund or WWF). Successful in my application, I then worked for WWF-International until 2004, after which I transferred to Plantlife International, another non-governmental organisation dedicated to conservation. My job in Plantlife was to help develop its international work under the direction of Jonathan Rudge who had recently been employed as its first International Director.

During my first four years at WWF-International, I supervised a large number of projects based in many parts of the world dealing with all sorts of aspects of plant conservation, but, after that, I concentrated my efforts mainly on the People and Plants Initiative (PPI, 1992–2005), a programme aimed at increasing global capacity in applied ethnobotany (see Chapter 10). While working for WWF, I realised that I needed to expand my knowledge of the humanities and organised a course for myself which eventually resulted in the publication of a book, *Human Nature and the Natural World* (Hamilton, 2001). In Plantlife, I was employed as its Plant Conservation and Livelihoods Officer, which, like the Plants Conservation job at WWF, was conceptually very wide-ranging. Realising that I need to concentrate my energies to achieve something practical, with Jonathan's encouragement, I launched the Medicinal Plants Conservation Initiative (MPCI, 2004–2008), the aim of which was to promote conservation and sustainable development, based on people's interests in medicinal plants (see Chapter 11). WWF did not continue with the People and Plants Initiative after my departure, nor did Plantlife continue with the Medicinal Plants and Conservation Initiative.

I retired from paid employment in 2008 and later on helped start and run a community group in my home town of Godalming in Britain. The aim of the group was to raise awareness about the danger of climate change and persuade people to take personal actions to moderate it. Among other activities, this included a campaign to rehabilitate neglected woodlands to serve as sources of sustainable wood fuel and help conserve biodiversity (see Section 5.5). I also continued my interests in conservation in Uganda and China. However, my main preoccupation has been the compilation, along with five other people, of a two-way Luganda–English dictionary (see Section 7.5.4) (Hamilton, 2020).

When I first went to Uganda, I had no idea about its politics, but after I had met my future wife and got to know her politically and culturally engaged family, I came to realise more about the historical context of the present political and economic problems facing the country and how these relate to conservation and development. For Buganda, I started to understand why its once effective administrative system no longer contributes to Uganda's system of governance and why there has not been a single constitutional and peaceful transition from one Ugandan ruler to another since 1962, which is when Uganda gained its political independence. My knowledge of the culture of Buganda and the history of Uganda has contributed to the accounts in this book of worldviews in Buganda and Uganda (see Chapter 4), and the histories in Uganda of forestry (Section 5.6), development aid (Section 5.7), Christianity, education, religion and politics (Section 6.8), and language policy (Section 7.5.4).

Acknowledgements

Alan Hamilton wishes to thank his wife Naomi and son Patrick for giving him essential guidance in the writing of this book. The understandings of Luganda which it contains owe much to Naomi, Keefa Ssentoogo, Phoebe Mukasa,

David Ssewanyana and Christine Kabuye. Pei Shengji acknowledges his debts to his mentor Tsai Xitao, one of the pioneer scientific botanists in China, and to his former research student and current assistant Yang Zhiwei. We wish to thank Ali Thompson, Ward Cooper, James Bishop, Helen Elliott, Rachel Bowen of CABI and Gillian Watling for their assistance with the production of this book.

References

Hamilton, A.C. (1984) *Deforestation in Uganda*. Oxford University Press, Nairobi.

Hamilton, A.C. (2001) *Human Nature and the Natural World*. New Millennium, London.

Hamilton, A.C. (2020) *Luganda–English and English–Luganda Dictionary*. Alan Hamilton, Godalming, UK.

Pei Shengji (1982) A preliminary study of the ethnobotany of Xishuangbanna (in Chinese). In: *Collected Research Papers on Tropical Botany*. Yunnan Publishing House, Kunming, China, pp. 16–30.

Pei Shengji (1985) Some effects of the Dai people's cultural beliefs and practices upon the plant environment of Xishuangbanna, Yunnan Province, southwest China. In: Hutterer, K.L., Rambo, A.T., *et al.* (eds) *Cultural Values and Human Ecology in Southeast Asia*. Michigan Papers on South and Southeast Asian Studies No. 27. University of Michigan Center for South and Southeast Asian Studies, Ann Arbor, Michigan, pp. 321–339.

Pei Shengji (1988) Plant products and ethnicity in the markets of Xishuangbanna, Yunnan Province, China. In: Rambo, A.T., Gillogly, K., *et al.* (eds) *Ethnic Diversity and The Control of Natural Resources in Southeast Asia*. Michigan Papers on South and Southeast Asia No. 32. University of Michigan Center for South and Southeast Asia, Ann Arbor, Michigan, pp. 119–142.

Pei Shengji, Chen Sanyang, *et al.* (1994–present) *Arecaceae (Palmae)*. In: Wu Zhengyi, Raven, P.H. and Hong Deyuan (eds) *Flora of China, Vol. 23 (Acoraceae through Cyperaceae)*. Science Press, Beijing, pp. 132–157.

Tsai Xitao and Pei Shengji (1959) On the integrated utilization of wild plant-resources in Yunnan. *Journal of Biology Bulletin* 7, 293–296.

Wang Xia (2020) Study on localization of Zitong Wenchang Culture in Japan. In: Mthembu, A. (ed.) *2020 2nd International Conference on Humanities, Cultures, Arts and Design, Sydney, Australia, 18–20 December 2020*. Francis Academic Press, London, pp. 78–84.

1

Sustainable Development: An Existential Challenge of Our Time (With a Note on the Luganda Language)

Abstract

During the past 50 years, it has become clear that the human has destabilised the ecosystem of the Earth and that it is now urgent to shift the direction of the human journey on to a more sustainable path. Whatever else is done in favour of sustainable development, little ultimately will be achieved unless plants receive targeted and concentrated attention. This is because of the fundamental roles that plants play in the functioning of both ecosystems and human economies, and the numerous ways in which people are culturally connected to plants. A note on the Luganda language is appended, to provide an introduction to references to it later in the book.

The human species, *Homo sapiens*, is known from the archaeological record to have existed for about 300,000 years and to have become widely distributed across Africa and Eurasia (Hublin *et al.*, 2017; Richter *et al.*, 2017; Reich, 2018). There have been several forms of *H. sapiens*, but genomic analysis shows that all modern people belong to the same form and that this was restricted to the African continent until about 50,000–80,000 ya (years ago), at which time it started to spread out to other places (Pääbo, 2003, 2014). Two other species of *Homo* have contributed genes to some modern populations of *H. sapiens*: the Neanderthals (*Homo neanderthalensis*), which formerly lived in Western Eurasia, and the Denisovans, which lived in Central Asia. *H. sapiens* is descended from *Homo erectus*, a long-lasting species that existed from about 2 million to 100,000 ya. By 400,000 ya, *H. erectus* knew how to make fire, a skill which *H. sapiens* also acquired once it had evolved (Gowlett, 2016). The ability to make fire enlarged the ecological niches of these species through enabling them to live in colder places (fire for warmth), eat an enlarged diet (fire for cooking), penetrate into dark places (fire for light) and transform habitats (through burning vegetation).

© Alan Hamilton and Pei Shengji 2024. *History and Future of Plants, Planet and People* (Alan Hamilton and Pei Shengji)
DOI: 10.1079/9781789248944.0001

The modern form of *H. sapiens* quickly achieved an extremely wide geographical distribution by the standards of large terrestrial animals. Established as a member of the local fauna in all the continents (except Antarctica) by 14,000 ya, its exceptional cultural ability enabled it to adapt to a wide range of habitats, from tundra to tropical forest. As it spread, its ecological forcefulness resulted in a wave of extinctions of large terrestrial animals (Sandom *et al.*, 2014). It was only in those places that it failed to reach until much later, such as Madagascar and Mauritius, that the large animal fauna remained intact. However, when humans did arrive, their large animals experienced the same fate.

The ability of humans to become so widely distributed was partly due to the invention of boats. People were in Australia and New Guinea by about 50,000 ya, which necessitated them taking a sea passage of at least 65 km (Kealy *et al.*, 2018). Later, the passage from Asia to America is thought to have required navigating by boat around a huge mass of ice that, at the time (the Last Glacial Maximum), was sitting on the northern part of North America. After arriving in ice-free land to the south of the ice sheet, people moved quickly down the western side of the Americas and were in southern Chile by 14,600 ya (Dillehay and Ocampo, 2015). Map 1.1 shows the routes taken by the human in its initial spread around the world. It also shows the locality where, 33,000 ya, it began a close association with the grey wolf (Wang Guo-Dong *et al.*, 2016). This was the beginning of a hunting partnership between people and the grey wolf (transitioning into the dog) that proved lethal for their prey (Lupo, 2011).

Major changes in the archaeological records of Africa and Eurasia show that, about 50,000 ya, *H. sapiens* adopted a more sophisticated way of life (Reich, 2018). Known as the cognitive revolution (Harari, 2011), it is thought that what fundamentally happened was the acquisition of open-ended language, which is the ability to form an almost unlimited number of sounds (i.e. words and sentences) from a limited number of vocalisations (Spranger *et al.*, 2010). The possession of this linguistic competency raised the potential for the human species to communicate and think in detailed and subtle ways. It enhanced people's abilities to organise their knowledge systematically and tell stories (see Section 7.1) (Armstrong, 2019). Among these stories are some, known as self-narratives, which people use to explain to themselves who they think they are, what purposes they see in life and how they envisage their futures (McAdams, 1996; Boyer and Bergstrom, 2008). Once acquired, the possession of open-ended language set modern *H. sapiens* on an evolutionary path unique to its species, resulting in an extraordinary proliferation in its cultural diversity.

Rare archaeological finds allow glimpses into the distant cultural past of *H. sapiens* and its relatives. Art, 46,000 years old in Australia and 44,000 years old in Indonesia, shows that people were then thinking mythologically, which is suggestive of the modern human capacity for thought (McDonald and Veth, 2008, 2013; Aubert *et al.*, 2019). Beliefs and practices that today would be considered to be religious in nature appear to be very ancient and not unique to our species. There is evidence that both Neanderthal man (*H. neanderthalensis*) and *Homo naledi* practised ritual burials, just as do modern humans. Neanderthal man had a brain even larger, on average, than that of *H. sapiens*. *H. naledi* was a species known to have lived in South Africa 335,000–236,000 ya that had the brain the size of a chimpanzee (Pomeroy *et al.*, 2020).

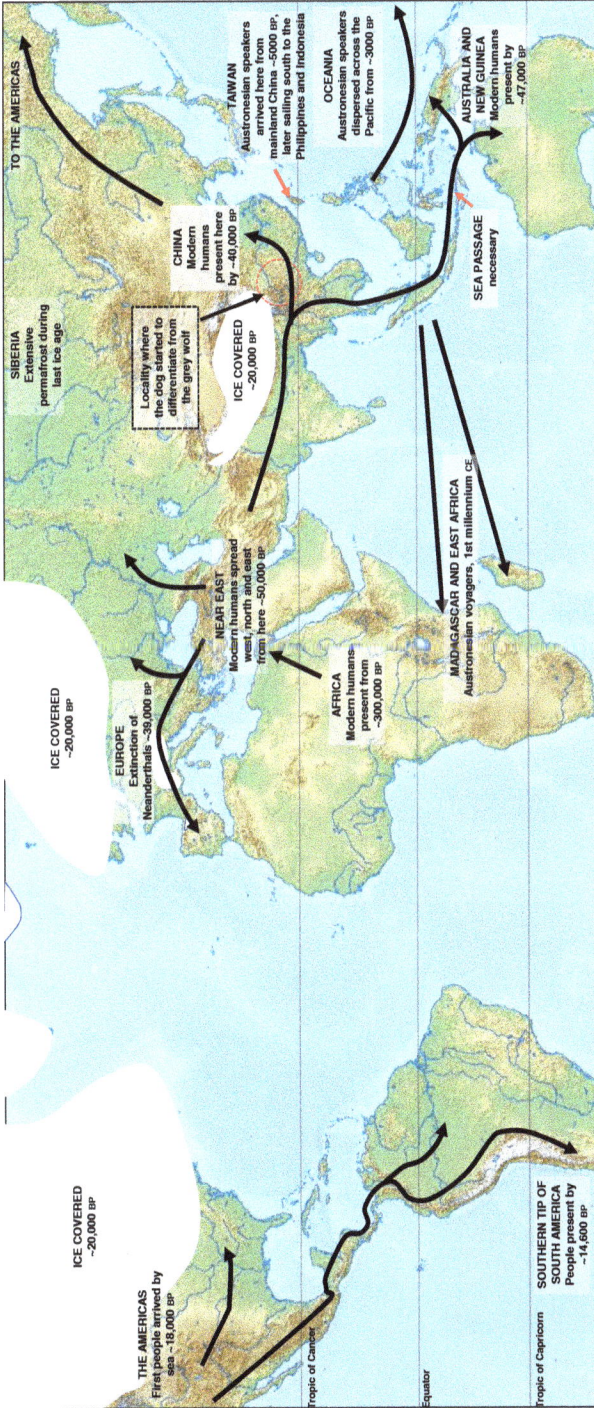

Map 1.1. Spread of modern *Homo sapiens*. Dates shown are those of first arrivals.

Depending on the locality, *H. sapiens* has acquired its food by gathering, hunting and sometimes fishing for most of its 300,000 years on Earth. The period since 10,000 BCE has been different, witnessing major changes in human cultural, political and socio-economic systems, a massive rise in the size of the human population and a much greater human influence on the Earth. Today, the whole world has become thoroughly infused with human influences and, because some of these influences are malign, the human has become, in effect, the first globally invasive species. It is only during the lifetimes of the present authors that scientists have come to realise the magnitude of the environmental transformations that *H. sapiens* has brought (Gore, 2006; Hails *et al.*, 2008). Some consider that we have now entered a new human-dominated geological era, the Anthropocene (or Homogenocene) (Zalasiewiczi *et al.*, 2011; Su Nan-Yao, 2023).

According to the Intergovernmental Panel on Climate Change (IPCC), which was established by the United Nations (UN) in 1988, the following are among the detrimental environmental and social trends that are now happening (IPCC, 2014): (i) the climate is warming and becoming less predictable; (ii) there are more frequent extreme weather events; (iii) changing patterns of precipitation and of snow and ice melting are altering hydrological systems (which, in turn, are causing changes to the quantities, qualities and the seasonal availability of fresh water); (iv) the geographical ranges of many terrestrial, freshwater and marine species are shifting; (v) sea levels are rising; (vi) the oceans are acidifying; (vii) coral reefs are being destroyed; (viii) crops in many regions have become subject to more negative than positive impacts; (ix) climate change is contributing to more violent human conflicts; and (x) the negative impacts of climate change are disproportionately affecting that proportion of the human population that is already marginalised. Most greenhouse gas emissions come from the burning of fossils fuels, but about 12% come from forest destruction and degradation, and about 5% from destruction of peatlands (IUCN, 2021a,b). Some physical and ecological systems are currently at risk of triggering abrupt and irreversible environmental change. For instance, if the unstable Thwaites Glacier in Antarctica was suddenly to collapse (which it might), sea levels worldwide would rise rapidly by 2–3 m, engulfing many coastal cities (Milillo *et al.*, 2019; Hogan *et al.*, 2020).

A mass extinction is now underway, similar in its potential severity to the earlier ones that have periodically punctuated geological time, such as the one at the Cretaceous/Tertiary boundary 66 million ya, which witnessed the extinction of all non-avian dinosaurs (Bacon and Swindles, 2016). Alarmingly for the integrity of ecosystems, insect populations are in worldwide decline (Harvey *et al.*, 2022). The rate of extinction of plants is now hundreds of times higher than the background rate that prevailed in more stable geological times. It has been estimated that 39% of the 325,000 species of vascular plants that currently exist are in imminent danger of extinction (Humphreys *et al.*, 2019; Antonelli *et al.*, 2020; Nic Lughadha, 2020).

Now that the danger is known, urgent attention is needed to shift the direction of human travel rapidly on to a more sustainable path. In the phraseology of the UN's Brundtland Report, humanity has to find a way to meet the needs of the present without compromising the abilities of future generations to do the same (WCED, 1987). Striving for sustainable development is a principle with which all states agree, judging by their official political positions. All those states that are parties to the UN and have ratified the Convention on Biological Diversity have agreed to conserve the biological diversity that lies within their borders and to use their natural resources sustainably (see Section 6.11). Every single member of the UN has signed up to Goal 7 of the Millennium Development Goals (which is 'Ensure environmental sustainability') and to Goals 14 and 15 of the Sustainable Development Goals (which are 'Conserve and sustainably use the oceans, seas and marine resources of the Earth' and 'Conserve and sustainably manage forests, combat desertification, halt and reverse land degradation, and halt biodiversity loss') (UNDP, 2000, 2015). A total of 189 states, plus the European Union, have ratified or acceded to the Paris Agreement (2016), the main aim of which is to strengthen the global response to climate change. Some theoretical economists in the West have woken up to the urgency of devising economic systems appropriate for modern environmental reality (Kneese, 1988; Raworth, 2017).

We, the authors, are botanists and our primary interests in conservation and sustainable development are those connected to plants. Whatever else is done in favour of sustainable development, little ultimately will be achieved unless plants receive targeted and concentrated attention. This is because of the fundamental roles that plants play in the functioning of ecosystems and economies, and their numerous cultural connections with people.

A major aspect of plant conservation is about saving plant species. Apart from its ethical merit, it is important because species are the biological systems that maintain the continuity of life and, if lost, are gone for ever. In practice, conservationists often have to make judgements about which species to try and save, but, at the same time, they should be mindful about the high degree of human ignorance about the properties and ecologies of almost all species. Scientific developments can spring surprises. Before the pioneering work of the Hungarian-born George P. Rédei (1921–2008), few botanists had much interest in thale cress (*Arabidopsis thaliana*), an inconspicuous temperate weed. Although thale cress is common in Britain, it is not a plant that many members of the public know. Yet today, Rédei's pioneering research has led to thale cress becoming a model species for studying many aspects of biology, from respiration to reproduction. Its contributions to improving food security are huge (Piquerez *et al.*, 2014). Thousands of experimental scientists around the world are engaged in researching its properties.

We have organised this book into four parts, the first to state the challenge of plant conservation and to outline the systems context (Chapters 1 and 2), the second to describe how the relationships between people and plants have changed over the course of history (Chapters 3–8), the third to suggest some ways in which plant conservation can be made more effective (Chapter 9) and the fourth to describe some of our own experiences in botany and conservation to back up these suggestions (Chapters 10–21). Map 1.2 shows the localities of the conservation experiences described in the final part.

Because we are dealing with such a big story, we have decided to concentrate on a few themes, which we see as being especially significant, and with three parts of the world with which we are relatively familiar. Two of these parts are China and the West (especially Britain), which are where we are from (Map 1.2).

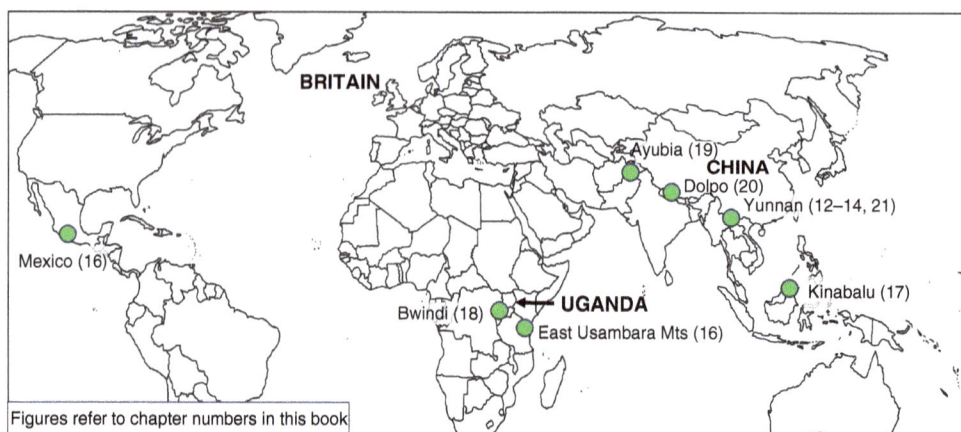

Map 1.2. Localities of China, Britain and Uganda and of the conservation experiences described in Chapters 12–21. Note regarding the positions of country boundaries: see Disclaimer.

The third is Uganda (especially Buganda), which is included because we wished to include an example of one of those many countries outside Europe which were subjected to European/Western expansionism during the last 500 years and were profoundly influenced by the experience. A.H. has lived on-and-off in Uganda since 1966 for a total of about eight years, researching its flora and environmental history, teaching at a university and being involved in a number of conservation projects. He has published many scientific papers and books about Uganda and also compiled a Luganda–English dictionary (Hamilton, 2020). He is married to a Ugandan (a Muganda).

Understanding the histories of China and Britain (as a part of the West) benefits from the availability of contemporary written records reaching back almost 3000 years. Table 1.1 summarises these histories, as relevant to the theme of this book. It shows connections between the two parts of the world from the 1st millennium BCE, including in human diseases and in trade, culture and politics. In contrast to China and the West, nothing was written about Uganda or the Baganda before the mid-19th century CE. To compensate for the lack of contemporaneous written information, we have placed added weight in this book on oral history and linguistic evidence when considering the history of relationships between people and plants in Buganda/Uganda. To help readers appreciate our references to the Baganda, we conclude this chapter with a note on Luganda, the language of the Baganda.

We find it interesting that, from the different histories of our two countries and of ourselves, we have arrived at much the same conclusions about what in plant conservation needs to be done. In this we have benefited from working together in two international conservation programmes that had a focus on people's relationships with plants, one the People and Plants Initiative (PPI – see Chapter 10) and the other the Medicinal Plants Conservation Initiative (MPCI – see Chapter 11). The purpose of PPI was to build global capacity in applied ethnobotany, ethnobotany being the interdisciplinary subject concerned with the relationships between people and plants, and 'applied' referring to

Table 1.1. Histories of China and Western Eurasia/Britain showing connections. Pandemics highlighted in orange. The histories denoted in the right hand column are for Western Eurasia below the dividing line and for Britain above.

China	Connections	Britain/Western Eurasia
Pandemic starts in China (2019).	COVID-19 pandemic.	Pandemic arrives in Britain (2020).
Belt and Road Initiative launched (2013).	Hong Kong returned to China (1997)	Britain withdraws from EU (2017) and joins Trans-Pacific Trade Block (2023).
China industrial superpower (from 1980s). Market liberalisation (1982).	Globalisation of trading systems (especially from the 1970s).	Rise of neoliberal political ideology (from the 1980s). British industry in decline (from 1945).
Cultural Revolution (1966–1976). Great Leap Forward (1958–1962). Communists take power (1949).	Cold War between USA and Soviet Union (1947–1991). Founding of United Nations (1945).	Decolonisation of empire (1945–1997).
Partial occupation of China by Japan (1937–1945).	Second World War (1939–1945).	Britain at war with Japan (1941–1945).
Civil war (1927–1949).Republic of China founded (1912).	Peak of European imperialism (1st half of 20th century).	British Empire at its peak (late 19th to early 20th C.).
Protestant missions operating freely in inland China (from 1858).	Protestant foreign missions active worldwide (from 19th C.).	Restrictions lifted on non-Anglicans holding professorships in Britain (1871).
Western plant hunters in China (mainly after 1860).	Worldwide botanical exploring by Western botanists (from late 18th C.).	Kew Gardens becomes botanical hub of British Empire (1840).
Opium Wars (1839–1842 & 1856–1860). Nanking/ Tientsin treaties.	British trading opium from Bengal to China (from late 18th C.).	Rise of humanitarian movement in Britain (during 19th C.).
Qing (Manchu) dynasty (1644) replaces Ming.	Jesuit (Catholic) missions active worldwide (late 16th to 18th C.).	Industrial revolution starts (c. 1800). Tea becomes national drink (18th C.).
East India Company trading with China (from 18th C.).	European chartered companies operating abroad (from 17th C.).	East India Company granted a royal charter (1600).
Portuguese trading station at Macau (from 1557). Ming dynasty begins (1368).	European global geographic and economic expansionism begins (c. 1500).	European mid-2nd millennium cultural revolution (from c. 1350), e.g., the enlightenment (1650–1800).
Pandemic contributes to demise of Yuan dynasty.	Plague pandemic (14th century).	Pandemic (the Black Death, 1346–1353) contributes to the demise of feudalism.
Yuan (Mongol) dynasty begins (1271). Song dynasty (960–1279).	Muslim Mongol Empire maximum extent (1300). Marco Polo in China (late 13th C.). Islamic Golden Age (c. 800–1300).	Ottomans seize Constantinople (1453). Muslim trading network around the Indian Ocean (from 8th C.).

Continued

Table 1.1. Continued.

China	Connections	Britain/Western Eurasia
Christianity arrives in China (635).	Smuggled silkworms from China launch Byzantine silk industry (6th C).	Life of Muhammad (c. 560–632).Christian reconversion of Britain (6th C).
First mention of bubonic plague in China 610 CE	Plague pandemic recorded in Europe to Central Asia (541–750 CE).	Plague pandemic contributes to fall of Western Roman Empire (Justinian Plague, 541–549 CE).
Plague outbreaks in in China during Eastern Han dynasty (25–220 CE).	Plague spread through Ancient Silk Roads?	Antonine Plague (165–180 CE)
Buddhism arrives in China during Han dynasty (206 BCE – CE 220). Life of Confucius (551–479 BCE). Life of Laozi (6th Century BCE).	Ancient Silk Roads opened by Emperor Wu of Han (141–87 BCE). Long-distance trade routes developing across Eurasia.	Life of Jesus Christ (c. 4 BCE – CE 30/33). Beginning of Roman Republic (509 BCE). Life of Aristotle (384–322 BCE).Classical Greece (c. 500–300 BCE).

its applications to conservation of biodiversity and the sustainable use of plant resources. One of the fields of capacity-building in which PPI was involved was the building up of capacities at national and international levels. P.S. followed up this theme in China by establishing the Chinese Association of Ethnobotany (CAE) in 2002 and then, in 2004, establishing the Asia-Pacific Forum on Ethnobotany (APFE). The two networks have met, mostly biennially, since 2002 in joint congresses held across China (Table 1.2). The themes chosen for these events, the presentations made at them and the discussions that they generated have been very helpful to us in developing the material for this book.

Addendum: A Note on the Luganda Language

Language shapes how people think (Boroditsky, 2018), a matter which we discuss in Section 7.5.4 in connection with making progress in conservation initiatives that involve people who think and speak in different languages. So familiar are people with looking at the world through their own eyes, that it can be difficult for them to accept that other people look out at the world in ways that are different from their own. Even with colours, which common sense might suggest are the same for all people except for the colour blind, the categories recognised and the words used can differ between cultures (Gibson *et al.*, 2017). For example, there appears to be no indigenous word for the colour 'blue' in Luganda, a feature which is not unusual in the world linguistically. Nowadays, the words used for 'blue' in Luganda are loan words from English (*bbulu* or *bbululu*).

Luganda is a member of the Bantu subgroup of the Niger–Congo language family, the dominant language family of sub-Saharan Africa. The 400 or so modern languages of the Bantu subgroup have their origin in a single language

Table 1.2. National Symposia on Ethnobotany organised by the Chinese Association of Ethnobotany (CAE) and meetings of the Asia-Pacific Forum on Ethnobotany (APFE).

Year	City	Province or Municipality	Themes	Meeting number CAE	APFE
2002	Hangzhou	Zhejiang	Ethnobotany of Natural Product Research and Development	1	
2004	Xishuangbanna	Yunnan	Ethnobotany and Sustainable Use of Botanical Resources	2	1
2006	Nanjing	Jiangsu	Ethnobotany and Medicinal Plants	3	2
2008	Urumqi	Xinjiang	Ethnobotany and Agrobiodiversity	4	3
2010	Beijing	Beijing	Development of Ethnobotany and Conservation of Traditional Knowledge	5	4
2012	Yinchuan	Ningxia	Ethnobotany and Ethnomedicine	6	5
2014	Guilin	Guangxi	Ethnobotany and Plant Germplasm Resources	7	6
2016	Huhhot	Inner Mongolia	Ethnobotany and Construction of Eco-civilisation	8	7
2018	Kunming	Yunnan	Ethnobotany and the Belt and Road Initiative	9	8
2021	Lishui	Zhejiang	Ethnobotany and Rural Revitalization	10	9
2023	Huaihua	Hunan	Ethnobotany and the Modernization of China	11	10

(known as Proto-Bantu) that was spoken in the Cameroon–Nigeria border area at about 3000 BCE (see Section 3.3.1). Earlier, the Niger–Congo language family is believed to have originated in the western half of what is now the Sahara Desert during the early part of the so-called Green Sahara, a period of higher rainfall that began at about 9000 BCE (see Section 3.2) (Blench, 2006).

Two characteristic features of Bantu languages are agglutination and inflexion. Agglutination refers to the property of a language whereby many words are composed of more fundamental parts (roots and stems) with components added to them. An example of a compound word in Luganda is *obuntubulamu*, the meaning of which is discussed in Section 4.4, and which is composed of two nouns, *obuntu* ('humanity') and *obulamu* ('health'). Inflexion refers to the property of a language whereby grammatical functions are indicated by changes *within* words, rather than (for instance) through the use of additional words or changes in their order.

Examples from Luganda may make this clearer. The Luganda word *omuntu*, which means 'person', is an agglutinated word composed of two parts, *omu-* (a prefix) and *-ntu* (a stem). Through inflexion, the prefix *omu-* can be substituted with other prefixes to give new words having different meanings, e.g. *abantu* ('people'), *ekintu* ('thing'), *ebintu* ('things'), *akantu* ('small thing'), *obuntu* ('small things' – also, 'humanity' and 'mankind') and *awantu* ('somewhere'). The first vowel of the prefix is not always present, depending on the context. The *obu-* prefix is often used to make abstract nouns (i.e. 'humanity' in this case).

The stem -*ntu* is not used on its own but can be said to have a meaning something like 'an existence', i.e. *omuntu* is a 'human existence', *ekintu* is a 'thing existence' and *awantu* is a 'place existence'. -*ntu* is a stem found in many Bantu languages, which suggests that it relates to something basic to the way that Bantu-speaking people have traditionally viewed the world. In Section 4.4, we follow this up by making reference to *obuntu* in an exploration of the commonalities in the worldviews of the Chinese and Baganda.

Another example of a stem in Luganda is -*ganda*, which, like -*ntu*, cannot be used on its own. It means something like 'a matter relating to the Baganda'. Nouns derived from -*ganda* through the addition of prefixes include *Muganda* ('a person of Buganda'), *Baganda* ('the people of Buganda'), *Luganda* ('the language of the Baganda') and *Buganda* ('the place of the Baganda'). The use of -*ganda* as an adjectival stem is seen in the phrase *omwenge omuganda* ('the beer of the Baganda' – which refers to banana beer, the Baganda's traditional brew). The abstract noun using the prefix *obu-* is not much used in everyday speech, but is found in sayings, for example '*Obuganda bwonna buli wano*' ('The whole of Buganda is here'), which refers (for instance) to a huge crowd of people who have come to see the *Kabaka* (the King of Buganda). Traditionally in Buganda, the use of sayings to convey fundamental truths in a pithy way was one of the hallmarks of a well-educated *Muganda* (see Section 4.4).

The botanical vocabularies of Luganda and Britain reflect differences in the floras of the homelands of the languages. Take, for example, the English word 'herb', which (in English) can refer to a small non-woody seed plant, a plant used for flavouring food or a plant used as a medicine (as in the phrase 'herbal medicine'). The word *omuddo* and the stem -*lagala* in Luganda cover aspects of these meanings. *Omuddo* can be translated into English as 'grass', 'herb(s)' or 'weed(s)'. -*lagala* is found in the words *olulagala* (pl. *endagala*), which means 'a banana leaf', *amalagala*, which means 'cuttings of sweet potatoes', and *eddagala*, which means 'medicine', 'dye', 'chemical' or 'drug'. As an adjective, -*lagala* means 'green', as in the phrase *olugoye olwa kiragala* ('a green garment'). It is noteworthy that, in Luganda, a word associated with leaves and greenness is used for traditional medicine (*eddagala ly'ekinnansi*), while, in South Africa, traditional medicine is called *muthi*, which means 'tree'. This reflects differences in herbal traditions in the two places relating to their climates and floras. In more ever-wet Buganda, many herbal medicines are prepared from fresh leaves, while in more seasonal South Africa, there is greater use of the bark or roots of trees or other parts of plants that can easily be stored.

Several names of plants in Luganda reflect their uses, as is also the case in English (see Section 7.3.4). Examples in Luganda are *akalunginsanvu* (*Syzygium guineense*), which means 'it cures seven diseases' and *mubajjangalabi* (*Alstonia boonei*), which means 'the tree that is used for making long thin drums'. Both of these species are rainforest trees. Botany, medicine and religion are closely linked traditionally in Buganda, as they are in many parts of the world (see Section 7.3). The Luganda names for the common small tree, *Shirakiopsis elliptica*, provide examples of related terminology. There are two names for this species in Luganda used in traditional medicine. One is *muzzaŋŋanda*, which is used when the intention is to bring reconciliation between people (this word is derived from the verb -*zza*, which means 'bring back', and *oluganda*, which means 'kinship'). The other name is *musasa*, which is used when the intention is to cause divisions between people (the word is derived from the verb *-sasa*, which means 'scatter'). Wood from this tree is preferred for the construction of traditional shrines in clan graveyards, but never for the construction of ordinary houses. The overlap between herbal medicine and traditional spiritual beliefs, as illustrated by this example, has resulted in some branches of Christianity rejecting all Ugandan traditional medicine out of hand, labelling it as 'witchcraft' (see Section 7.5.3).

The vocabulary of Luganda reflects their traditional ecological knowledge (TEK), as it has evolved during the history of the Baganda (Kyebogola *et al.*, 2020). In Section 7.2, we describe how the vocabulary of Luganda matches the Baganda's traditional plant-resource system, as it has evolved in their present-day homeland near Lake Victoria. To give an example relating to clay-rich soils, the Baganda recognise *ettosi* (wet clay in valleys), *obudongo* (used as mortar in building), *olufuufu* (used as plaster) and *ebbumba* (used in pottery). Apart from mortar, *ebbumba* is used to make *emmumba*, which are sticks of clay containing herbal ingredients and which are ground into powder when taken as medicine (*ebbumba* and *emmumba* are derived from the stem *-bumba*, which means 'to make (with clay)').

Over the course of time, the linguistic ancestors of the Baganda acquired new words, as they encountered new environments and people, and discovered new technologies. As discussed in Section 3.3.1, some words were borrowed from speakers of Nilo-Saharan languages when the Baganda encountered them in the Lake Victoria region during the 1st millennium BCE. Many words have come into Luganda from Swahili and English. Examples of Luganda words that have been borrowed from Swahili and, before that, Arabic are *ekitabo* ('book', from Swahili *kitabu* and Arabic *kitab*), *Bulaaya* ('Europe', from Swahili *ulaya* and Arabic *uwrabaa*), *eddiini* ('religion', from Swahili *dini* and Arabic *din*) and *malayika* ('angel', from Swahili *malaika* and Arabic *malak*). The Luganda word for 'tea' is *caayi*, which was introduced as a drink in the 19th century by the British or, possibly, Swahili or Arab traders. *Caayi* may be derived from 'cha', a colloquial word used for tea by the British, or otherwise possibly from *chai* (Swahili), *shay* (Arabic), *chaay* (Hindi) or *tē* (Urdu), but anyway ultimately from *chá* (Chinese).

The Luganda name for the Irish potato (*Solanum tuberosum*) is either *akamonde akazungu* (meaning 'the little potato of the white man') or *obumonde*, which is the plural. These words are adaptations of *lumonde*, which means 'sweet potato' (*Ipomoea batatas*). The Irish potato is a crop of South American origin that was introduced into Uganda at about the turn of the 20th century (Table 3.1). In contrast, the sweet potato is a crop, also of South American origin, that has been part of the sub-Saharan food system for several hundred years (Low *et al.*, 2009). According to Keefa Ssentoogo, an expert on the traditional culture of the Baganda and a member of a team that collated a Luganda/English dictionary (Hamilton, 2020), red varieties of sweet potatoes (*lumonde omumyufu*), but not white varieties (*lumonde omweru*), have been in Buganda for a very long time. The impression given is that it was being grown even earlier than the Columbian discovery of America (in 1492).

Evidence from historical linguistics shows that several crops of South American origin were present in Uganda from about 1800 CE, which was before the arrival of any Western explorer (Ehret, 2011). These crops included tobacco (*Nicotiana tabacum*), peanut (*Arachis hypogaea*) and the American beans (*Phaseolus*), all of which (the linguistic evidence suggests) spread overland to Uganda from the *west* (not the east) coast of Africa. The apparent earlier presence of red varieties of the sweet potato in Uganda suggests that they may have been among the suite of crops brought from Asia by Austronesian-speaking seafarers during the early centuries CE and which subsequently spread into Uganda from the *east*. It is known that the Austronesians picked up the sweet potato in South America about 2000 ya and spread it to islands in the Western Pacific (see Section 3.3.3).

The European introduction of Christianity to Uganda in the 1870s was soon followed by conflict breaking out between the Baganda who converted to Protestantism and those who converted to Catholicism (see Section 6.8.2). The two factions came to use different words for some fundamental aspects of Christianity, such as for the name of their religion (*Obukulisitaayo* for Protestants, *Obukirisitu* for Catholics) and the name of Jesus (*Yesu* for Protestants, *Yezu* for Catholics).

The traditional calendar in Buganda is based on a combination of the agricultural calendar, lunar months and two pairs of wet and dry seasons in each sidereal year (Roscoe, 1921). The following are examples of the agricultural calendar: *amasiga* (time of sowing), *amasimba* (time of planting), *amakoola* (time of weeding) and *amakungula* (time of harvesting). The appearance of the new moon divided up the year into lunar months. There was a rest of at least one day at the appearance of the new moon, while, at the king's court and one or two temples, special ceremonies of a religious nature were carried out and there was a cessation of work for nine days. The rainy season, occurring between March and May, was termed *ttoggo mukazi* and that between September and November *ddumbi musajja*. The words *mukazi* and *musajja* refer to maleness and femininity, which is said to be because the March–May rainy season (known as the 'long rains' in English) were characterised by rain falling without much thunder and the September–November rainy season (known in English as the 'short rains') were accompanied by much thunder and frequent deaths from lightening (Roscoe, 1911).

There are three names for each of the 12 months in modern Luganda, all of them new and seemingly invented by the British. One of the names refers to the order of a month in the year, another is a loan word from English and the third often makes reference to the agricultural calendar. For example, the names for January are *Omwezi ogw'olubereberye* (which literally means 'the first month'), *Janwali* (a loan word from English) and *Gatonnya*, which is a reference to the many ripe fruits which fall off cooking banana plants in this month. (The verb *-tonnya* means 'fall' and the noun of reference is *amenvu*, which, in this case, means 'ripe cooking bananas'.) Other examples are *Museenene* for November and *Ntenvu* for December. These are references to the swarming of the edible grasshopper, which is called *enseenene* in Luganda, and to a type of beetle larva called *entenvu*, which is found in the stems of banana plants. References to cash crops can be seen in the words *Mugulansigo* for March and *Mutunda* for September. The first makes reference to the verb *-gula* and the noun *ensigo*, which mean 'buy' and 'seeds', and the latter to *-tunda*, which means 'sell'. This fits with the commercial agricultural calendar for cotton, which was the main commercial crop promoted by the colonial government for growing by indigenous farmers (see Section 5.4.3).

2 Plants, Planet and People – An Overview

Abstract

Plants helped to create a world habitable by people and remain essential for its continuing maintenance. Their present diversity and geographical distribution are the products of a long history of evolution. Since first appearing 50,000 years ago, the modern, linguistically competent, form of the human species has discovered increasingly efficient ways of acquiring food, which has resulted in an increasing number of people and increasing disruption of the Earth's ecosystem. The larger human population, combined with the increasing ability of people to rapidly travel long distances, has resulted in the spread of types of organisms detrimental to human interests. Some are pests; others cause diseases of people, plants or animals; and others have become invasive and disrupted native ecosystems. Two case studies of the spread of organisms detrimental to human interests are presented, both concerning Uganda. One is the introduction of pests and diseases of the banana and the other the introduction of the Nile perch into Lake Victoria.

2.1 Plants in Planetary History

The histories of the Earth and its plants have been greatly clarified by advances in science made during the last 500 years. It is now known that the Earth is a planet encircling the Sun and that the distribution of land on its surface has changed over time, driven by the movements of convection currents in its mantle and crust. All life currently existing on the Earth has evolved from one or a few simple forms of life that lived about 3.7 billion ya (years ago). Plants were instrumental in the creation of a world that is habitable by people and remain essential for its continuing maintenance.

Plants are eukaryotic green photosynthetic organisms. Their photosynthetic ability enables them to make organic from inorganic matter, in doing so producing the organic materials and the oxygen that animals require to survive. In the early years of the planet's existence, plants, together with a number of other

© Alan Hamilton and Pei Shengji 2024. *History and Future of Plants, Planet and People* (Alan Hamilton and Pei Shengji)
DOI: 10.1079/9781789248944.0002

early-evolved types of organisms (notably the blue-green algae), gradually oxygenated the atmosphere and created an ozone-rich layer. A critical moment in the history of life occurred when sufficient ozone had accumulated in the upper atmosphere to block enough of the damaging ultraviolet radiation reaching the Earth from destroying any organism that tried to venture out of water on to land.

Once able to live on land, plants were instrumental in creating the first soils and, from then on, have been fundamental parts of the ecosystems that maintain them. Plants, together with the soils which they help create and in which they grow, form an important regulator of the ways that water moves through the upper layers of the Earth. They help to determine the moistness of regional and local climates, seasonal variations in river flows, the quality of the water available for human consumption, the rates and types of soil erosion, the incidence of landslides and, over the longer term, the shape of the land.

One species of green alga, living approximately 1 billion ya, fanned out along many evolutionary lines and into a great variety of morphological forms, both aquatic and terrestrial (Essig, 2015). This species was the ancestor of nearly all of the organisms that people commonly think of as plants (though not the red and brown seaweeds, and the fungi, which lie on different evolutionary lineages). Large trees were in existence by the late Devonian (383–360 million ya). All modern bryophytes have evolved from a single ancestral species that lived ~500 million ya, all modern vascular plants from another species living ~425 million ya, all modern seed plants from another species living before the end of the Devonian (360 million ya) and all modern flowering plants (Angiosperms) from another species that lived between 200 and 140 million ya. Much of the world's coal dates to the Carboniferous Period (360–299 million ya). It originated through the incomplete decomposition of plants growing in swamp forests.

The ancestors of all plant species that are indigenous to isolated oceanic islands that have never been connected to the continents must have arrived on them by means of long-distance dispersal, such as through the carriage of their seeds on the outsides or insides of birds (Viana, 2016). Elsewhere, the present-day diversity and distribution of terrestrial plants owes much to how their ancestors reacted to the past movements of the continents and the coming and going of the land bridges that, at times, formed connections between them. The many similarities that exist between the floras of temperate North America and Eurasia owe much to the interchanges of plants that took place between them at the times when land bridges connected the two across the place where the Bering Straits now stands (Jiang Decohun *et al.*, 2019). Similarities between the floras of Australasia and the southern tips of Africa and South America are believed to be relics of a time when they were physically connected in a giant landmass called Gondwana.

The Quaternary, the most recent era of geologic time (2.6 million ya to now), has been marked by pronounced oscillations between cooler and warmer climates (typically, drier and wetter in the tropics). These oscillations have strongly influenced the evolution and distributions of plants and animals (see Chapter 15) (Hamilton and Taylor, 1991; Hewitt, 2000; Fjeldså *et al.*, 2020). Prior to the Quaternary, a steady drying of the African climate from the beginning of the Miocene (23 million ya) is believed to have contributed to the present floristic poverty of its tropical rainforests, compared with those of Southeast Asia and Amazonia (Plana, 2004; deMenocal, 2014; Muscarella *et al.*, 2020).

The Earth is periodically impacted by large extra-terrestrial objects causing large-scale extinctions. One such object, an asteroid that fell on the Yucatan Peninsula in Mexico 66 million ya (the K–T boundary), is believed to have caused the extinction of about 57% of all plant species living in North America (Askin and Jacobson, 1996; Nichols and Johnson, 2009). It also exterminated all non-avian dinosaurs, everywhere. Plants reacted quickly and soon the same general types of vegetation were restored as before (Vajda and Bercovici, 2014). However, it took much longer for the vegetation to recover in terms of its diversity of species (Jablinski, 2001).

Other causes of periodic biological catastrophes have been huge volcanic eruptions. The last such global event was the Campanian Ignimbrite super-eruption that occurred close to Mt Vesuvius in Italy about 40,000 ya (see Section 15.4.1). It helped to trigger the start of a long, cool and dry climatic phase in Western Eurasia and tropical Africa that lasted until the first stirrings of the present postglacial period about 14,500 ya. It is believed to have been the decisive factor that led, in Western Eurasia, to the replacement of the Neanderthals by modern *Homo sapiens*.

2.2 The Human in Planetary History

2.2.1 The influence of methods of food acquisition

Changes in the ways that people have acquired their food have been major drivers of global environmental change during the last few thousand years. The seminal event was the transition from hunting and gathering to farming and animal husbandry, which, at the earliest, was at about 10,000 BCE (Romer, 2012; Reich, 2018). Exactly why and how this transition happened is a matter of continuing debate, but it is surely significant that it occurred independently in a number of places at about the time (or soon thereafter) when the global climate switched decisively to an interglacial climatic mode at the end of the last ice age.

One theory about the origin of agriculture is that there were certain 'primer' crops that stimulated people into becoming farmers. For instance, the broad bean (*Vicia faba*) has been suggested as a primer crop for agriculture in the Near East, on the basis of it being a conspicuous plant with large seeds which can not only be eaten, but which also allow a clear observation of a plant germinating from a seed – perhaps stimulating people to think about the causal links between growing and sowing (Kosterin, 2014). The invention of fire-hardened pottery, which is useful for storing and processing food, is likely to have been a further stimulus. Pottery of this type was invented in two places in the Old World independently of one another, in both cases a little before the local beginnings of agriculture (Dunne *et al.*, 2016). One was in East Asia, dating to *c*.14,000 BCE, and the other in North Africa, dating to *c*.10,000 BCE.

The acquisition of domesticated plants and animals increased the amount of food available to people and caused the human population to rise. The first agriculture was by hand and the crops were either dependent for water on the rain that fell directly where they grew or they were grown in seasonally inundated places by the sides of rivers or lakes – factors which limited, to an extent, the amount of food produced and the rate of population increase. Then, from *c.*3000 BCE, more intensive agricultural methods started to be developed and crop yields increased. Irrigation was introduced in places, as was the use of draught animals to pull carts and ploughs. Then, in *c.*1800 CE, further intensification occurred through the use of fossil fuels to drive machinery; then again, in *c.*1900, when artificial fertilisers and other agrichemicals began to be applied; and then again, in *c.*1950, when the green revolution brought industrial agriculture to the Global South (see Section 5.4). The demographic response of the human has been to increase its numbers from an estimated 4 million people in 10,000 BCE, to 1 billion in 1800 CE and then to 7.6 billion in 2018. It is predicted that, if present trends continue, the human population will reach 10 billion by 2050. However, this is unlikely to happen, since, it has been calculated, the sum total of all the plants on Earth will be incapable physiologically of meeting the demands for food of such a huge population (Kraussmann, 2013).

A consequence of the expanding area of land devoted to producing food and other biological commodities for human use is that the remaining area of more natural habitat, containing wild plants and animals, has been shrinking. The total area of cropland in the world rose from zero in 10,000 BCE to 272 million ha in 1600 CE, and stood at 1.59 billion ha in 2016, at which time four crops (wheat, maize, rice and soybean) each covered more than 100 million ha (Ritchie and Roser, 2024). It is estimated that, of the world's habitable land, 11% is under crops, 38% is under pasture and 38% is under forest, the last two being very extensively used to produce products for people (Ritchie and Roser, 2024). Currently, livestock constitute roughly 60% of all mammalian biomass, humans 30% and wild animals just 4% (Bar-On *et al.*, 2018).

Subsequent to attaining its modern capacities for language and thought about 50,000 ya, the story of the human species has been characterised by much movement and mixing of its populations (Reich, 2018). Males have tended to be more mobile than females and population mixing has been disproportionately by males belonging to migrating populations interbreeding with resident females. The acquisition of certain new technologies has spurred population movements. For instance, those populations that acquired domesticated plants or animals tended to spread into the areas of those that did not. Later, the same pattern repeated for those populations that acquired new, more productive, ways of growing crops or keeping livestock, compared to those that stuck conservatively to their familiar methods.

As human populations have spread and fanned out historically, contacts between their spreading subpopulations have tended to reduce as the distances between them have increased. Their subpopulations have tended to diverge genetically and culturally, including linguistically. Genetic diversification also occurred within the domesticated plants and animals that the people took along with them on their journeys. Countering these trends towards increasing genetic and cultural diversification of human populations have been ones towards greater homogenisation, as certain human societies proved to be particularly successful at expanding their territories at the expense of others. Beginning with the formation of the first states in prehistoric times (see Section 3.4), the pace of this globalising trend picked up in the mid-2nd millennium CE when Europe and subsequently the West expanded their influences to the global level, eventually resulting in today's global economic and cultural systems.

2.2.2 Human influences on the global carbon cycle

Three elements (carbon, oxygen and hydrogen) are required by living things in large amounts, three others (nitrogen, phosphorus and potassium) in medium amounts and a number of others in trace quantities. The stores of these elements within individual organisms and their flows into and out of them form parts of larger systems that include other organisms as well as non-living components of the environment. It is known that, on longer time scales, major changes in some of these larger systems have been closely associated with the history of life, for instance the changes in the oxygen cycle that eventually led to the accumulation of ozone in the atmosphere and life being able to move on to dry land (see Section 2.1). In recent years, humans have caused severe disruptions to some of the cycles, for instance the nitrogen cycle, as discussed in Section 7.4.2.

The two key chemical reactions in plants that involve carbon are photosynthesis and respiration. The former is an endothermic (energy-requiring) reaction that uses solar energy to convert carbon dioxide (CO_2) and water (H_2O) into glucose ($C_6H_{12}O_6$) and oxygen (O_2). The latter is an exothermic (energy-releasing) reaction that converts the energy contained within glucose into forms useful to plants, at the same time releasing carbon dioxide and water back into the environment. In a steady-state ecosystem, plants absorb as much carbon through photosynthesis as they lose through respiration, the net result of photosynthesis and respiration being a zero-sum game:

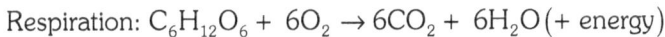

Photosynthesis: $6CO_2 + 6H_2O (+ \text{energy}) \rightarrow C_6H_{12}O_6 + 6O_2$

Respiration: $C_6H_{12}O_6 + 6O_2 \rightarrow 6CO_2 + 6H_2O (+ \text{energy})$

If the organic matter synthesised from glucose by plants is not completely decomposed, it can accumulate in the form of carbon-rich deposits, such as peat and organic-rich mud. Later in geological time, such deposits may become metamorphosised and become incorporated into the geological column as layers of lignite and coal, or else as accumulations of liquid petroleum and natural gas (which is mainly methane, CH_4). Carbon dioxide and, even more potently, methane are greenhouse gases, and it is their accumulation in the atmosphere since the time when the burning of fossil fuels began at scale at about 1800 CE that has been the principal driver of anthropogenic climate change (IPCC, 2021).

The carbon cycle is complex and new discoveries continue to be made. For instance, it has been found recently that, until the late 20th century, the increasing amount of carbon dioxide in the atmosphere has acted as a fertiliser for tropical forest, causing the trees to grow faster and sequester more carbon (Lewis *et al.*, 2009). However, since the 1990s in Amazonia and 2015 in Africa, intact tropical forest has started to become a carbon source, rather than a carbon sink. This is because, under the higher temperatures, more carbon is being lost through respiration than is being absorbed by photosynthesis (Hubau *et al.*, 2020; Sullivan, 2020).

According to current scientific understanding, it was the human ancestor, *Homo erectus*, that discovered how to make fire (see Chapter 1). The deliberate burning of vegetation to create new habitats may have been invented by *H. erectus*. Nowadays, there are several reasons why people burn vegetation, among them to stimulate the growth of pasture grasses, to clear vegetation for agriculture, to destroy brash after harvesting timber and to create fire-protection corridors to limit the spread of wildfires. Extreme wildfire events, some started maliciously, are becoming increasingly common (Bowman *et al.*, 2017). Mediterranean-type ecosystems, one of the richest types of ecosystems floristically, are particularly vulnerable to damage by over-burning. Such ecosystems are found in places with Mediterranean-type climates – that is, in Southern California, Central Chile, the Cape Floristic Region of South Africa and in South Australia – in addition to around the Mediterranean.

2.3   Pests, Diseases and Invasive Species

A number of organisms, unwelcome to humans, have been advantaged by the increasing numbers and mobility of the human species. Some of these species are pathogens that have jumped to humans from other species – events that have become increasingly common with the passage of history (McNeill, 1976). The arrival of such species in human populations that have not acquired genetic immunity can be devastating.

One infectious disease that has had catastrophic effects on human populations is bubonic plague, which is caused by a bacterium, *Yersinia pestis*. Originally, this pathogen is believed to have lived in colonies of underground rodents inhabiting the steppes of central Eurasia. It is transmitted between humans via the intermediary of fleas carried on black rats, a species well-adapted to living in human settlements and to hitching transoceanic rides on ships. The first known cases of bubonic plague (identified through palaeopathology) were in the Samara region of Russia (north of the Caspian Sea), dating to about 1800 BCE (Sprou *et al.*, 2018). The Justinian plague pandemic in the 6th century CE is believed to have contributed to the fall of the Western Roman Empire (see Table 1.1) (Harper, 2017). The next plague pandemic, the Black Death, is estimated to have killed 75 million to 200 million people in Eurasia between 1330 and 1350.

In England, where it killed about one-third of the population, the Black Death is believed to have helped bring about the end of feudalism. The reason for this causal association was because it killed such large numbers of serfs (the agricultural labourers of the day) that those who survived became a scarce resource, which enabled them to increase their bargaining power and obtain better conditions of employment. In this way, the Black Death served as an early stimulant to the mid-2nd millennium European cultural revolution discussed in Section 6.1. In China, the Black Death arrived at a time of political confusion and contributed to the replacement of the Yuan dynasty by the Ming (Wood, 2020).

Another infectious disease which has changed the course of history is falciparum malaria, the deadliest type of malaria. This originated when *Plasmodium falciparum*, the pathogen which causes the disease, jumped from the gorilla to the human in Africa 40,000–60,000 ya (Liu Weimin *et al.*, 2010; Otto *et al.*, 2018). Very roughly 22,000 ya, the ancestors of the Bantu-speaking people acquired a mutation (the sickle-cell mutation) that provided them with a measure of protection against malaria (Laval *et al.*, 2019). Much later, when they penetrated into the African rainforest, which was after 3000 BCE, the Bantu-speakers spread the sickle-cell mutation into the hunter-gatherer 'pygmy' people who were already living there. Falciparum malaria was one of several infectious diseases that were introduced from the Old World into the New World, following its Columbian discovery in 1492 (see Section 5.1) (Rodrigues *et al.*, 2018). It was the absence of genetic resistance to malaria that prevented Europeans from being able to enter the African interior in force prior to the late 19th century (see Section 5.2).

It is not only diseases of humans that have changed the course of history. So, too, have diseases of plants and animals. An example is the arrival in 1845 in Ireland of blight, a disease caused by a fungus-like organism (*Phytophthora infestans*) which is fatal to the potato (*Solanum tuberosum*). The failure of the potato crop was disastrous for the Irish poor, who depended on it as their staple food. An estimated 1 million of them died of starvation between 1846 and 1851 and about 2 million more were forced to emigrate.

Additional to disease-carrying organisms, there are other organisms, damaging to human interests, that have been advantaged by the increasing numbers and mobility of humans. Among them are alien invasive species – plants and animals transported to new places by people, deliberately or otherwise – and which have moved into native ecosystems and become threats to the survival of their floras and faunas. The European geographic and economic expansionism that started in the mid-2nd millennium CE galvanised their spread. The scale of alien invasions is now so great that it is a major cause of worldwide biodiversity decline (Rejmánek, 2000; Cronk, 2016; Paini *et al.*, 2016). The biotas of long-isolated oceanic islands, such as Mauritius, are particularly susceptible to biological invasions (see Section 5.3.1). 'Inland islands' too are vulnerable, such as the archipelago-like isolated forests on the Eastern Arc Mountains, such as the East Usambaras (see Section 16.4).

Below, we present two case studies of alien invasive species, both concerning Uganda and both responsible for causing severe disruptions to biodiversity and society. They demonstrate the interconnectedness of the natural world and the limited human understanding of how it works.

2.3.1   Introduction of pests and diseases of the banana into Uganda

The banana, a food crop originally domesticated in New Guinea, was brought to the coast of East Africa by people during the early centuries CE (see Section 3.3.3). From the coast, the ancestors of two genome groups, the Plantain Genome Group (AAB) and the East African Highland Genome Group

(AAA-EA), were carried inland and diversified into a number of local landraces (Table 2.1 – see legend for an explanation of the codes used for the genome groups). Plantains are roasting bananas, which are common staple foods in West and Central Africa, but in Uganda are regarded as snacks. East African Highland bananas are cooking and brewing bananas. They are the preferred staple food (*matooke*) of millions of people in Uganda and fermented to prepare a traditionally favoured alcoholic drink (*omwenge* – banana beer).

Cultivated bananas rarely reproduce sexually and the normal way in which they spread geographically is through the excision of suckers by farmers and their planting elsewhere. Genetic diversification in cultivated bananas is by somatic mutation and this, combined with farmer selection, has resulted, over time, in the creation of a large number of indigenous cultivars of AAA-EA bananas in Uganda. Today, the country is recognised as a major secondary centre of genetic diversity for AAA-EA bananas. Two separate surveys of on-farm banana diversity in Uganda identified 82 and 120 locally named types of AAA-EA bananas, respectively (Gold *et al.*, 2002; Edmeades *et al.*, 2007). Figure 2.1 shows a fruit bunch of *mbwazirume*, one of the common types

Table 2.1. Types of bananas present on farms in Uganda in 2004–2005 (Edmeades and Karamura, 2007). The wild species that have contributed to the genomes are *Musa acuminata*, which is native to Southeast Asia (AA diploid), and *Musa balbisiana* (BB diploid), which is native from Northeast India to China (Perrier *et al.*, 2011). The earliest record of cultivation of bananas globally is from New Guinea, where their cultivation dates back to at least 5000–4500 BCE (Denham *et al.*, 2003). Adapted from: Hamilton *et al.* (2016).

Genome group or modern cultivar	Genome	No. of varieties identified	Examples of local names	Uses	Origin and history
Modern cultivars					
Tetraploid hybrids	AAAA	4	*Kawanda*	Mainly cooking; not considered tasty	Bred in research laboratories
Recently introduced genome groups (from 1900 CE onwards)					
Gros Michel	AAA	2	*Bbogoya*	Dessert	Introduced after
Kamaramasenge	AAB	1	*Ndiizi*	Dessert	1900 CE, at first
Ney Poovan	AB	1	*Kisubi*	Brewing	into Entebbe
Bluggoe	ABB	2	*Kivuuvu*	Multi-use	Botanical Garden, later through agricultural research stations
Indigenous genome groups (present in Uganda for over 1000 years)					
East African Highland	AAA-EA	82	*Mbwazirume*	Cooking and brewing	Locally evolved from types of bananas
Plantain	AAB	3	*Gonja*	Roasting	introduced into Africa during the early centuries CE

Fig. 2.1. Fruit bunch of a cooking banana. Photo: A.H. (2015).

of cooking bananas. It was photographed on a plant in the Regional Musa Germplasm Collection of the National Agricultural Research Organisation (NARO) at Mbarara. The original name of *mbwazirume* is said to have been *mwanakufe*, but it was changed after Kabaka Mwanga II, a king of Buganda in the late 19th century, commented that it was so delicious that people would not notice being bitten by flies while eating it (information from Keefa Ssentoogo). *Mbwazirume* combines the noun *embwa* ('biting fly' – *Simulium*) and the verb *-luma* ('be painful').

Further details about how banana diversity in Uganda has been created and maintained have been revealed through ethnobotanical research. It has been found that, despite political, socio-economic and cultural change, there are still smallholding farmers in the country who have a particular interest in traditional banana varieties and have large collections on their farms. It has also been found that it is common for families and friends to gift suckers of favoured varieties to one another (Karamura and Mgenzi, 2004; Karamura *et al.*, 2004). On the scientific side, the main method that has been used to conserve banana diversity has been through the establishment of field collections (see Section 8.6). However, these have proved vulnerable to loss at times of funding shortages, political strife and changes in government policy (Table 2.2). Suggestions have been made about ways of linking together traditional conservation and the field collections, so as to make a more robust system (Hamilton *et al.*, 2016).

Table 2.2. Field collections of bananas in Uganda (1898–2016). Modified from: Hamilton *et al.* (2016).

Institute	History
Entebbe Botanic Gardens	Founded 1898, especially to trial the local suitability of potential economic crops. Banana varieties were introduced from many countries. Work on bananas was transferred to agricultural stations from 1910
Kampala Plantation	Banana varieties accumulated from Buganda and Ankole (1919–1925) and trials of cooking bananas started (1927). The collection no longer exists, related to expansion of Kampala City
Bukalasa Agricultural College (formerly Bukalasa Substation)	Trials on banana varieties started 1927 using materials obtained from Kampala Plantation. The work was largely transferred to KARI in about 1940. A banana collection was re-established in the 1960s, eventually containing 600 accessions, but lost by 1985
Kawanda Agricultural Research Institute (KARI)	Banana collection started 1940, but later lost. It was restarted in 1989, triggered by concern caused by the discovery of Black Sikokota in Uganda in 1988. When Uganda started a banana breeding programme (2003), the Kawanda collection was transferred to MZARDI. Kawanda then became developed as a collection of breeding lines
Makerere University Agricultural Research Institute – Kabanyolo	Banana collection established 1989 but lost by 2004. A few representative samples exist in tissue culture
Namulonge Agricultural Research Institute	Collection established in 2006–2008 as part of a plan to move all research on crops from Kawanda to Namulonge. It contains about 100 accessions. A banana collection belonging to the International Institute of Tropical Agriculture (IITA) was started at Sendusu, near Namulonge, in the 1990s
Mbarara Zonal Agricultural Research and Development Institute (MZARDI)	Banana collection started in 1998 as a reference collection and to duplicate the collection at Kawanda. The collection became a Regional Musa Banana Collection in 2008 (covering DR Congo, Kenya, Tanzania and Uganda). It currently has 450 accessions. It is run by the National Banana Programme (based in KARI) of NARO, backstopped by Bioversity International

A decline in the productivity of AAA-EA bananas started to be noticed in Uganda in the 1950s (Wrigley, 1989). Initially attributed to exhausted soils (see Section 7.2), subsequent observations brought to light an additional problem – the ravaging of the banana plants by pests and diseases (Karamura and Mgenzi, 2004; Mulumba *et al.*, 2004). On investigation, it turned out that many, perhaps all, of the pests and diseases had been introduced attached to banana stock brought into the country for trials by agricultural scientists (Blomme *et al.*, 2012).

The introduced pests and diseases of the banana (and times of arrival, if known) include: the banana weevil (*Cosmopolites sordidus*), which was present by 1908; some damaging nematodes (notably *Radopholus similis*); the pathogens responsible for Sigatoka leaf spot (first seen, 1938); *Fusarium* wilt (first reported, 1953); Black Sigatoka (first recorded, 1988); and banana

Fig. 2.2. The disease Black Sigatoka on the leaf of a cooking banana. Photo: A.H. (2015).

Xanthomonas wilt (BXW, first reported 2001) (Hamilton *et al.*, 2016). Figure 2.2 is a picture of a leaf of *mbwazirume* diseased with Black Sigatoka. It was photographed in the Regional Musa Germplasm Collection at Mbarara, Uganda.

All banana varieties in Uganda succumb to BXW, while AAA-EA bananas are especially susceptible to *R. similis*, weevils and Black Sigatoka. The latter is a disease caused by a fungus *Mycosphaerella musicola*, first reported in Honduras in 1972 and which today is the greatest threat to the production of bananas worldwide. The *Fusarium* wilt that started to infect AAA-EA bananas in Uganda in the 1950s also affected an introduced variety called Gros Michel. Gros Michel (known as *bbogoya* in Luganda) was once the variety of dessert banana that filled the shelves of supermarkets in the Western world. Its collapse in Uganda due to *Fusarium* wilt was paralleled in commercial plantations in other parts of the world, a calamity that led to its worldwide replacement by another variety, Cavendish, which today is the common type of dessert banana seen in supermarkets. Scientists have responded to the decline in the availability of traditional varieties of cooking bananas in Uganda by breeding disease-resistant hybrid replacements. However, the *matooke* prepared from these hybrids is regarded as tasteless by *matooke* connoisseurs.

An unfortunate consequence of the problems bedevilling AAA-EA bananas in Uganda has been to raise the cost of buying *matooke* to levels that make it out of reach of the day-by-day budgets of those ordinary Ugandans who have to buy their food, for example the millions who live and work in cities. Another is to add to the precariousness of life for smallholding farmers, who rely on their surplus agricultural produce to raise cash. There is an extensive traditional banana-related vocabulary in Luganda (Hamilton, 2020), but this is becoming less well known to the younger generation (see Section 7.5.4).

2.3.2   Introduction of the Nile perch into Lake Victoria

The extinction of an estimated two-thirds of an original total of ~300 species of small haplochromid fish (*enkejje*) in Lake Victoria during the last 60 years has been blamed principally on disruption to the aquatic ecosystem caused by the introduction of the Nile perch (*Lates niloticus*), which is a large predatory fish (Jansen, 1997; Pringle, 2005; Goudswaard *et al.*, 2007). Before this introduction, the fishing industry of Lake Victoria was multi-species based, but, with the disruption, it narrowed down to dependency on just three species – the Nile perch (known locally as *empuuta*), a species of introduced tilapia (*Oreochromis niloticus – engege*) and a tiny silvery zooplanktonic fish (*Rastrineobola argentea – mukene*). Before the lake's ecological transformation, the haplochromid fish had constituted 83% of the total biomass of all fish in the lake (Jansen, 1997).

Research into the knowledge of local fisherman by Keefa Ssentoogo, undertaken for a new Luganda–English dictionary (Hamilton, 2020), revealed that village fishermen believe that the introduction of the Nile perch caused reductions in other types of fish, additional to the haplochromids. They blame the Nile perch for the disappearance of a fish known locally as *enningu* (*Labeo victorianus*), which then became replaced by a fish known as *nnunguli* (not scientifically identified), which itself subsequently declined. Also gone are two types of elephant snout fish, one of which is known to science as *Mormyrus kannume* and the other (*nsuma*) which was slightly larger and seems not to be known scientifically. The elephant snout fish has three names in Luganda – *kasulubbana, kasulu* (a diminutive) and *somatulye*. The last is an affectionate nickname meaning 'pray and let's eat', referring to the fact that it was regarded as especially tasty. Said by the elder among the fishermen, it gave them permission to eat *somatulye* while they were still out on a fishing trip – the humorous implication being that, otherwise, they would have had to take it back to the village, where it would have to be shared.

British colonial officials in Uganda began expressing concern in the 1920s about the declining catches of an economically significant species of tilapia that was indigenous to Lake Victoria (*Oreochromis esculentus*) and debate ensued as to whether attempts should be made to boost the lake's fisheries by introducing other, relatively large, species of fish. Some of the colonial officials were disdainful towards the abundant *enkejje*, considering them useless for the types of economic development that they had in mind. One British officer in the Uganda Game and Fisheries Department (UGFD) suggested that the best use that could be made of *enkejje* was to turn it into fishmeal for livestock feed (Anderson, 1961).

Eventually, in 1953, the pro-introduction big-fish lobby succeeded in receiving political authorisation to release into the lake four, non-indigenous, species of tilapia (including *O. niloticus*) and debate continued about whether or not the Nile perch should follow. British personnel attached to a number of territorial and interterritorial organisations contributed to the debate, the interterritorial organisations being included because, at the time, Britain controlled all three of the political territories that bordered the lake (i.e. Kenya, Uganda and Tanganyika – now Tanzania) and was considering amalgamating them into a single political unit (see Section 16.2). One of the interterritorial organisations was the East African Fisheries Research Organisation (EAFRO, founded 1947), which was a scientific research agency, and another the Lake Victoria Fisheries Service (LVFS, founded 1948), which had legal responsibility for administering the lake's fisheries. EAFRO and LVFS reported directly to the East African High Commission, which was the interterritorial political organisation responsible for administering the common affairs of the three territories.

It was British staff working for the UGFD who actually introduced the Nile perch into Lake Victoria, which they did on several occasions (Pringle, 2005). The first introduction was in 1954, when it was done without official permission, and this was soon followed by several other (officially sanctioned) introductions. At the time, moving species of fish from one African water system to another was considered quite normal by European colonial officials. Many such transfers were made between the 1920s and 1970s, the peak of transfers being the 1950s, when a total of 67 species of fish were transferred (Goudswaard et al., 2007).

A striking degree of self-confidence is apparent in the way that species of fish were being moved liberally around the African continent during the latter part of the colonial period, indicating a self-belief among those responsible that they knew what they were doing and that what they were doing was right. The characters of the two British game wardens who successively led UGFD during the first 35 years of its existence are revealing of the worldviews of senior British colonial officials at the time. Charles Pitman (in charge, 1925–1951) was a first-class field naturalist with a profound knowledge of Ugandan natural history, while Bruce Kinloch (in charge, 1951–1960) was a devoted big-game hunter and sports fisherman (Kinloch, 1957; Benson, 1976; Pringle, 2005). Many of the other senior British officials responsible for administering the colonial territories shared these cultural interests in natural history, big-game hunting and sports fishing, forming a powerful political group that succeeded in creating many Game Reserves during the colonial era (see Chapter 8.2.3). No contradiction was seen at the time between the pursuit of natural history and hunting game (Thompson, 2010).

Scientists at EAFRO and in the Natural History Museum in London were fascinated by the diversity of the haplochromid fishes of Lake Victoria, which, from a scientific perspective, constitutes an extraordinary evolutionary phenomenon (Greenwood, 1966). However, they were careful to place little weight on this in presenting their opposition to the proposed introduction of the Nile perch, wary of being accused by political officials and fisheries officers of putting too much effort into researching matters of 'esoteric scientific interest' and not enough into subjects that might yield economic benefits. This cultural gap between the colonial scientific and political officers was mirrored in other parts of the British Empire. For instance, in India, colonial botanists were fascinated by the uses of plants in local medicine, but this exasperated their political bosses, who considered that such research was not what they were being paid for (Chatterjee, 1948). For its part, the UGFD had a straightforward argument for introducing the Nile perch into Lake Victoria, namely that it would allow the creation of a game-fishing tourist industry and boost the territory's income.

Wary of showing too much enthusiasm for matters of 'pure' scientific interest in political circles, the case made by the fish scientists who opposed the introduction of the Nile perch into Lake Victoria stressed the economic harm that this might cause through disruption of the lake's ecosystem (Pringle, 2005). They predicted that tilapia, the most important commercial fish and one that is largely herbivorous, might prove to be easy prey. As it turned out, the scientists were proved right about the ecological disruption, but wrong in predicting that the economic value of Lake Victoria fisheries would decline – at least in the way that economic value is calculated in conventional economics (i.e. with social and environmental costs excluded) (Hall and Klitgaard, 2012). As the population of the Nile perch exploded, the total value of the commercial fishery haul increased, from ~100,000 tons annually in the 1960s–1970s to ~500,000 tons annually in the late 1980s–1990s (Jansen, 1997).

However, as this happened, the nature of the fishery became transformed from being a small-scale artisanal industry, in which most of the boats and fishing gear belonged to owner-operators, into a technologically modern fishery, involving stronger, more expensive gillnets and controlled by businesspeople with access to capital. Today, much of the Nile perch catch in Lake Victoria is processed in factories and the bulk of the choice cuts (such as fish fillets) frozen and either exported to Europe or sold to more affluent people in Uganda. A report by the Food and Agricultural Organization (FAO) of the United Nations has proposed even further integration of the Lake Victoria fisheries into the global economy, pointing out that there is a good potential to develop exports for the foreign pet food industry (Josupeit, 2006). If this happened it would result in an even larger share of Lake Victoria's biological production being exported for use in the Global North. Keeping pets is popular in the Global North. In the United States, it is estimated that the consumption of animal-derived energy by pet dogs and cats is equivalent to 33% of that of humans, with equivalent environmental impacts on the Earth's ecosystem (Okin, 2017).

The changes made to the fish fauna of Lake Victoria described above have disadvantaged ordinary Ugandans in the same three ways that they became disadvantaged by the decline in indigenous banana diversity. The availability of a favoured and nutritious food has declined, opportunities for monetary income have been lost and their culture has been eroded. During the 1960s, when *enkejje* were still abundant in the lake, many villages situated within about 30 km of its shore were regularly visited by traders on bicycles selling sun-dried *enkejje* caught by artisanal fishermen. The fish were sold at locally affordable prices. These days, *enkejje* are no longer carried around by bicycle traders and they are rarely seen in markets.

Ordinary Ugandans cannot afford to buy the choice cuts of frozen Nile perch that are sold in supermarkets. The Nile perch is too large and oily to be easily sun-dried. What happens to less-choice parts of the fish is illustrated by Figs 2.3 and 2.4, which are photographs taken in the fish landing site of Gaba near Kampala. The part of the landing site shown in Fig. 2.3 is devoted to heat-drying the fish using wood fuel, stacks of which can be seen in Fig. 2.4. The heap in the foreground of Fig. 2.3 consists of rolled-up dried skins of Nile perch. According to the lady who was attending these skins, they are destined for sale in Eastern DR Congo, where, she said, they are considered a delicacy. This use of wood to dry the fish has introduced yet another way in which ordinary Ugandans have become disadvantaged by the introduction of the Nile perch into Lake Victoria. It has added another pressure on the tree resources of the country, which is already suffering from a high rate of deforestation (see Section 5.6).

Fig. 2.3. Fish smoking section of the landing site at Gaba, Lake Victoria. Photo: A.H. (2010).

Fig. 2.4. Stacks of firewood at the fish landing site at Gaba, Lake Victoria. Photo: A.H. (2010).

3

Plants in Human History to 1500 CE

Abstract
A very large number of plant species have been used by people for one purpose or another. One type of use (for food) and one category of food-producing plants (those with starch-storing organs) are used to illustrate how people have used plants. Plants were domesticated and farming invented in several parts of the world independently, starting at about 10,000 BCE. Later, more intensive forms of farming were independently invented in several places. As well as domesticating plants, people have domesticated herbivores with the ability to digest cellulose, which is not something that the human can do. By eating and otherwise making use of these animals, people have further increased the range of habitats they have been able to exploit. Over history, those human populations that acquired more efficient ways of obtaining food fanned out into new places at the expense of those that did not. New political, socio-economic and cultural systems arose along with these developments. The most impressive spreaders of crops prior to 1500 CE were Austronesian-speaking people whose original homeland was in mainland South China. From there, some moved to Taiwan and then Southeast Asia, from where they launched blue-water expeditions that took them and their domesticated plants and animals across the Pacific and Indian oceans.

3.1 People's Uses of Plants

People have collectively made use of a huge variety of plants. A recent compilation for China has revealed that 10,525 (27%) out of the 39,174 species (including infraspecies) of vascular plants considered to be native to the country have been used at some time for one purpose or another (Zhuang *et al.*, 2021). The largest categories of use are medicine (9620 species), food (1647), ornamental (826), industrial raw materials (760), construction (760), fibre (663), oilseed (547), and a category that covers both fodder and veterinary medicine (535). The figure of 27% for the total percentage of the national flora recorded as being used in China is similar to the figures obtained for some other countries in which full inventories have been attempted (Díaz-Forestier *et al.*, 2019). Many studies confirm the large number of plant species that have been used medicinally.

We have chosen one category of use (food) and one subcategory of plants within this – those having starch-storing organs – to provide a little more detail in this overview of human uses of plants. Starch is a carbohydrate, and carbohydrates form one of the three categories of macronutrients present in the human diet, the others being fats and proteins. Starch is the main chemical used by plants to store the energy that they capture from the sun. Therefore, it is not surprising that starch-storing plants have often provided people with their staple foods.

One type of starch-storing body produced by plants is the grain, a small hard seed, physiologically capable of prolonged dormancy. Two categories of grains are recognised, one is the cereals, produced by plants belonging to the grass family (*Poaceae*), and the other, the pseudo-cereals, is produced by other types of plants. (A grass grain is actually a special type of fruit known technically as a caryopsis.) Ecologically, having seeds in the form of grains is a good survival strategy for annual plants living in places that are seasonally too cold or too dry for plants to grow. It enables them to survive the unfavourable seasons in dormant, protected states. Several prehistoric groups of humans, inhabiting different parts of the world, came to realise, independently of one another, that grain-producing plants constitute a very useful resource for their own survival, grains being both nutritious and storable. However, before they could fully profit from these realisations, first they had to develop ways to harvest the seeds in bulk and store them in dry states out of the reach of rodents.

The history of human use of grain-producing plants is relatively well-known because grains are frequently preserved in a carbonised state at archaeological sites and leave characteristic impressions on pottery. The following are among the most prominent types of cereals and their places of domestication (Map 3.1): maize (*Zea mays*) – domesticated in the Mexican Highlands; fonio (*Digitaria exilis*), pearl millet (*Cenchrus americanus*), African rice (*Oryza glaberrima*) and sorghum (*Sorghum bicolor*) – domesticated in the northern savannah zone of Africa; finger millet (*Eleusine coracana*) and teff (*Eragrostis tef*) – domesticated in the Ethiopian Highlands; wheat (*Triticum*), barley (*Hordeum*), oats (*Avena sativa*) and rye (*Secale cereale*) – domesticated in the Fertile Crescent; broomcorn millet (*Panicum miliaceum*) and foxtail millet (*Setaria italica*) – domesticated in the lower catchment of the Yellow River in China; and Asian rice (*Oryza sativa*) – domesticated in the lower catchment of the Yangtze River in China. (The Fertile Crescent is a crescent-shaped region in the Middle East, extending from Israel/Jordan to Western Iran.) The pseudo-cereals, quinoa (*Chenopodium quinoa*) and buckwheat (*Fagopyrum*), were domesticated in the Andes and on the Qinghai–Tibet Plateau, respectively.

It is known that, both in the Fertile Crescent and China, cereals were fermented to make alcoholic drinks from very early on in agricultural history and even slightly before. Breweries, dating back to 11,000 BCE, are known from the Fertile Crescent, wheat and barley being the cereals fermented, and, back to 7000 BCE in China, where the fermented cereal was broomcorn millet (Liu Li *et al.*, 2018, 2019). Alcoholic drinks, brewed in traditional ways, are drunk on a daily basis in some modern agricultural societies. These drinks

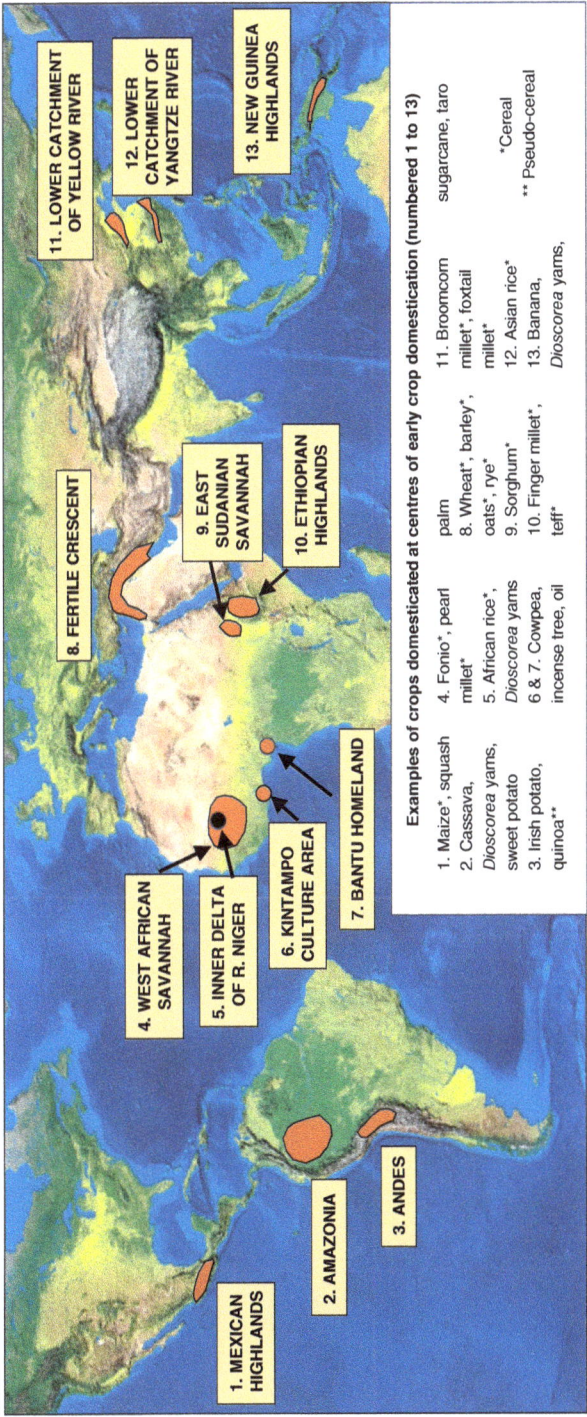

Examples of crops domesticated at centres of early crop domestication (numbered 1 to 13)

1. Maize*, squash
2. Cassava, *Dioscorea* yams, sweet potato
3. Irish potato, quinoa**

4. Fonio*, pearl millet*
5. African rice*, *Dioscorea* yams
6 & 7. Cowpea, incense tree, oil palm

8. Wheat*, barley*, oats*, rye*
9. Sorghum*
10. Finger millet*, teff*

11. Broomcorn millet*, foxtail millet*
12. Asian rice*
13. Banana, *Dioscorea* yams,

sugarcane, taro

*Cereal
** Pseudo-cereal

Map 3.1. Centres of early crop domestication.

differ from modern industrially produced wines and beers in being rich in plant residues. They are nutritious (Rawat et al., 2021) and can have the additional health-promoting benefit of sometimes containing lower loads of pathogenic organisms than the local drinking water. When drunk in company, they can help lubricate social intercourse through providing regular meeting points where (typically) men assemble at the end of the day to discuss local affairs and cement relationships.

Apart from grains, there are other types of starch-containing seeds and fruits with decay-resistant coats that are commonly found at archaeological sites. One of them is hazel (*Corylus avellana*), a small forest tree native to temperate Western Eurasia. Hazel shells can be common at Mesolithic habitation sites in the British Isles. The Mesolithic people, who were hunter-gatherers and occupied the British Isles during the earlier part of the postglacial period (9600–4300 BCE), were preceded by the Palaeolithic people, who were present during the last ice age and earlier, and were succeeded by the Neolithic people, who introduced agriculture. At Mountsandel near Coleraine in Northern Ireland, there is a very early Mesolithic habitation site dating to c.7000 BCE, which A.H. has investigated as a palynologist (Hamilton, 1985). Hazel shells were common at the site, which also contained abundant evidence of the rest of their diet, which included wild boar and two types of fish, salmon and trout, both of which live mostly in the sea, but return to freshwater rivers to breed (Woodman, 1985).

The Mountsandel site dates to the time during the postglacial when the first, faster-spreading, trees (such as hazel) had arrived in Ireland from their ice age refugia in Southern Europe, but before slower-spreading trees (such as oak – *Quercus petraea*) had managed to do so. In Europe, there is much evidence that the acorns (fruits) of oaks were a regular feature of the Mesolithic diet. One difference between consuming hazel nuts and consuming acorns is that acorns need to be detoxified before they are safe for people to eat. *Quercus*, the oak genus, is very widely distributed around the North Temperate climatic zone. There is only one native species in Ireland (*Q. petraea*) and only two in the British Isles as a whole (also *Quercus robur*), but altogether there are 600 species of oaks in the world, these varying in stature from shrubs to large trees. The two principal centres of global diversity for oaks are China and Mexico/ the United States. In both, pre-agricultural or very agricultural people included acorns in their diet.

In China, flour made from acorns (and perhaps also from the nuts of the related genera, *Cyclobalanopsis* and *Lithocarpus*) formed a major ingredient in the diet of the Peiligang, a people living in the upper catchment of the Yellow River at 7000–5000 BCE (Liu Li et al., 2010). The Peiligang had a culture that was transitioning from hunter-gathering to farming and, as with the Mesolithic people at Mountsandel, the people were also eating wild boar. In the United States, the Ahwahnechee, the people who were living at Yosemite when the California gold rush began in the mid-19th century CE, were also grinding up acorns to make flour. The oak in question was the California black oak (*Quercus kelloggii*) (Fig. 8.5). The grinding was done in depressions on exposed rocks which, over time, became enlarged into pits (Fig. 3.1). The Ahwahnechee lacked agriculture but manipulated populations of the black oak using fire to

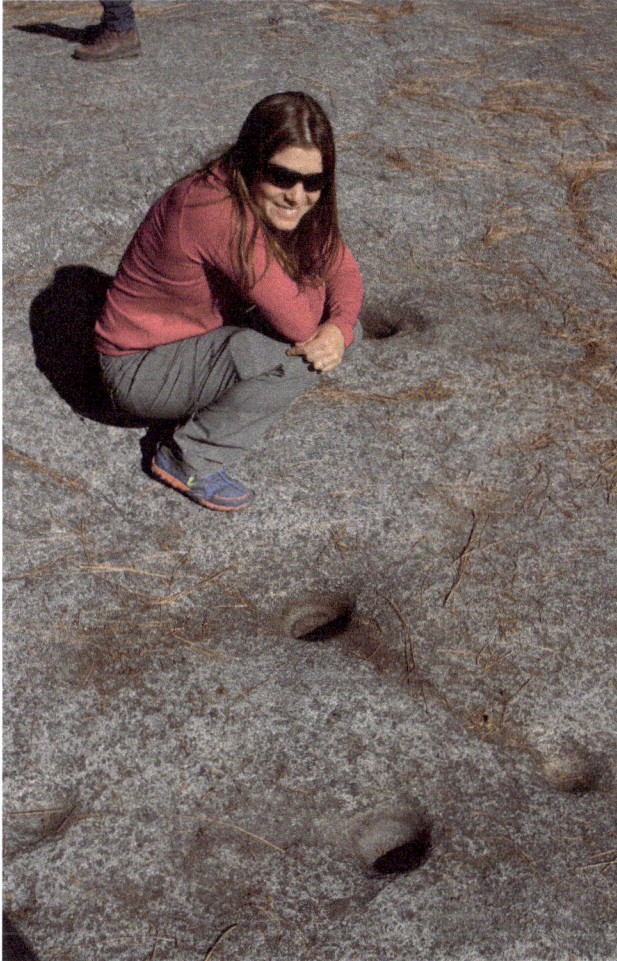

Fig. 3.1. Pits created by grinding up acorns, Yosemite. Photo: A.H. (2013).

increase the yields of acorns (see Section 8.2.2). The Ahwahnechee ate many species of wild animals, but not pigs, which were absent from the American fauna until introduced by Europeans in the early 16th century.

For the most part, the histories of human use of plants that store starch in ways other than as grains or nuts are poorly known, having lower visibility in the archaeological record. However, a better understanding is starting to emerge through studying the starch grains, phytoliths and chemical markers that some of these plants leave behind (Fullagar *et al.*, 2006). Classified according to their types of starch-storing organs, some of the most important of these plants are: using underground parts to store starch – yams (*Dioscorea*), cassava (*Manihot esculenta*), sweet potato (*Ipomoea batatas*), Irish potato (*Solanum tuberosum*), Livingstone potato (*Coleus esculentus*), arrowroot (*Maranta arundinacea*), tannia (*Xanthosoma sagittifolium*) and taro (*Colocasia esculenta*); using stems to store starch – sugarcane (*Saccharum officinarum*), the sago palm (*Metroxylon sago*) and a cycad (*Cycas revoluta*); using young shoots to store starch – some species of bamboos (which belong to the grass family); using leaf sheaths to store starch – ensete (*Ensete ventricosum*) and the golden lotus banana (*Musella lasiocarpa*); and using fruits to store starch – squashes (*Cucurbita*), the breadfruit (*Artocarpus altilis*), the jackfruit (*Artocarpus heterophyllus*) and the banana (*Musa*).

3.2   The Transition to Agriculture

During the last 12,000 years, human societies in several parts of the world have, independently of one another, domesticated food plants and transitioned from hunter-gathering to farming (Map 3.1). The places where this happened are sometimes known as Vavilov Centres, in honour of the Russian geneticist and conservationist, Nicolai Vavilov (see Section 6.2.2). The transition to farming was probably sometimes triggered by climatic change. For Western Eurasia and North Africa, this is suggested by two sets of coincidences. One coincidence is that the date of about 10,000 BCE, which is when the first farming (anywhere) started (in the Fertile Crescent), is also the date when the global climate switched from a glacial to an interglacial climatic mode. The other coincidence is that the date of about 3000 BCE, which is when the first farming started in sub-Saharan Africa, is also the date when the climate of Northern and tropical Africa became drier (see Section 15.4.1). The domestication of plants is not something that only happened in the past. The home gardens of the Baganda contain a range of species lying along the spectrum from fully domesticated to fully wild (see Section 7.2).

Apart from consuming plants directly as food, people also consume them indirectly through their consumption of herbivorous animals. Among the species of herbivorous animals that people have domesticated during the last 10,000 years are a number of ruminant and pseudo-ruminant mammals. The ruminants, which are characterised by having an extra stomach to assist in the digestive process, include cattle, the yak, goats, sheep and water buffalo, all of which were first domesticated in the Old World. The pseudo-ruminants, which lack the extra stomach, include (in the Old World) two species of camel, the ass and the horse, and (in South America) the llama and alpaca. The domestication of cattle from their wild ancestor (the aurochs) occurred in two places independently of one another, namely the Greater Indus Valley Region and the Fertile Crescent. Later, cattle domesticated in the Middle East interbred with wild aurochs in Northeast Africa (Pérez-Pardal *et al.*, 2010; Upadhyay *et al.*, 2016).

Both ruminants and pseudo-ruminants possess the ability to digest cellulose, a type of carbohydrate which is abundant in plants, but cannot be digested by humans. For this reason, owning domesticated ruminants and pseudo-ruminants can be tremendously advantageous to people who live in ecosystems where food for these animals is abundant, but scarce for humans. The food products provided to humans by domesticated ruminants and pseudo-ruminants include meat, fat, milk and blood. Products manufactured from these animals include coverings and leather (made from hides), straps and strings (made from strips of hide or guts), textiles (made from wool) and glue (made from bones and other parts).

People living in the Fertile Crescent acquired four typical features of a farming way of life during the 3000-year period between 10,000 and 7000 BCE. These were domesticated plants, domesticated animals, pottery and permanent settlements (Gibbs, 2015). The cereals that they grew at first were wheat and barley, their legumes included the broad bean (*Vicia faba*), chickpea (*Cicer arietinum*) and lentil (*Vicia lens*), and their domesticated ruminants were cattle, goats and sheep.

To the west of the Fertile Crescent, in North Africa, a much wetter climate than now prevailed during the earlier part of the postglacial (9000–3000 BCE), a period that archaeologists refer to as the Green Sahara (Tierney *et al.*, 2011, 2017). During the first part of the Green Sahara, hunter-gatherers preyed on the wildlife that abounded there and ate a variety of wild plant foods, among them the grains of various savannah grasses, including species of *Echinochloa*, *Panicum* and *Setaria* (Dunne *et al.*, 2016). Later, pastoralists with herds of cattle and flocks of sheep moved into the area where the Sahara Desert is now situated. However, the people continued to depend on wild plant foods, still lacking cultivated crops. Pots were used to prepare meals from plants, the first place in the world where people are known to have done so (Dunne *et al.*, 2016).

The transition to a drier climate in North Africa at *c*.3000 BCE caused the Sahara Desert to expand and pushed the pastoralists south. It is suspected that this caused the people to start exploiting some types of plants more intensively, leading eventually to their domestication (Dunne *et al.*, 2016; Grillo *et al.*, 2022). It has been proposed that, in the case of one of these species, the cowpea (*Vigna unguiculata*), the process of domestication began when herders started to uproot entire plants to feed to their cattle (Ng, 1995). Through selecting certain variants of the cowpea for this purpose, it is thought that, over time, the result was the evolution of the fully domesticated cowpea, which differs from the wild version in having non-shattering pods, large seed size, large pod size and a lengthened period of seed dormancy.

It is interesting that one of the clans of the Baganda has the domesticated cowpea (*mpindi*) as its totem and the wild cowpea (*kiyindiru*) as its secondary totem (see Section 8.1). Traditionally, the members of the cowpea clan were responsible for taking care of the king's cattle and, uniquely, their clan is the only one whose birth rites do not involve a role for the banana plant (see Section 7.2) (Roscoe, 1911). Instead of the ritual of placing the afterbirth at the foot of a banana plant, the members of this clan bury it within their houses close to their doors. This suggests the retention of a cultural feature of great antiquity, a hangover from before the time of arrival of the banana in Africa, which was at about 500 CE (see Table 3.1).

Early centres of crop domestication are known from both sides of North Africa (Map 3.1). The crops domesticated in the Niger River Basin included the African yam (*Dioscorea cayenensis*), which was domesticated from a forest species (*Dioscorea praehensilis*), African rice, pearl millet, African

Table 3.1. Origins and dates of introduction of crops into Uganda. Main sources of information: Ehret (2011); Hamilton *et al.* (2016); Neumann *et al.* (2022).

Origins of crops	Approximate date of first arrival in Uganda	Crop types
Possibly domesticated locally		Ensete, robusta coffee. The incense tree can be regarded as semi-domesticated in Uganda; it may be native. It is possible that the oil palm is indigenous in the Bwamba area of Western Uganda
Crops domesticated in Africa (introduced by Nilo-Saharan speakers)	Second millennium BCE	Sorghum, finger millet, Livingstone potato, Bambara groundnut, cowpea, sesame; probably pearl millet
Crops domesticated in Africa (introduced by Bantu speakers)	Early 1st millennium BCE	*Dioscorea* yam, castor bean, edible gourd, bottle gourd; possibly cowpea and pearl millet
Asian crops brought to the east coast of Africa by Austronesian speakers	About 500 CE	Bananas of East African Highland and Plantain genome groups, Asian rice, sugarcane, taro, bud yam; possibly red types of sweet potato
American crops transported across the Atlantic Ocean by Europeans; present in Uganda prior to the arrival of the first Europeans	About 1800 CE (tobacco possibly earlier)	Tobacco, kidney bean, maize, peanut, sweet potato (white types, at least), cocoyam
Crops introduced during the colonial era or bred locally	From 1894	A large number of crops were introduced during the colonial era, including dessert bananas, the Irish potato and cassava. Cassava was probably introduced by the British as a famine crop

potato (*Coleus esculentus*) and cowpea (D'Andrea *et al.*, 2007; Burgarella *et al.*, 2018; Cubry *et al.*, 2018; Scarcelli *et al.*, 2019). Taxonomically related species of *Coleus* with edible tubers are known from other parts of Africa and are known respectively as Ethiopian, Sudan and Madagascar potato (Titus *et al.*, 2023). The process of domesticating African rice resulted in the cultivars diverging along two evolutionary lines, one leading to the modern wetland types and the other to the dryland. An archaeological excavation of a site in Ghana, dating to about 1600 BCE, has shown that the Kintampo, the people who lived there at the time, were also eating the fruits of the oil palm (*Elaeis guineensis*) and the incense tree (*Canarium schweinfurthii*) (D'Andrea *et al.*, 2007). On the other side of Africa, the crops domesticated post-3000 BCE in the East Sudanian/ Ethiopian region included the cereals teff, finger millet and sorghum

(Winchell *et al.*, 2017). Wheat, barley and lentil had been introduced from the Fertile Crescent into North Ethiopia by the 1st millennium BCE (Harrower *et al.*, 2010).

Agriculture within a tropical forest setting has been practised for ~7000 years in New Guinea, ~6000 years in Amazonia and ~2500 years in Africa (Fullagar *et al.*, 2006; Neumann *et al.*, 2012, 2022; McMichael *et al.*, 2015). The agricultural practices of these prehistoric societies have been reconstructed from archaeological research combined with ethnographic studies of people who are believed to be following similar ways of life (Rappaport, 1984). According to reconstructions, their agricultural techniques included leaving trees of valued species standing when clearings were made by slash-and-burn in the forests, planting crops in the clearings, enriching areas of forest with valued types of trees and leaving some areas of virgin forest untouched as intact stands.

Prehistoric agriculturalists in tropical forests have left enduring marks on modern forest ecosystems. Studies in Amazonia have revealed that pre-Columbian earthworks are extensive throughout the Amazonian rainforest and that the activities of the people associated with them have left a lasting impact on the floristic composition of the forests (Peripato *et al.*, 2023). Fertile *terra preta* soils are widespread in Amazonia, the creations of generations of agriculturalists (McMichael *et al.*, 2015). At Ituri in Eastern DR Congo, studies of fossil charcoal contained in soils under rainforest have revealed that the floristic composition of the forests became transformed from *c.*350 BCE by the activities of agriculturalists (Hart *et al.*, 1996). (The locality of Ituri is shown on Map 3.2.) Fourteen of the 36 woody species identified in the charcoal are no longer present in this part of the Congolese forest. One of the species that has replaced them is the large tree *Gilbertiodendron dewevrei*, which is common in the forests today, but is completely unrepresented in the charcoal. Sediment analyses have revealed that early agriculturalists also caused substantial alterations to the floristic composition of forests in Uganda, as well as increasing the rates of soil erosion (Lejju *et al.*, 2005; Hamilton *et al.*, 2016). The siltation encouraged the growth of the giant swamp herbaceous plants, bullrush (*Typha*) and papyrus (*Cyperus papyrus*).

3.3 The Spread of Crops

3.3.1 Africa

Historically, the substitution of a hunter-gathering with an agricultural way of life has been followed by the numbers of people multiplying and expanding their areas of occupation. One such episode of population growth and spread began in the borderland between Nigeria and Cameroon at *c.*3000 BCE among a people who spoke Proto-Bantu, the ancestral language of the 400 or so modern languages belonging to the Bantu subgroup of the Niger–Congo language family (Nurse and Philippson, 2003). According to linguistic and archaeological evidence, the crops grown by the early Bantu speakers included *Dioscorea* yam, cowpea (*Vigna unguiculata*), Bambara groundnut (*V. subterranea*), the castor bean (*Ricinus communis*), the edible gourd (*Cucumeropsis mannii*) and the bottle gourd (*Lagenaria siceraria*) (Ehret, 1998; Oslisly and White, 2007). They also ate the fruits of the oil palm (*Elaius guineensis*) and the incense tree (*Canarium schweinfurthii*), both of which may have been wild or perhaps semi-cultivated. They knew how to make pottery, drums, fishhooks and boats, and possessed the domesticated goat.

Some of the Bantu speakers spread out from their Nigeria/Cameroon homeland into the Congo rainforest and established commensal relationships with the hunter-gatherer 'pigmy' people who lived there (Turnbull, 1961). They intermarried with them and exchanged their agricultural produce with products from the forest. Research carried out during the 1990s in Bwindi Impenetrable Forest, an eastern extension of the Congolese rainforest in Uganda, revealed the depth of knowledge about the plants and animals in the forest that these hunter-gatherers still possess, including about the *Dioscorea* yams that they eat and the stingless bees that provide them with honey (see Chapter 18).

Some descendants of the Bantu speakers who entered the Congolese forest emerged into the savannah belt on its southern side and fanned out, eventually becoming the dominant ethno-linguistic group over much of Equatorial and Southern Africa (Map 3.2). As this great wave of expansion occurred, the languages of their geographically diverging subpopulations became increasingly different from one another. Bantu-speaking people arrived with their crops at *c.*500 CE in South Africa, where they encountered another people who had settled there 500 years earlier and who possessed cattle and sheep which their distant ancestors had brought down with them from Northeast Africa (Henn *et al.*, 2008).

Bantu-speaking people arrived in the neighbourhood of Lake Victoria during the period 1000–500 BCE. On their arrival, they encountered another agricultural people already living there, but speaking languages of the Nilo-Saharan language family. Loan words in Luganda, a modern Bantu language spoken in the northwestern hinterland of Lake Victoria, show that the Baganda's linguistic ancestors acquired the knowledge of how to cultivate sorghum and finger millet, keep cattle and how to smelt iron from these Nilo-Saharan speakers (Ehret, 2011).

Within the western part of East Africa, Rwanda is known to have been a major centre of iron-smelting during the 1st millennium BCE (Van Grunderbeek *et al.*, 1982; Van Grunderbeek and Roche, 2007). For Uganda, pre-colonial iron-smelting centres are known from the Masindi and Mwenge areas of the Kingdom of Bunyoro-Kitara, and in the Kyaggwe and the Rakai/Masaka areas close to Lake Victoria (Map 15.1) (Iles, 2009). It is of interest to note, in the context of the present book, that the Iron Age began during the 1st millennium BCE in all three of the countries that we have chosen for special attention (China, Britain and Uganda). It is possible that the discovery of how to smelt iron was first made in sub-Saharan Africa (Killick, 2015; Humphris and Scheibner, 2017).

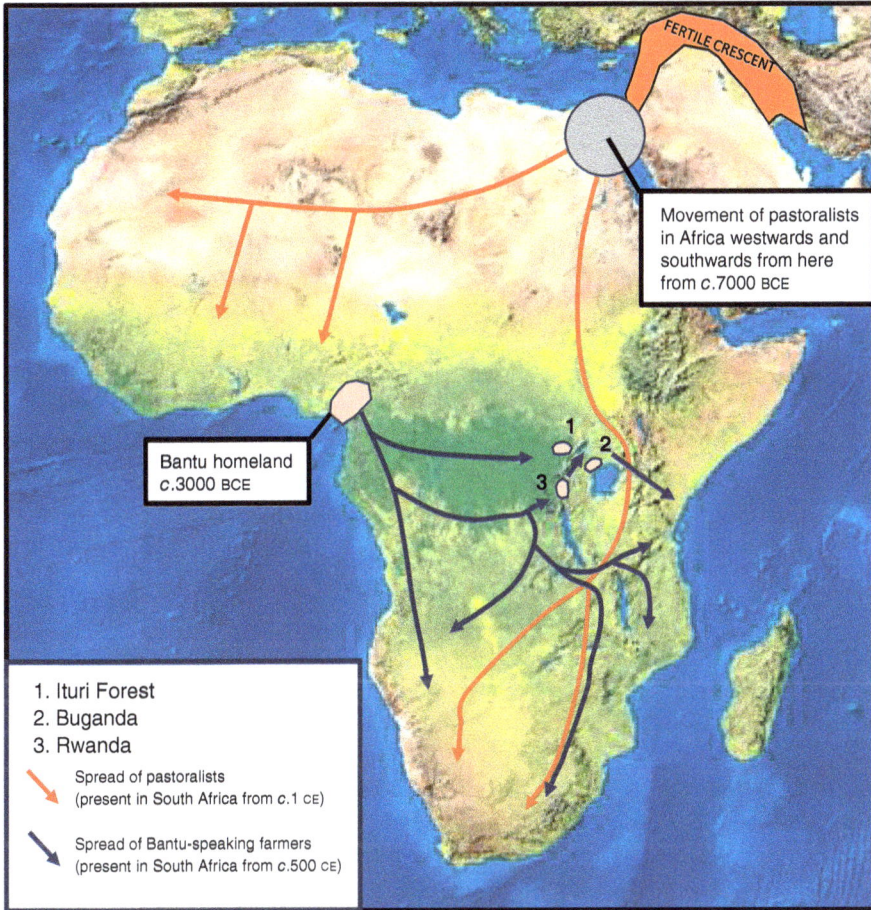

Map 3.2. Expansion of pastoralism and of Bantu agriculturalists in Africa.

Research into the dates of acquisition of crops by the ancestors of the Baganda reveals that the overall pattern of their plant-resource system has existed for a long time, although sometimes with one species being substituted by another (Table 3.1). Two crops that may have been domesticated locally (as well as elsewhere) are ensete (*Ensete ventricosum*) and robusta coffee (*Coffea canephora*), both of which are thought to be indigenous to Uganda. Both of them have deep symbolic significance in traditional Kiganda culture (see Section 7.2). Pollen grains of sorghum and finger millet have been found in Early Iron Age archaeological contexts in Rwanda, dating to 100–500 CE. This shows that, by that time, people were growing these cereals in the Great Lakes equatorial region (Roche, 1996; Van Grunderbeek and Roche, 2007). Both were originally domesticated further north in Africa.

When crops of Asian origin arrived in about 500 CE, plantains and cooking bananas extensively replaced African yams (*Dioscorea*) in the diet and, possibly, Asian rice replaced African rice. When crops of American origin arrived from about 1800, they extensively replaced crops of African origin as follows: American beans (*Phaseolus*) replaced the cowpea; the peanut replaced the Bambara groundnut; and maize replaced sorghum and finger millet. During the colonial era, *akafumbo* (*Gomphocarpus physocarpus*) became extensively replaced by *ppamba* (cotton) as a source of hairs for stuffing pillows and the seeds of ensete (*ekitembe – E. ventricosum*) by the seeds of an American species (*omupiki – Sapindus saponaria*) as the preferred counters for playing the ancient African board game of *omweso* (Hamilton, 2020).

Studies of the pollen contents and other properties of sediments in Uganda have revealed signs of localised clearings being made in the rainforest, presumably for agriculture, from about 1700 BCE (see Section 15.4.2). However, much more prominent in the sediment record is a major environmental transition, dating to about 1000 CE, marked by large-scale forest clearing and increased soil erosion. Archaeological research has revealed other things that happened at about 1000 CE. They were the replacement of a pottery type known as Urewe ware by one known as rouletted ware, the replacement of the Early Iron Age by the Later Iron Age and the construction of large, earth-banked enclosures in the savannah zone of Western Uganda.

According to oral tradition, the ruling dynasties of the pre-colonial kingdoms of Bunyoro-Kitara and Buganda can be traced back to about 1000 CE (Roscoe, 1911; Dunbar, 1965; Nuwagaba, 2014), which strongly suggests that the environmental and cultural transitions at that time, as seen in the archaeological record, were related to the formation of more hierarchical societies. One of the things that is thought to have happened is the adoption of new, economically more productive forms of food production (Schoenbrun, 1993a,b, 1998). For the incipient Kingdom of Bunyoro-Kitara, whose homeland is in the savannah zone of Western Uganda, the sedimentary record suggests that the boost to the economy was due to the adoption of large-scale cereal-growing and cattle-keeping (Lejju *et al.*, 2005, 2006). For the incipient Kingdom of Buganda, which originated in a wetter, naturally forested, area near Lake Victoria, two factors have been suggested. One was the adoption of a perennially productive banana-dominated type of home garden, known as the *lusuku* (see Section 7.2). The other was the ability to access resources and control trade along an extensive part of the shoreline of Lake Victoria, due to the people having acquired the art of constructing fast, paddle-driven, boats (Schoenbrun, 2021). The same canoe transport was used by the first British Christian missionaries to travel to Uganda in the 19th century (see Section 6.8.1). The traditional religious practices of the Baganda have strong connections with the lake (Roscoe, 1909).

The closeness of Lake Victoria to the traditional identity of the Baganda is apparent in the name given to the lake in Luganda, which is *Nnalubaale*, which can be translated into English as 'mother of the gods'. (*Nnalubaale*, is a compound word, composed of the noun, *lubaale*, and the prefix, *nna-*; *lubaale* means 'god' or 'deity'; *nna-* is an honorific prefix given to women.) The deity most associated with the lake is Mukasa, who is considered to be the life-giver and bringer of crops and children. There are a number of important ritual sites on the Ssese Islands, which are situated in the lake close to its shore and which the Baganda consider to be part of their heartland. They include a forest containing the tree that is used to fashion the mace (*ddamula*) that gives authority to the Baganda's traditional prime minister (*katikkiro*). (The verb *-ddamula* is concordant with the verb *-lamula*, which means 'judge' or 'arbitrate'.) Several of the Kiganda clans have connections with the lake, including the *nvuma* clan, the totem of which is the fruit of the water chestnut (*Trapa natans*) (see Section 8.1).

Both Bunyoro-Kitara and Buganda engaged in trade. Glass beads and a copper bangle, dating to about 900–1200 CE, have been recorded during archaeological research at Munsa, one of the large enclosure sites in the savannah zone of Western Uganda (Robertshaw *et al.*, 1997). Chemical analysis of the beads indicates an Asian origin, and the copper is suspected to have come from Katanga, which is situated 2500 km to the south. The Baganda are known to have been engaged in trade within their region, for instance importing iron tools and salt from places that were under the control of the Kingdom of Bunyoro-Kitara (see Map 15.1) (Schoenbrun, 1993a,b, 2021; Iles, 2009). There is a large pre-colonial kaolinite mine at Ttanda in Buganda, whose size suggests that some of its produce may have been destined for external trade. It is possible that the kaolinite was used for making tuyères (pipes of heat-resistant material used for blowing air into iron-smelting furnaces). Ttanda is an important place mythologically for the Baganda and a pilgrimage site for modern Baganda interested in learning more about their religious heritage (see Section 6.8.4).

3.3.2 Eurasia

Analysis of ancient human DNA shows that, at 10,000–7000 BCE, two populations, situated on either side of the Fertile Crescent, in Israel/Jordan and in Western Iran, were involved in the domestication of plants and animals, and became the first farmers (Reich, 2018). The two populations were as genetically distinct from one another as modern-day Europeans are from modern-day East Asians – findings that form part of the evidence that has led to the general conclusion that population movement and population mixing have been two of the outstanding features of human evolution during the last 50,000 years (see Section 2.2.1). Those farming people who spread from the western end of the Fertile Crescent across Europe arrived in the British Isles in about 4300 BCE, where their arrival was archaeologically associated with the arrivals of pottery-making, the building of monumental architecture and the making of compound (Neolithic) tools. The fact that these new technologies all arrived together gave rise to the idea among early British archaeologists that the transition from hunter-gathering to farming always involved the simultaneous adoption of a set of new technologies (known as 'the Neolithic package'). However, elsewhere in the world, these technologies did not necessarily arrive together. Any one of them could arrive first.

The story of the introduction of agriculture into large parts of sub-Saharan Africa by Bantu-speaking people appears to have been comparatively straightforward, in that farming and speaking Bantu languages spread together in a single cultural wave. In contrast, the stories of Western Eurasia and of Southwestern to Southern Asia are complicated by the intrusion, at *c*.3000 BCE, of another people, the Yamnaya, whose ancestral home was in the steppe region to the north of the Black and Caspian seas (Reich, 2018). These people, who spoke languages belonging to the Indo-European language family, had more socially stratified societies than did the established farmers that they encountered as they spread. They also knew how to work with bronze, how to use oxen to draw ploughs and carts, and had the domesticated horse (which was used for riding, drawing chariots, milking and eating). Today, languages belonging to the Indo-European language family are the most widely spoken languages in Europe and the northern part of the Indian subcontinent (both English and Hindi are Indo-European languages). When the descendants of the Yamnaya arrived in Britain at about 2500 BCE, they initiated its Bronze Age, which lasted until the start of the Iron Age at about 800 BCE.

Farming was invented independently in two places in China, both at about 6000 BCE (Map 3.1). One was in the north, in the lower catchment of the Yellow River, where the cereals foxtail millet and broomcorn millet were domesticated by people speaking languages of the Sino-Tibetan language family. The other was in the south, in the lower catchment of the Yangtze River, where Asian rice was the cereal domesticated. As was the case with African rice, the ancestor of Asian rice (*O. sativa*) was a swamp grass living in seasonally flooded marshlands along the sides of rivers. Also, as with African rice, the process of domesticating Asian rice resulted in the cultivars diverging along two evolutionary lines, one leading to the modern wetland types and the other to the dryland (Gong *et al.*, 2007; Li Chunhai *et al.*, 2012). Figure 3.2 is a photograph of paddy fields on hillsides in Honghe Hani and Yi Autonomous Prefecture, South Yunnan, China, where, it is thought, this complex system of ponds and channels has been in existence for over 1300 years. It is recognised by UNESCO as a World Heritage Site. Many landraces of Asian rice are grown in these paddies, including common rice, and red, black and sticky rice. The growing of rice by the Hani forms part of an integrated farming system that involves keeping buffalo, cattle, duck and fish. Palms (*Trachycarpus fortunei*) were traditionally planted between the paddy fields, their hollowed-out trunks being used to store rice straw for feeding to buffalo.

With population growth and the passage of time, the people associated with the development of early farming in North China spread towards the south and those associated with the development of early farming in South China spread towards the north. Where they met, they intermarried, the result being that, as of today, a longitudinal gradient of genetic variation can be detected among the Han people of China (Reich, 2018). As well as spreading south, the early farmers in North China spread west on to the lower slopes of the Qinghai–Tibet Plateau, where, at about 3200 BCE and at altitudes of up to ~2500 m, they were growing foxtail and broomcorn millet and keeping pigs (Guedes *et al.*, 2013; Chen *et al.*, 2015). Subsequently, at about 1500 BCE, Sino-Tibetan-speaking populations introduced farming on to the Qinghai–Tibet

Fig. 3.2. Hillside paddy fields in South Yunnan, China. Photo: Mrs Xu Yinfeng (2019). Reproduced with permission.

Plateau, which, with an average elevation of 4000 m, presented a new set of livelihood challenges.

The response of the incoming farmers from China to living on the Qinghai–Tibet Plateau was to switch to a radically different type of agricultural economy, based on buckwheat, the yak and a suite of plants and animals that had originally been domesticated in the Fertile Crescent. Buckwheat was domesticated locally and so too, most likely, was the yak, a long-haired relative of the American buffalo (Wu *et al.*, 2010; Qiu *et al.*, 2015). The domesticates introduced from the Fertile Crescent included barley, wheat, the goat and the sheep. The invention of this unique farming system laid the foundation for subsequent political and cultural developments in Tibet, which included the development of a Tibetan Empire, which once extended into the lowlands, well beyond the plateau. King Songtsen Gampo (ruled *c.*618–650 CE) is credited with expanding the Tibetan Empire, as well as with introducing Buddhism into Tibet. He may have been the person responsible for conquering the Kingdom of Zhangzhung (see Section 3.4).

3.3.3   Southeast Asia and the Pacific and Indian oceans

The farmers who moved from the Yellow River southwards carried along with them their Sino-Tibetan languages, which eventually displaced the Austronesian languages that are believed to have been spoken by the people responsible for domesticating Asian rice (Reich, 2018). The Austronesian language family is no longer represented in China. However, Austronesian speakers did not disappear, because, the evidence suggests, they moved to Taiwan at about 3000 BCE, where they mastered the art of building true ocean-going ships – the first people in the world to do so – and then journeyed on, arriving first in the Philippines and then in Indonesia, out of which they launched blue-water expeditions into the Pacific and Indian oceans (Map 3.3). The technological accomplishments of these early Austronesian-speaking people made them by far the most impressive spreaders of crops around the world prior to the exploits of Europeans after 1500 CE.

Austronesian speakers were the first people to inhabit islands in the Pacific Ocean, such as the Marquesas, Hawaii, Easter Island and New Zealand. It is known from ethnographic evidence that Austronesian sailors took domesticated plants and animals with them on their voyages, enabling them to found new well-provisioned colonies (Suggs, 1951). Their main food plants included yams (*Dioscorea alata* and *D. esculenta*), taro (*Colocasia esculenta*), breadfruit (*Artocarpus altilis*), sugarcane (*Saccharum officinarum*) and Asian rice (*Oryza sativa*). Their domesticated animals were the pig, chicken and dog. At one stage, these people touched on the coast of South America, where they picked up the sweet potato (*Ipomoea batatas*), which they then proceeded to introduce to the Marquesas and other islands in the Pacific (Rull, 2021; Ioannidis *et al.*, 2020). There is a possibility that, from the western fringes of the Pacific, they carried sweet potatoes further west and introduced them to Africa (see Addendum to Chapter 1).

During the first centuries CE, the westward expansion of the Austronesian-speaking people took them to Madagascar, where they became the first people ever to live there. They introduced crops of Asian origin, such as dryland rice, and caused the extinction of many larger species of terrestrial animals (see Chapter 1). Figure 3.3 is a photograph of dryland Asian rice growing among felled trees on land recently cleared of forest through slash-and-burn near Ambodiscakoana Village, Manongarivo, Madagascar. Landing on the east coast of tropical Africa at about the same time, the crops they introduced are believed to have included Asian rice, sugarcane, taro, probably red sweet potatoes and ancestral stock of the Plantain and East African Highland genome groups, both of which have evolved within Africa (see Section 2.3.1).

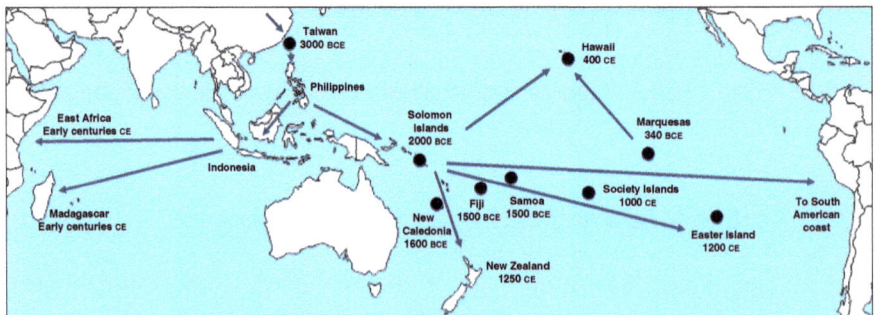

Map 3.3. Spread of Austronesian speakers. The dates shown are those of first arrivals. Note regarding the positions of country boundaries: see Disclaimer.

Fig. 3.3. Dryland Asian rice in Madagascar. Photo: A.H. (1993).

There is an ancient ritual involving sugarcane in Buganda that is performed as part of the ceremony marking the accession of a new king. This is a mock battle fought between the prince who has been designated as the next monarch and the spiritual guardians of Buddo Hill, which is where the coronation ceremonies are performed. The thin stems of a special type of sugarcane (known as *ekirumbirumbi*) are used in the battle, the thinness of its stem being reminiscent of that of the stem of the New Guinean swamp grass that was the ancestor of the modern thick-stemmed varieties used for the production of crystalline sugar (see Section 5.3.1) (Daniels and Daniels, 1993). (The word, *ekirumbirumbi*, is composed of the reduplicated verb, *-lumba*, which means 'attack' or 'assault', plus the prefix, *eki-*, which means 'the thing of'. Reduplication of a stem in Luganda indicates that it has a new or modified meaning.) This linguistic evidence supports the proposition that sugarcane has been a long-standing member of the plant-resource system of the Baganda.

3.4 Agricultural Intensification, Social Stratification and Urban Civilisations

Archaeological research, backed up by ethnographic studies of modern hunter-gatherers, suggests that, except in the most bountiful places, prehistoric hunter-gatherers led mobile lives, moving in small bands around their territories and tapping into food and other natural resources as they became seasonally available (Chaseling, 1957; Leacock and Lee, 1982). It was customary for the bands to come together occasionally in larger gatherings – events that allowed them to strengthen their wider social networks, such as through arranging marriages. Social roles were ascribed to people on the basis of characteristics over which individuals have no control – biological gender, age and position within families. Apart from this, societies were egalitarian, since all in a band faced much the same challenges of survival together.

Early farmers lived in societies that were similarly small in scale and, likewise, required their members to have a good knowledge of the types of plants found within their territories. However, there was now a need for more permanent settlements, since a continuing guard had to be maintained on their crops, food stores and homes. Greater occupational specialisation developed, along with firmer social stratification and an increased emphasis on private property (Price and Fienman, 2010; Bowles and Choi, 2013; Perret *et al.*, 2017).

People's ways of life continued to change as methods of acquiring food intensified, populations grew and settlements expanded. Along with this came more complex types of social organisation – states and empires – characterised by having even greater occupational specialisation and social stratification, and in being multi-ethnic and multilingual (Crone, 1989). The affairs of the states and empire were now the responsibilities of elites, some of whom specialised in matters such as ruling, administration, engineering and record-keeping. Meanwhile, the lives of the ordinary people were little different in some ways from those of their forebears who lived in small-scale societies. Most worked in agriculture and had identities rooted in local communities. On occasion, they could be called upon to work for the affairs of the state, such as to construct monuments to honour the gods and rulers, maintain roads and serve in the army. New faith traditions emerged, attuned to meeting the psychological needs of people under their new political and socio-economic circumstances. Rulers were identified as the representatives of the deities and charged with maintaining justice and social order, while the ordinary people were required to play out the social roles to which they had been assigned by destiny.

Taking 'civilisation' to be a populous, socially stratified and urban-centred society (*civitas* = 'city', in Latin), the earliest known civilisations in the Old World were the Sumerian, Egyptian and Indus Valley civilisations, all of which came into existence at about 3000 BCE. The earliest civilisation in South America is thought to have been the Norte Chico Civilisation in north coastal Peru (in existence, *c.*3700–1800 BCE) (Solis *et al.*, 2001) and the earliest in Central America that of the Olmec in Southern Mexico (*c.*1200–400 BCE). The civilisations existing in Eurasia and North Africa at the beginning of the 1st century CE included (from east to west) the empire of the Han Dynasty in China, the Zhangzhung Kingdom in central to west

Tibet, the Satavahana Empire in India, the Kushan Empire centred in Afghanistan, the Parthian Empire in Persia, the Kingdoms of Axum and Kush in Northeast Africa and the Roman Empire, which ringed the Mediterranean (Map 3.4).

The development of states and empires was possible only because sufficient surplus food was produced to feed the urban populations. The first civilisations in the Old World (the Sumerian, Egyptian and Indus Valley civilisations) achieved this partly through being efficiently organised and partly through proficiency in hydroengineering. All three originated in the lower reaches of large, seasonally-flooding rivers (the Tigris/Euphrates, Nile and Indus) and, in all three, the people found ways to channel the high flows of the rivers, with their nutrient-rich sediments, into riverside fields. In the New World, the predecessors of the Inca developed irrigation systems based on tapping into the rivers that flowed westwards from the Andes, creating islands of agricultural prosperity at intervals within the arid Pacific coastal belt.

Tanks are a feature of monsoonal South Asia, these being artificial pools which, prior to the neglect of many during recent years, were administered efficiently to supply water to agricultural fields. Their systems of management involved farmers cooperating with one another to share the water and oversight by religious officials (Sakthivadivel *et al.*, 2004). Various civilisations have built aqueducts to transfer water across or between catchments, most famously the Romans, whose main phase of aqueduct construction was between 312 BCE and 226 CE. In Karnataka in southern India, an impressive aqueduct was constructed at Hampi, mostly during the time of the Vijayanagara Empire (1336–1646 CE).

It was probably during the time of the Nasca Civilisation (100 BCE–800 CE) that the people living in the hyper-arid coastal region of Southern Peru and Northern Chile constructed *puquios* – subterranean aqueducts, sometimes many kilometres long, that tapped into aquifers and fed water into oases (Ponce 2015). Similar systems were built in the Old World, where they are

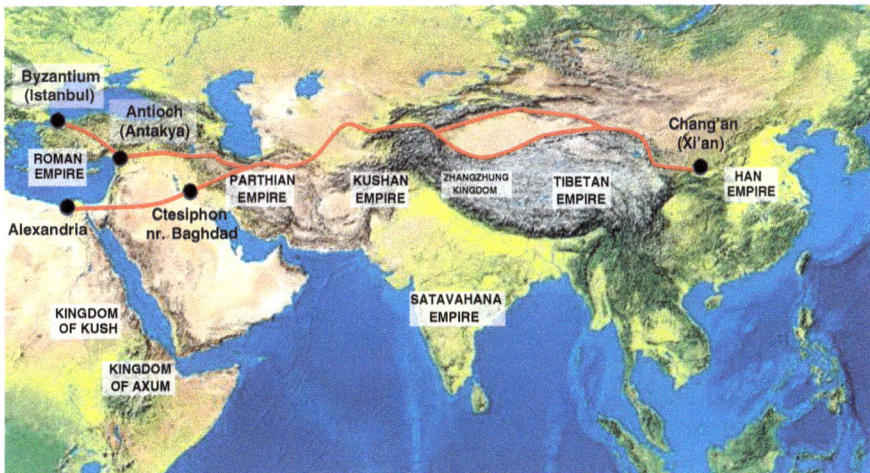

Map 3.4. Civilisations in Eurasia and Northeast Africa at 1 CE emphasising land connections between the Han and Roman empires.

called *qanats*, *foggaras* or *karez*. They were constructed at several places within the sub-subtropical arid zone that extends from the Atlantic to Mongolia, for instance in Saharan Algeria, Iran and Turban in Xinjiang Province in West China (Remini *et al.*, 2014; Barbaix *et al.*, 2020). They served as critical watering and resting spots enabling trading caravans to travel across arid sections of the Ancient Silk Roads.

In places, lakes lay at the hearts of complex systems of crop production, as at Tonle Sap in Cambodia and Lake Texcoco in Mexico (now the site of Mexico City). The Aztecs started to construct their capital city of Tenochtitlan on two islands in Lake Texcoco in the early 15th century CE, later extending its area using *chinampas*. These were artificial islands created by enclosing areas of the lake by building underwater fences and then filling them up with soil and plant materials until their surfaces were raised above the water level. The agriculture that was developed on these platforms was highly productive, giving the Aztecs a crucial edge when they engaged in campaigns of conquest against their enemies.

One of the places in the Americas where people in pre-Hispanic times developed lake-based agriculture was Lake Titicaca, which is situated at the extremely high elevation of 3812 m in the Peruvian Andes (Erickson, 2000). The agriculture was practised on platforms constructed from bundles of reeds harvested from the swamp sedge *totora* (*Schoenoplectus californicus*). The same sedge and a related species, *tule* (*Schoenoplectus acutus*), were used to build boats for fishing along the Pacific coast. Pre-Hispanic people in South America also had larger boats (*balsas*), driven by sails and constructed of logs of lightweight wood, which they used for travelling and trading up and down the western seaboard of South and Central America (Prescott, 1847). Interestingly, *S. californicus* was present in pre-Columbian times on Easter Island, which is 3500 km west of Chile in the open Pacific Ocean – one of the facts that led the Norwegian ethnobotanist, Thor Heyerdahl, to speculate that it, together with the sweet potato, may have been introduced there from South America by pre-Columbian seafarers (Heyerdahl, 1961). Heyerdahl's hypothesis, that the sweet potato was transported to islands in the Pacific Ocean during pre-Columbian times, has subsequently been verified, although the people who did the transporting were probably not indigenous Americans travelling from the east, but people speaking Austronesian languages who came from the west (see Section 3.3.3).

3.5   Developments in Communications, Exchanges and Trade

3.5.1   Improvements in communications

Communication systems required adjustment as societies became more urban-centred and extensive. One reason for this was to expedite the transport of food and other commodities from the countryside to the towns and cities. Another was to ensure that the knowledge systems that bound the states and empires together penetrated everywhere, for instance to ensure that information passed in a timely way from their centres to their peripheries and vice versa. A third was to enable troops to be deployed in short order to suppress internal unrest, keep tributary polities under control and stave off invasions.

The roads built in western Europe during the Roman Empire (31 BCE–476 CE) were so well-constructed that some continued to be used for over a thousand years after the empire fell, no subsequent political power having the competency to make roads of similar quality. In the central Andean region of South America, the Inca and their predecessors built at least 40,000 km of roads to form a huge and efficient communication system extending from Chile to Colombia. There were three trunk roads, one running within the submontane forest zone along the eastern flanks of the Andes, another high in the mountains, using suspension bridges with cables woven from an altiplano grass (*Jarava ichu*) to cross ravines and gorges, and the third close to the Pacific in the arid coastal plain. Relays of runners were stationed at intervals along the roads to ensure the rapid transmission of messages. The Mongol Empire, which, at its peak (1300 CE), was the largest contiguous land empire that the world has ever known, had a rapid transport system involving spaced-out staging posts along the routes that the messengers used. Rested horses were stationed at the staging posts, enabling orders to be transmitted over huge distances at impressive speed.

Prior to their contact with Europe, the Baganda possessed roads (*enguudo*) and associated messengers (*ababaka*), whose job it was to transmit messages (*obubaka*) between the king (*kabaka*) and the chiefs (*abaami*). The highways were maintained by the communities living along their lengths. This is an example of the traditional custom in Buganda, known today as *bulungibwansi*, which is undertaking work for the community good. Lacking writing, the messengers were expected to remember and deliver the messages exactly as received, on pain of severe punishment. The Baganda also had 'talking drums', described as follows (Roscoe, 1921):

> There were literally several hundred different beats for drums, and each rhythm was known by the people, as the waves of sound do to the wireless telegraphist. … In the case of any urgent call or claim, it was the duty of the first person at a distance who heard the rhythm to repeat the message, and thus in a few minutes a claim or call was carried hundreds of miles.

3.5.2 From local tea trade to the Ancient Silk Roads

Tea is an herbal beverage with medicinal properties prepared from the young leaves and shoot tips of a type of small forest tree, *Camellia sinensis*, which is native to the moist submontane forests of Southwest China, Northern Myanmar and Assam (India). For a long time, the Hani and other ethnic minorities living in the uplands of Xishuangbanna (a prefecture in the south of Yunnan Province, China) have picked the leaves of the tea tree to make drinks. They have also long exchanged crude tea and other commodities found in the uplands with rice and other commodities available to the Dai and the Han Chinese who lived in the lowlands (Fig. 3.4) (Pei Shengji, 1988).

The Han Chinese traders, who purchased the crude tea, transported it to Pu'er City (a prefecture to the north of Xishuangbanna), where it was

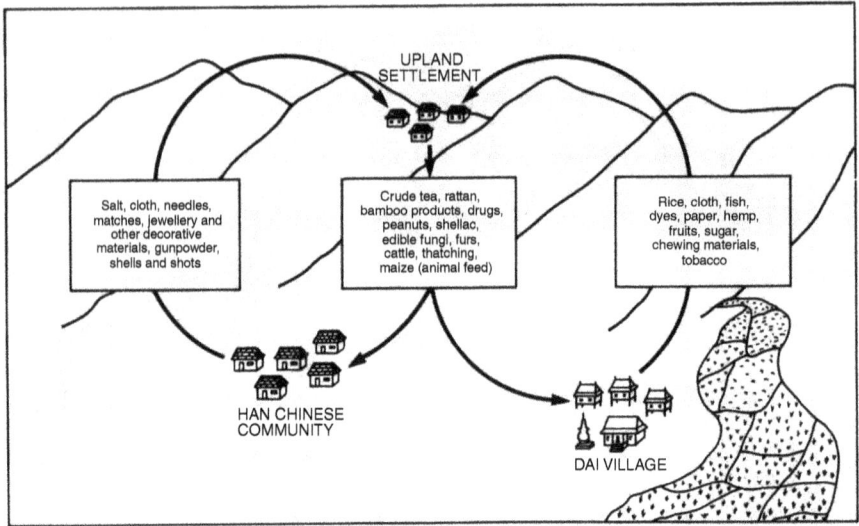

Fig. 3.4. Traditional commodity exchange in Xishuangbanna Prefecture, Yunnan, China. Reproduced from: Pei Shengji (1988).

processed in family workshops into a variety of end products, which subsequently were carried by porters and mule trains to destinations across China. The earliest find of tea in an archaeological context is from the royal mausoleum in Xi'an, which was built for Liu Qi, the Emperor Jing of Han (188–141 BCE) – a find that shows that, by then, Chinese emperors were drinking tea (Lu et al., 2016). During the Tang dynasty (618–906 CE), tea drinking spread to all parts of China and tea houses sprang up across the country. Tea houses were places where, amid rituals and ceremonies, friends met to pass the time and business deals were done. The drinking of tea in tea houses has close connections with the developments of philosophy and the arts in China.

One place to which tea from China has long been exported from the South China provinces is Tibet, where, ever since the 1st millennium BCE, it has been customary to drink tea as a nutritious brew. Known as milk tea or salted-butter tea, this is prepared by adding salt and yak butter to a well-boiled infusion of tea leaves that have been scraped off 'bricks' of tea. These bricks were manufactured in workshops in the Chinese lowlands by fermenting tea leaves under carefully controlled conditions and then compressing them into compact, dry, tablets. The reduced volume and high durability of tea in tablet form are advantageous when it is to be taken on long and protracted journeys. The launching-off points for the tea trade from China to Tibet were Dali in Yunnan and Chengdu in Sichuan, out of which mule trains carrying tea and other commodities once regularly departed on the long winding tracks that eventually arrived on the Qinghai–Tibet Plateau. One of these routes passed through the Powo Valley, which is mentioned in Section 20.5. Hardy Tibetan ponies, much prized in China as warhorses, were among the commodities traded in exchange.

The tea–horse exchange routes connecting China and Tibet eventually became plugged into the other long-distance trade routes that were being developed in Asia (Map 3.5). The way that the trading was done along these routes was by relays of merchants using mules, camels and other animals to carry their goods, and meeting at marketplaces to bargain and exchange commodities with one another (Kocka, 2016). The presence of tea dating to the 2nd to 3rd centuries CE in a cemetery in the capital of the Zhangzhung Kingdom, plus other archaeological finds, shows that, by then, trade from Southwest China extended right across the Qinghai–Tibet Plateau, from east to west (Lu et al., 2016). This vast terrestrial trading network, which became known as the Ancient Silk Roads, was made even more extensive when it became connected with the maritime trading networks that were developing in the Mediterranean and the Indian Ocean. By the 2nd century BCE, silk cloth from China was being traded in Rome, where it was highly prized by the aristocracy. The Silk Roads were formally declared open for business in 130 BCE by the Chinese Government, which at that time was based at Chang'an (near Xi'an) in Shaanxi Province at the eastern terminus of the Silk Roads. Subsequently, trading networks became extended from the Black Sea northwards up the Don and Dnieper rivers, and then, with portage, on to ports on the Baltic Sea coast and so to other parts of Northwest Europe.

An early case of botanically connected espionage occurred during the 6th century CE, when the Byzantine emperor, Justinian 1, who was based in Constantinople (Istanbul), secretly charged two Christian monks travelling to China on a diplomatic mission to return with the secrets of silk (Hunt, 2011). By then, silk was a commodity that had been known and highly prized in Europe for over 600 years, but how silk originated remained a closely guarded Chinese secret. Silk is actually a fibre produced by a caterpillar (the silkworm)

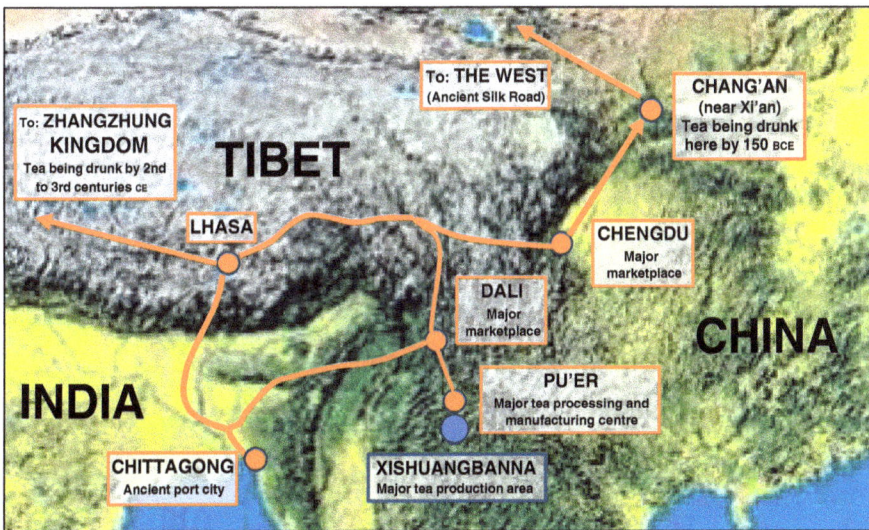

Map 3.5. Tea trading routes in China and their onward connections.

when feeding on the leaves of its food plant (the white mulberry, *Morus alba*). In the event, the monks proved successful in their commission, returning with silkworm larvae secreted in their canes. Thus, the Chinese monopoly over the production of silk was broken and the Byzantine silk industry was born.

3.5.3   From gifting to currency

The ways that goods and commodities have been transferred between people have evolved over time (Kocka, 2016). The earliest form of transfer is believed to be reciprocal giving, which is common in small-scale societies. A more anonymous version of this is bargaining, which involves exchanging goods based on their relative values to each party. Bargaining is a cumbersome and inflexible way of doing business when trading networks become long or complicated. Historically, a solution to this problem was found by inventing money, this consisting of small, easily carried tokens that have values accepted by all involved parties. A further development was the invention of currency, which involved states guaranteeing the worth of money, for instance by stamping their seals on coins.

Worldwide, the first known use of money was in China during the Zhou dynasty (1046–256 BCE) and the first certification of the value of coins was in the Greek city states of Asia Minor in the 7th century BCE. The Tang dynasty in China (7th century CE) was responsible for issuing the world's first banknotes. A consequence of the invention of money was that having access to financial capital became an important adjunct to engaging in trade – hence the origin of financiers and banks. Even more useful than money for long-distance trade are documents (promissory notes) whose value is trusted by all involved, because this avoids the risk of being waylaid by thieves and having one's money stolen.

Money, in the form of cowrie shells (*ensimbi*), was being used in Buganda when it was first visited by Europeans in the mid-19th century (Roscoe, 1911). The value of cowrie shells was a consequence of the difficulty of their acquisition, since cowrie shells are produced by marine molluscs and the nearest sea to Buganda is 1000 km away. Cowrie shells are said to have been introduced as currency during the reign of King Ssemakookiro (1797–1814 CE). Prior to this, the tokens used for currency were small blue beads (*ensinda*) and, prior to them, small ivory discs (*amasanga*) (Roscoe, 1911). It is interesting to note that glass beads of Asian origin have been found at the archaeological site of Munsa in Western Uganda, where they have been dated to about 900–1200 CE (see Section 3.3.1).

3.5.4   Sea travel, maritime trade and waterways

Maritime trade, compared with land trade, is advantageous when trading networks are extensive and the commodities to be transported are heavy or bulky. The Austronesian-speaking people, mentioned in Section 3.3.3, were using large cargo ships to trade between Indonesia, India and China by the 1st millennium BCE (Manguin, 2016). By the 1st century CE, merchants from the Roman Empire were trading along the Malabar Coast of Southwest India and down the east coast of Africa, possibly as far as Zanzibar. The rapid expansion of Islam following the death of the Prophet Muhammad in 632 CE led to a network of Islamic trading stations becoming established around the shores of the Indian Ocean and in Southeast Asia (Map 3.6).

At times, China became directly engaged in trading in the Indian Ocean, for example during the Song dynasty (960–1279 CE) and, again, during the early 15th century, when Yongle Emperor of the Ming dynasty dispatched a fleet of huge ships under the command of Admiral Zheng He on diplomatic missions to destinations around the Indian Ocean (Mote and Twitchett, 1988; Kocka, 2016). The wooden hulls of these ships were treated yearly with oil extracted from the seeds of the tung oil tree (*Vernicia fordii*), which maintained them in good condition, thanks to the oil's anti-rotting properties. (Tung oil seeds were collected by Pei Shengji when he was a boy to have the oil extracted for use in the lamps in his home – see Preface.) After the death of Yongle Emperor in 1424, China ceased its long-distance maritime voyages and the government concentrated instead on resisting the threats of invasion that were developing on its northern frontier. Eventually, the order was given to have the fleet destroyed.

The idea of improving water transportation by digging canals occurred to a number of people in antiquity. For example, the idea of building a canal between the Nile and the Red Sea had been mooted for centuries before one was actually constructed, which may have been during the reign of Pharoah Necho II (610–595 BCE) or possibly Darius the Great (c.550–486 BCE). The canal is said to have remained in use during the time when Egypt was ruled by Rome (30 BCE to c.640 CE), but, like the Roman roads of Western Europe, was not maintained after the Roman Empire collapsed and it fell into a state of disrepair. It was not until 1869 CE that a new waterway, the Suez Canal, was constructed connecting the Mediterranean and Red seas. Subsequently, it played an instrumental role in changing the nature of British imperial rule in India (see Section 5.3.3). In China, a notable engineering achievement was the construction of the Grand Canal (built, 486 BCE–609 CE). This canal connects the Yellow and Yangtze rivers, also the cities of Beijing and Hangzhou (one of the capital cities of Ancient China). It proved politically important in history for maintaining the unity of China, because it enabled the government to move large quantities of grain between northern and southern China, as and when

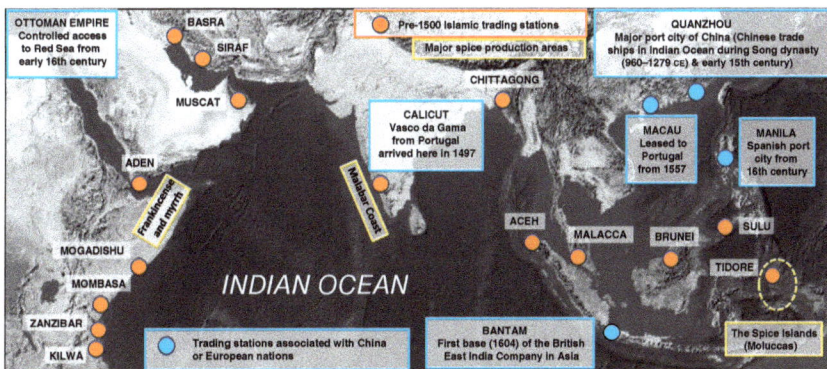

Map 3.6. Pre-1500 Islamic trading stations around the Indian Ocean and in Southeast Asia.

needed, to feed the people in times of shortage and supply the army. At 1770 km, it was, when completed, and still is the longest artificial waterway in the world.

Spices and medicinal plants were among the principal items traded around the Indian Ocean prior to 1500 CE. The main spices traded, and the places at which they were acquired, included (see Map 3.6): nutmeg (*Myristica fragrans*) and cloves (*Syzygium aromaticum*) – obtained from the Spice Islands (the Moluccas in East Indonesia); camphor (*Cinnamomum camphora*) – obtained from East Asia; sandalwood (*Santalum album*) – obtained from India; cinnamon (*Cinnamomum verum*) – obtained from South Asia; cardamom (*Elettaria cardamomum*) and black pepper (*Piper nigrum*) – obtained from south India; frankincense (*Boswellia*) – obtained from Arabia; and myrrh (*Commiphora myrrha*) – obtained from Arabia and Northeast Africa. In Section 16.1, mention is made of cardamon as being a crop that has been cultivated for some time on the East Usambara Mountains of Tanzania. These mountains are visible from the Indian Ocean coast, which has been visited by Islamic traders for hundreds of years. While it has been reported that cardamom was introduced to the East Usambaras in *c.*1960 (see Section 16.1), there is a possibility that it may have been introduced much earlier. Cardamom is native to the Western Ghats of South India – the area known as the Malabar Coast, where the Romans also traded (Gaikwad *et al.*, 2023).

Some of the earliest records of the use of spices in China are contained in ancient documents concerned with Kai-bao (Southern) Materia Medica (Pei Shengji and Youkai, 2020). One of these books is *Nan-fang ts'ao-mu chuang* dating to *c.*304 CE and written by the scholar and botanist Ji Han (263–307 CE) during the Western Jin dynasty. It describes the flora of present-day South China and Northern Vietnam and is regarded as the world's first regional flora and ethno-flora (Li Hui-Lin, 1979). The first mention of cloves in Kai-bao Materia Medica is for the year 973 CE, although, in other documents specifically concerned with Traditional Chinese Medicine (TCM), its use is recorded as early as the Chin dynasty (221–207 BCE). Kai-bao Materia Medica mentions that cloves were imported into China through two routes, one overland from Myanmar and the other via a maritime and land route that terminated in the province of Guizhou.

At the western end of the Ancient Silk Roads, in Italy, the medieval maritime states of Genoa and Venice competed to dominate the lucrative trade in spices entering Europe from Asia and Africa, one consequence of which was that the two were more or less continuously at war between 1256 and 1381 (Crowley, 2011). When Constantinople, situated at a strategic location at the eastern end of the Mediterranean, was captured by the Ottomans in 1453, the terrestrial trading routes that linked Christian Europe to sources of spices in the East became blocked, which stimulated rulers and traders to wonder whether there might be other ways to reach the East – the spur that stimulated the voyages of Christopher Columbus and Vasco da Gama and the expansion of the European powers into the Americas and Asia (see Section 5.1).

3.6   Writing and Paper

Writing was invented independently by the Egyptians, Sumerians and people of the Indus Valley Civilisation at about 3000 BCE. One early use was to post inscriptions on monuments to proclaim the linked power of the gods and rulers – warnings to potential invaders and subjects alike not to disturb the established order. Another was to keep economic records, such as of the amounts of crops grown and traded, and taxes paid. Only from about 2000 BCE, at the earliest, did writing come to be used to convey people's thoughts and feelings, and to record stories (Armstrong, 2019). In China, writing dates back to the time of the Shang state (1550s to 1045 BCE), later becoming standardised under the Qin (221–206 BCE), which was the dynasty that founded the Chinese Empire. Several pre-Columbian civilisations in the Americas also possessed writing, the best developed being that of the Maya (300–1500 CE). The Wari people, living in the Andes and coastal Peru at 500–1000 CE, used a portable device known as the *quipu* for recording information. This consisted of a string of knotted threads tied along a line. Originally used for storing economic data, such as the amounts of crops grown, *quipus* were developed by the Inca and other civilisations into a system of recording information – not yet fully understood – that is said to be capable of recording stories, poems and songs.

The Sumerians wrote on tablets of clay, a material abundant in the alluvial plains where they lived. Elsewhere, organic materials were employed, including parchment (made from prepared animal skins) and materials obtained from plants. The Egyptians and, later, the Ancient Greeks and Romans wrote on papyrus, the Ancient Indians and Sri Lankans on *ola* leaves (from the palm *Corypha umbraculifera*) and the Chinese on wooden tablets, boards and slips of bamboo; also, on cloth made of silk. In Central America, the Maya and Aztecs wrote on *huun* and *amatl* paper, made from the inner bark of certain trees. However, rather little is known about these materials, because the priests who accompanied the Spanish conquistadors regarded the people as heathens and their writings as blasphemous and burnt all the Maya and Aztec writings that they found (von Hagen, 1944).

Paper made from plants is comprised of cellulose fibres that have been extracted by mechanical and chemical means and then deposited in thin layers. The invention of paper-making (in the Old World) is traditionally attributed to Cai Lun, an official of the Chinese Han dynasty (202 BCE–220 CE), although it is known, archaeologically, that paper was being manufactured in China earlier. A number of plant materials have been used for making paper in China, among them bamboo, sugarcane residues, rice straw and cotton rags. Paper made from particular species of plants was used for special purposes, for example fibres extracted from the inner bark of the paper mulberry (*Broussonetia papyrifera*) were traditionally preferred for the manufacture of superior quality paper. Buddhist scriptures in Xishuangbanna in Southwest China were traditionally written on paper made from the bark of the *snay* tree (*Streblus asper*).

From China, the art of paper-making spread to Baghdad (*c.*793 CE), Morocco (*c.*1125), Venice (*c.*1276) and Britain (*c.*1494). In Britain, until about 1800, most paper was manufactured from used textiles. Then, from around the mid-19th century, two other types of raw materials became popular. One was esparto, which is derived from the grasses *Stipa tenacissima* and *Lygeum spartum*, which are native to Southern Europe and North Africa; the other was wood, preferably that of conifers, which was preferred because of its abundance of long fibres.

4 Cultural Influences on People/Plant Relationships, Part 1 – Worldviews

Abstract

Collaboration between people to create a more ecologically sustainable world benefits if they have some understanding of one another's worldviews. A worldview is a property of the individual, but, because people who share similar experiences are exposed to similar formative influences, it can also be considered to be a property of a social group. Here, a comparative approach is used to understand the worldviews of people in three parts of the world – China, the West (particularly Britain) and sub-Saharan Africa (as represented by Buganda). These comparisons show that China and Buganda have more collectivistic ways of viewing the world than is typical of individualistic Britain. The family provides a useful point of focus in the context of this book for understanding how worldviews have changed in Britain and Uganda during the last 150 years.

4.1 The Concept of the Worldview

We pause our account of the history of relationships between people and plants in the mid-2nd millennium CE to take a look at the concept of the worldview, a feature of human culture that fundamentally influences the ways in which people interact with one another and with the natural world. We pick up our historical narrative again in Chapter 5 and then, in Chapter 6, discuss some aspects of European and Western culture that have been prominent influences over the ways that people have interacted with the natural world during the last 500 years. Chapter 7, entitled 'Human Knowledge of Plants', takes another angle on trying to understand people's relationships with plants. In Chapter 8, we consider the history of plant conservation, which opens the way in Chapter 9 for some suggestions about how to make plant conservation more effective.

The German philosopher Immanuel Kant (1724–1804) used the analogy of each person viewing the world through her or his individually tinted glasses to try to convey something of the meaning of the concept of the worldview. According to a more recent attempt (Sire, 2004):

> A worldview is a commitment, a fundamental orientation of the heart, that can be expressed as a story or in a set of presuppositions (assumptions which may be true, partially true or entirely false) which we hold (consciously or subconsciously, consistently or inconsistently) about the basic constitution of reality, and that provides the foundations on which we live and move and have our being.

The concept of the worldview is a useful one for plant conservationists to bear in mind when they engage with people having cultural backgrounds different from their own.

A worldview is a property of an individual, but, because people who share similar experiences tend to acquire similarities of outlook, it can also be considered to be a property of a human group. This form of a worldview (sometimes known as a cultural worldview) is one aspect of the knowledge systems that influence how the members of a society interact with one another and with the natural world (see Section 7.1).

In this chapter, we first give an account of the worldviews that people in earlier times are believed to have had and then concentrate on trying to understand the worldviews associated with the three sample parts of the world that we have selected for special attention in this book – China, the West (especially Britain) and Uganda (especially Buganda). Our approach is comparative, which is similar to that taken by Julian Baggini in his book *How the World Thinks* (Baggini, 2018). One advantage of taking this approach is that it forces the analyst to take a look at her or his own worldview, which is helpful for overcoming the tendency to imagine that the ways that other people view the world are the same as oneself.

4.2 Worldviews in Earlier Times

The evidence suggests that the worldviews of the people who lived in small-scale prehistoric societies of hunter-gatherers, early pastoralists and subsistence farmers were less partitioned into different spheres of perceived reality than is typically the case with the inhabitants of today's urban-industrial societies (Tylor, 1871; Peoples *et al.*, 2016). One aspect of this holism was inward-looking – the feeling of a high degree of unity between the different parts of the perceived self (body, mind and soul). Another was outward-looking – a sense of the self being firmly embedded within the settings of local society, the local natural world and the realm of the gods and spirits. The belief that particular objects, places and creatures possess their own distinct spiritual essences is believed to have been universal (a belief known as animism).

Within this general condition, certain modifications in worldview are thought to have occurred with the transition from hunter-gathering to farming. The domination of the agricultural calendar raised the imperative to try and influence how the deities behaved, these now conceived as forces more active in the day-to-day running of the universe. Pleas and offerings would have been made for the rains to begin and thanksgiving festivals held when harvests were received. Harvest festivals are ancient traditions in China and Britain and still persist to varying extents. In China, the harvest festival, which is known as the Mid-Autumn Festival, is a national holiday. One of its traditions is the eating of mooncakes, one of the products that Professor Pei's father used to make when he worked as a baker (see Preface).

In Britain, harvest festivals have deep roots in pre-Christian history, but they were not incorporated into the calendar of the Anglican Church until the 19th century.

4.3   The Chinese and Western Worldviews Considered

Analysis of the worldviews associated with China and the West is relatively straight-forward in that the roots of their worldviews can be researched by studying historical documents dating as far back as the 1st millennium BCE. A useful concept for analysing the worldviews in these and other parts of Eurasia and Northern Africa for which ancient written records exist is that of the Axial Age (Jaspers, 1953). This was a period, dating very approximately to the 1st millennium BCE, when a number of faith traditions (religions and philosophies) appeared that are still influential in the modern world. Sometimes collectively referred to as world religions, they include Judaism, Christianity, Buddhism, Daoism, Confucianism and Hinduism. The first two of these are Abrahamic religions. They belong to a historical series that also includes the later religions of Islam and Bahai. Not all the faith traditions that arose during the Axial Age have survived (Russell, 2014). Those that did tended to be those that found favour with the rulers of states and empires. The most successful of them in terms of early spread were Buddhism and Christianity (Map 4.1). Psychologically, the transition from tribal religions, as found in small-scale societies, to world religions can be considered as necessary to enable people to continue to function effectively as integrated beings of body, mind and soul in the new political and socio-economic circumstances (see Section 3.4).

4.3.1   The Chinese worldview

China is a country with a civilisation unmatched worldwide in terms of its longevity and high technical and artistic accomplishments, sustained through

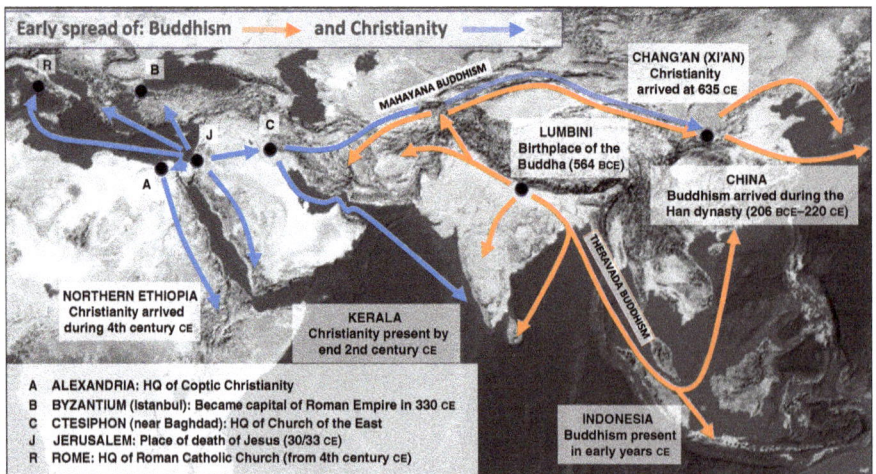

Early spread of: Buddhism ➡ and Christianity ➡

CHANG'AN (XI'AN)
Christianity
arrived at 635 CE

MAHAYANA BUDDHISM

LUMBINI
Birthplace of the
Buddha (564 BCE)

CHINA
Buddhism arrived during the
Han dynasty (206 BCE–220 CE)

THERAVADA BUDDHISM

NORTHERN ETHIOPIA
Christianity arrived
during 4th century CE

KERALA
Christianity present by
end 2nd century CE

A   ALEXANDRIA: HQ of Coptic Christianity
B   BYZANTIUM (Istanbul): Became capital of Roman Empire in 330 CE
C   CTESIPHON (near Baghdad): HQ of Church of the East
J   JERUSALEM: Place of death of Jesus (30/33 CE)
R   ROME: HQ of Roman Catholic Church (from 4th century CE)

INDONESIA
Buddhism present
in early years CE

Map 4.1. Early spread of Buddhism and Christianity.

the millennia. To this day, the deep sense of historical connectedness felt by many Chinese people is demonstrated by their participation in two ancient annual rituals, evidence of a strong sense of allegiance to family and respect for the ancestors. One is the Spring Festival, a time when hundreds of millions of urban Chinese travel back to their traditional family homes – the biggest annual human migration on Earth. The other is the Qingming Festival (Tomb-Sweeping Day) in April (15th day after the March equinox in the Chinese luni-solar calendar), when the graves of ancestors are visited and rituals performed.

Three Axial Age roots have contributed to the development of the modern Chinese worldview. Two are indigenous, one associated with the philosopher Confucius (551–479 BCE) and the other with the legendary figure of Laozi (6th century BCE). The third is Buddhism, a religion founded on the teachings of Siddhartha Gautama, better known as the Buddha, a philosopher and teacher who was born in India sometime around 500 BCE. Buddhism first came to China during the Han dynasty (206 BCE–220 CE) and was subsequently promoted during the Sui dynasty (581–618 CE). Buddhism, along with all other religions, was banned during the Cultural Revolution (1966–1976). However, since 1980, Buddhism has enjoyed a dramatic revival, as, too, have some other religions, including Christianity (Stark and Liu, 2011; Wood, 2020).

Three prominent, interrelated concepts are foremost in the Chinese worldview – harmony, the Way and the dynasty. The virtuous person is traditionally seen as one who looks to the past for lessons on how to achieve a harmonious Way along the journey of life. Education is commended, especially through self-study, learning-by-doing and engaging in a pupil-to-master relationship. Figure 4.1 is a photograph of Pei Shengji standing between the statues of two

Fig. 4.1. Pei Shengji standing between statues of Sun Simiao (on left) and Laozi. Photo: Mr Gao Chao (2016).

famous sages in Chinese history, Sun Simiao (on left) and Laozhi. It was taken in Fuleshan Memorial Park in Mianyang City, the prefectural capital of Pei's hometown in Sichuan Province (see Preface). Sun Simiao (581–682 CE), who was a physician during the Tang dynasty (618–907 CE), is commonly referred to as the 'King of Medicine' in China. He was responsible for two works, *Beiji Qianjin Yao Fang* and *Qianjin Yifang*, which set out his medicinal knowledge (Acosta Güemes and Cusumano, 2022). As well as giving many herbal formulations, he took a strong interest in public health and disease prevention. Sometimes called the Chinese Hippocrates, he insisted that patients be treated equally and warned against physicians being influenced by a desire for rewards, including financial rewards, fame or favours.

As particularly emphasised by Confucius, who lived at a turbulent time in Chinese history, there is a core belief that the highest good comes when members of a family, of families within a community, and of communities within a nation, stand in the right relationships one to another – for instance, in the roles of father and son, and ruler and subject. The relevance of Confucian thought to modern times was summed up by a group of Korean Confucian scholars at a special ceremony at the cemetery of Confucius in Qufu in 2016, as quoted by Wood (2020): 'He defined our collective values of hard work, duty and benevolence; his belief in universal brotherhood in this age of individualism makes his message still true after 2500 years, for all the world.' Maintaining harmonious relations with nature is particularly commended in Daoism.

Traceable back to the overthrow of the Shang state by the Zhou Kingdom in 1045 BCE and idealised by Confucian philosophers, the dynasty is a key marker used by Chinese people when they contemplate their country's past (Wood, 2020). It is somewhat akin to the way that the British think about their national history in convenient slices of time, such as the Roman Period and the Victorian Era, but has more of a recurring than replacing feel. An ideal dynasty for the Chinese represents an historical phase when all China is in order, with both political power and moral authority flowing out from a core embodied in the person of a just and wise ruler. Interrupting the succession of orderly dynasties are periods of disorder, marked by rulers neglecting their duties or becoming tyrannical – periods characterised by political fragmentation, conflicts between warlords and foreign invasions. Fortunately, so the idea goes, eventually the fundamental civilisational strength of China reasserts itself and order is restored. New dynasties emerge. This cyclical view of their country's history lies behind the dread that many Chinese feel about their country descending into disorder.

The imperial civil service is an example of the sophistication of Chinese culture (Wood, 2020). Ending only when the last dynasty fell (1911) and China became a republic (1912), its origins extend far back into Chinese history. A foundational event was when Emperor Jing of the Han dynasty (ruled, 157–141 BCE) accepted a suggestion made by scholar Dong Zhongshu that formal education in China should draw on both Confucian humanism and Yin Yang cosmology (an aspect of Daoism) (Lou Yulie, 2017).

Restructured on several occasions, the imperial civil service was reorganised during the Song dynasty (960–1279 CE) into a form that thereafter persisted into the 20th century. Four basic principles underlay the way that it was structured and functioned, these being that: (i) recruitment to its ranks should be based on merit determined through competitive examination; (ii) candidates should demonstrate that they have received a good general education; (iii) successful recruits should be graded into a hierarchy; and (iv) promotion should be based on achievement. The highly exacting examination for senior posts in the civil service was designed to ensure that all successful candidates possessed a common knowledge of how to write properly (including expertise in calligraphy) and had mastery of the Confucian principles that determined how an ideal society should be run. The common possession of these competences among senior civil servants helped to ensure good communication between government departments and the ease of transfer of officials between provinces. The Chinese imperial civil service so impressed the British that, when they learnt of its existence, they used it as a model when they reformed their own (corrupt) civil service in 1854 (see Section 6.6). 'Mandarin', which is the Chinese word for a senior civil servant, passed into the English language and is still in use today.

4.3.2 The Western worldview

Two Axial Age roots have contributed to the development of the modern Western worldview. One was Christianity, in the forms that it has taken in Europe and the West, plus some contributions from medieval Islam; the other was composed of the writings of the philosophers of Classical Greece (c.500–300 BCE), plus some contributions from Classical Rome. However, highly pertinent for understanding the modern Western worldview are certain cultural developments, themselves nourished by these ancient roots, that started to unfold in Europe from about the middle of the 2nd millennium CE (see Section 6.1). This was the time when a phase of European geographical and economic expansionism began (see Section 5.1) and, it seems likely, both the internal cultural revolution within Europe and Europe's foreign expansionism were propelled by the same underlying force, namely rivalries for supremacy between the rulers of the many small states into which Europe had become divided. One way that the rulers could increase their power was through finding new sources of wealth abroad. Another was to attract the leading creatives of the day to their states to adorn their courts. For innovative thinkers, this had the advantage that, if they fell out of favour with one ruler, they might be able to find patronage from another.

Roman Catholicism was the particular form of Christianity present in Western Europe when the mid-2nd millennium cultural revolution started. Its worldview was a medieval feudal one, the universe being seen as divided into heavenly and earthly tiers. God occupied the upper tier, different classes of heavenly and spiritual beings the second, and different classes of people currently alive on Earth the third. Underneath was the fiery furnace of hell. Christianity, like Judaism and Islam, is a salvation religion, its particular doctrine holding that those who believe and follow the way of Jesus Christ open themselves to the possibility of salvation after their Earthly lives are over. If they do not accept Christ's offering of salvation, then, according to the medieval worldview, they are condemned to endure an eternity of suffering in hell. The association of this salvationist worldview with post-1500 European geographic expansionism added to its forcefulness, since the expansionism could be justified on the basis of providing opportunities to preach to the heathen and give them an opportunity to save their souls.

Ordinary people in medieval Europe were discouraged from questioning this scholastic construct. For truth about religion, it was deemed sufficient for them to follow the teachings of the church, as recounted to them by the clergy, and, for truth about the natural world, it was deemed enough for them to consult the writings of the Ancients, particularly the Greek philosopher Aristotle (384–322 BCE).

Three powerful ideas rooted in the Greco-Romano Classical Age contributed to the way that Europe and the West developed after 1500. One was the belief that the universe is an orderly place, its workings potentially capable of being understood by people using their powers of observation and reason. The increasing hold of this belief on the Western mind has resulted in a continually shrinking space being available for the supernatural. The second powerful idea was that, in some fundamental sense, all people are due the same fundamental human rights (see Section 6.1). The third powerful idea is one that is common in faith traditions with origins dating back to the Axial Age. This is that life is a journey in which one is trying to discover the meaning of one's life (or expand its meaning). It is vividly described by John Bunyan (1628–1699) in his novel *The Pilgrim's Progress*, one of the most influential books in the history of English-speaking Protestantism (Bunyan, 1678). Pilgrimage as an actual journey is a common feature of Axial Age religions. They involve visiting places of special significance in the religion. In Islam, going on Hajj at least once during an adult lifetime is a mandatory duty for Muslims who are physically and economically able. When the 'must visit' places of different religions coincide, it can be a recipe for conflict, as seen in the disputes that arise between Christians, Jews and Muslims over access to holy sites in Jerusalem and between Hindus and Muslims to the site of the Babri Masjid Mosque in Ayodhya, India.

In the West, faith in the power of science as a way of investigating the order in the universe and of answering human problems, plus a mounting emphasis on individual human rights, has led to the concept of progress being extended to other facets of human living, in addition to (or replacing) individual spiritual development. It has become more or less axiomatic for politicians to propose to their electorates that they know better than their rivals how to deliver higher material standards of living. Even candidates from green parties can be wary of pointing out that what is fundamentally needed is a change of basic values, fearing that this would be electoral suicide. This forward-looking worldview of the West with its emphasis on constant progress contrasts with the worldview of traditional China, which encourages people to look to the past to know how things should be done – one aspect of which is the great respect that the Chinese traditionally give to the elderly (Laidlawa *et al.*, 2010).

The Western expansionist view of human potential was being developed in Europe as it was expanding its geographic influence. The relative ease with which Europe, later the West, was able to impose its political and economic will on the rest of the world has fed into a narrative that holds that what the West knows, thinks and does stands at the vanguard of human progress. In colonial Africa, this attitude is exemplified by the smug self-confidence shown by those colonial officials who liberally transferred species of fish from one water body to another (see Section 2.3.2). In modern Britain, it can take the form of nostalgia for empire, a sentiment that contributed to the decision of Britain in 2016 to leave the European Union (Koegler et al., 2020). In the United States, it underlies the idea of American exceptionalism, which is the doctrine that the United States is inherently different from all other nations, being destined, and entitled, to be the world leader (Ceaser, 2012).

4.4 Commonalities in the Worldviews of the Chinese and the Baganda

In his book, *How the World Thinks*, Baggini faced the problem of trying to explain the worldview of the only people included in his book who have not long possessed literacy. He calls these people the 'Southern Africans', which is a very broad category since it contains a huge variety of people differing in types of society and ways of life (Iliffe, 2017). However, a clue to the people he has in mind is given by his repeated stress on the word *obuntu*, which, he maintains, contains the essence of their worldview. In his opinion, this essence is to perceive the self is being strongly linked to humanity (in general) and family (in particular). As explained in the Addendum to Chapter 1, *obuntu* is an abstract noun based on the stem *-ntu*, which is a root deeply embedded in the languages of the Bantu subgroup of the Niger–Congo language family (Schadeberg, 2003).

In Luganda, which is a Bantu language, the translations of *obuntu* given in a recent Luganda/English bilingual dictionary include 'humanity' and 'mankind' (Hamilton, 2020). This ties in well with Baggini's thesis. Further light on the meaning of *obuntu* in Luganda is shed by considering the agglutinated Luganda word, *obuntubulamu*. This is composed of two nouns, *obuntu* and *obulamu*, the latter on its own meaning 'life' or 'health'. According to the same dictionary, the agglutinated word *obuntubalumu* has a meaning circumscribed in English by the words 'politeness', 'courtesy', 'civility', 'decency' and 'good manners'. In Buganda, *obuntubulamu* is considered to be a core characteristic of a person of good character (*omuntubulamu*) – in other words, a key feature of the way that one person is expected to treat another.

We detect a commonality between the Chinese and Baganda worldviews in the great significance of the word *ren* in the traditional Chinese way of thinking and which, like *obuntu* in Luganda, can similarly be translated into English as 'humanity'. *Ren* expresses the foundational virtue recognised in Confucianism – the bearing and behaviour that a pragmatic human should exhibit in order to promote a flourishing society. Aspects of *ren* include filial piety, loyalty, uprightness, altruism, sincerity and modesty (Boedicker and Boedicker, 2009). *Ren* is seen as part of the nature of people, but it must be developed through education and guidance.

The Chinese idea that an ideal dynasty is a time of settled social order, when both political power and moral authority flow out from a just and wise ruler, is matched in Buganda by the meanings embedded in the word *omulembe* and its plural form *emirembe* (Roscoe, 1921). *Emirembe* means 'peace'. The phrase *emirembe n'emirembe* is frequently said by the clergy during Anglican Church services in Buganda to convey blessings. Two of the meanings of *omulembe* are 'period' and 'era'. Referring to the times of the former kings of Buganda, as in the phrase *omulembe gwa Kintu* ('in the era of Kintu'), the implication is that this was a time when all Buganda was in good order.

In China, one of the cultural traits that enabled its civilisation to last so long was the possession of an efficient civil service (see Section 4.3.1). In Buganda, there were several such traits that helped to ensure the unity of the nation and the quality of its leadership (Roscoe, 1911). One was the custom of the king being the only person in the kingdom who belonged to the lineage of his mother (*not* his father) and the tradition of the king being chosen by the senior chiefs when they met as an assembly (*lukiiko*) (see Section 6.8.1). These chiefs were also responsible for administering the day-to-day affairs of the kingdom, which made them well-aware of the challenges that it was currently facing and therefore the qualities required of a new monarch. There was generally a long list of eligible candidates, because, although the king had to be a son of the previous king, the kings had many wives.

Another mechanism that gave adhesion to the kingdom was the openness of the government to the people. To quote one of the examples of this in Roscoe (1921):

> [In public meetings of the royal court] every kind of business was discussed, and the meetings were enlivened by amusing accounts of current events, or stories of huntsmen who had escaped death from some animal in the chase, and also by the exhibition of any peculiar growth of vegetables. …
> In these meetings, the king learnt about his kingdom, and gained such accurate information of the country that he could describe distant places with minute detail; this enabled him to rule with great wisdom, and to decide matters concerning places which he had never visited; it also gave him an intelligent interest in the country, advantageous to the needs of the people. He also learnt in these meetings to gauge the feeling of the country accurately.

Traditionally, and still today, the Baganda have placed a high value on education – a national trait that it shares with China, although with the difference that, prior to the second half of the 19th century, education in Buganda had no connection with literacy. Instead, the education of children traditionally in Buganda involved their participation in many of the everyday activities of the family and spending time with their elders, who passed on to them their knowledge and wisdom in the form of advice and stories (see Section 7.1). Like the Chinese, the Baganda had a traditional system of higher education to train those who administered the country, which persisted well into the second half of the 20th century. The current *Nnaabagereka* (Queen of Buganda) has been trying to revive it in an attempt to help the youth know about and value their cultural roots.

The traditional system of higher education in Buganda involved families sending youths who showed exceptional promise to go and live in the courts of the chiefs or the king to assist with administration and acquire the art of governance. This tradition is known in Luganda as '*mu kisaakate*' in the case of the courts of chiefs and '*mu lubiri*' in the case of the court of the king. These courts were surrounded by palisades made of the stems of elephant grass (*ekisagazi* – *Cenchrus purpureus*). The tradition in Buganda is for the palisade around the courts of chief to be constructed of interwoven reeds set diagonally and for that of the king to be constructed of reeds placed vertically.

According to comparative psychologists, the ways that people behave owe much to where their societies are positioned along a spectrum of cultural variation from individualistic to collectivistic (Kim *et al.*, 1994). The definitions of individualism and collectivism have been given as follows (Hofstede, 1980):

> Individualism pertains to societies in which the ties between individuals are loose: everyone is expected to look after himself or herself and his or her immediate family. Collectivism as its opposite pertains to societies in which people from birth onwards are integrated into strong, cohesive groups, which throughout people's lifetime continue to protect them in exchange for unquestioning loyalty.

In pre-modern times, it was normal for societies to be positioned towards the collectivistic end of this spectrum, but in Europe, as its mid-2nd cultural millennium unfolded, societies started to become more individualistic. Nowadays, all societies have been touched to some extent by individualism.

Today, the Baganda and Chinese alike typically retain more of the traditional collectivistic features of their societies than is the case with the West. This cultural conservatism is fortunate for the Baganda, many of whom are financially very poor. With little in the way of social services available from the state, most people are heavily dependent on their families when they become ill, old or financially broke.

The deep-rooted cultural beliefs of long-established social groups are often conveyed in the forms of pithy sayings, which are widely regarded as formulations of time-honoured wisdom (Ho and Chiu, 1994). Chinese is well-known for its aphorisms and Luganda for its proverbs, examples being:

> Chinese: 授人以鱼不如授人以渔 (Literally: 'Giving a man a fish is not equal to teaching a man to fish.' Meaning: 'Doing things for people can satisfy their immediate needs. Teaching people how to do things means that they can look after themselves.') This saying illustrates the high value traditionally placed on education by the Chinese.

> Chinese: 吃一堑，长一智 (Literally: 'Fall into the moat and you'll be wiser next time.' Metaphorically: 'One learns from one's mistakes'.) This illustrates the pragmatism associated with traditional Chinese culture and the emphasis placed on learning through experience.

> Luganda: '*Omulya mmamba aba omu, n'avumaganya ekika.*' (Literally: 'He who eats the lungfish brings the whole clan into disrepute.') This refers to the fact that the lungfish is the totem of one of the clans of the Baganda and it should not be eaten by clan members. This proverb can be translated into English as follows (Lule, 2006): 'If one person in a group breaks the rules, or does something outrageous, the entire group is blamed for his misdeeds and is disgraced.'

Luganda: '*Omwami tafuga ttaka, afuga abantu.*' ('A chief does not rule land, he rules people.') This way of looking at human authority and power illustrates one reason why tropical Africa proved to be so vulnerable to European colonialisation. The traditional African way of thinking found it hard to envisage how a single individual could have exclusive rights of ownership of land.

4.5   Commonalities and Divergences in Worldviews in Britain/the West and Buganda/Uganda

The concept of 'the family' provides a useful focus in the context of this book for considering the commonalities and divergences in worldviews between Britain/the West and Buganda/Uganda since 1862, which was the year when they first came into direct contact. It is appropriate because of the key role that the family has traditionally played in societies for passing on knowledge about plants down the generations (see Section 7.1).

In the 19th century, the family was regarded as a pillar of society both in Britain/the West and Buganda/Uganda but was conceived differently. In 19th century Britain (as an example of the West), an ideal family was seen as a small 'nuclear' social unit consisting of a man (regarded as the head of the family), his wife and their children. The family of Queen Victoria, the monarch who reigned for most of the 19th century (1837–1901), was considered exemplary (see Section 6.8.1). In contrast, the family in Buganda (as an example of Uganda) was a much larger 'extended' unit. If a family is defined as a social group within which sexual relations and marriage are taboo, then a family in Buganda was (and is) equivalent to a clan (*ekika*) and some clans have tens of thousands of members. With the exception of the Kiganda king, males in both Britain and Buganda were favoured in matters of inheritance. In Buganda, the children of men belonged to their clans (not to those of their wives, as with the king) and in Britain, until 1870, the property of a woman and any money she earned became the property of her husband on marriage.

An advantage for the Baganda in belonging to extended families in the 19th century was that every person had the security of a recognised social identity within the collectives of the kingdom and the clan. This advantage was described by John Roscoe as follows (Roscoe, 1921):

> Whatever the origin may have been, the social benefits [of the clan family system] were great, for not only were the marriage relations regulated by means of the totemic beliefs, but the numerous calls made by a member of a family upon others who bore the same totems, the financial help, the sympathetic assistance in sickness, and the communal rites were a great boon to the family.

Roscoe (1861–1932) was one of the first Anglican missionaries to work in Uganda and a trained anthropologist.

When Britain assumed full charge of Buganda in 1900 and the rest of Uganda soon thereafter, the colonial authorities were confronted by the question of what to do legally about the family. Pressure was mounting on it to act, because the creation of Uganda had brought together, or was soon to do so, over 40 indigenous ethnic groups and several world religions, each with its own understanding of the meaning of a family. Particularly pressing were questions about what to do about marriage and divorce between individuals whose cultural backgrounds with respect to the family were different (Hansen, 1984). The world religions represented in Uganda by the early decades of the 20th century included the Bahá'í Faith, Christianity (in the forms of Anglicanism, Catholicism, the Orthodox Church and Pentecostalism), Hinduism, Islam (in the forms of Shia, Sunni and others), Judaism and Sikhism (see Section 6.8.4) (Riggs, 2015).

Responding to this pressure, the colonial authorities passed a series of laws about marriage and divorce during 1902–1906 which applied to everyone except for Muslims and Hindus, who were allowed their own dispensations. With one significant addition, the laws passed at this time remained unchanged until well after Uganda became a politically independent state, which was in 1962. The significant addition was a little-noticed part of a revision of the Ugandan Legal Code published in 1950. Uncontroversial at the time, it made same-sex sexual acts illegal in Uganda, a provision that had existed in British law since 1885.

In addition to trying to create a common legal base for the family, the colonial authorities had the ambition of turning Uganda into a Protestant Anglican state and making the Victorian concept of an ideal family the cultural norm. As mentioned in Section 6.8.2, one move that they took to do so was to break with Kiganda tradition and install a 1-year-old child, Daudi Chwa II, as the *kabaka* ('king') of Buganda in 1897 and then follow this up by providing him and his successor, Kabaka Muteesa II, with upper-class Anglicised upbringings. Roscoe, who was head of the theological college established by the Protestant Church Missionary Society (CMS), was hesitant about the wisdom of promoting the Victorian concept of an ideal marriage in Uganda, being more aware than the political authorities about the realities of people's lives. That his hesitancy was justified is suggested by the fact that it became a common practice for Ugandan men to choose one of their wives to be their official wife and pay less attention to the others. (The term used for an official wife in Luganda is *omukyala w'empeta*, which means 'the wife with the ring'.) The effects of this on society were described by Roscoe as follows (Roscoe, 1921):

> When Christianity introduced monogamy and broke down the old social customs, hundreds of women were rendered husbandless, without the former rigid restrictions to protect them against their sexual desires; and when the new hut taxes imposed by the British Government made it impossible for chiefs to provide homes for their clan relatives, hundreds of women were left to face the problems of life without any special guardian.

(See Section 5.4.3 for an explanation of hut tax.) Another tactic adopted by some Christian converts was to reconvert to Islam so as avail themselves of provisions within Islamic law that allowed for polygamous marriages and made it easier for men to divorce their wives (Hansen, 1984).

Nowadays, the Anglican Church of Uganda acknowledges the psychological importance of the clan system to the identity of the Baganda and has reconciled it with its theology. Wilson Mutebi, a retired bishop of the Church of Uganda, has pointed out that there are commonalities in the social structures of the clan and the church, and that the positions within them remain constant even while the individuals who occupy them change with time (Mutebi, 2005). In both the clan and the church, everyone has a recognised social position and, along with it, certain benefits and responsibilities in relation to others. He likens his own position in the church (that of a bishop) to that of a chief and describes how both bishops and chiefs are responsible for looking after 'their' people. This perspective allows the ancestors and the yet-to-be-born to be seen just as much as members of the clans and the church as the living, which fits in well with the respect that the Baganda feel for their ancestors and their interest in having children.

As mentioned in Section 5.4.3, there are numerous disputes in Buganda today between the customary owners of land, as represented by the ordinary people and their chiefs, and those who have come into possession of legal titles to land – disputes which are particularly bitter if the land contains graveyards, as is often the case (see Section 7.2). One expression of this tension is in the difference in meaning of the concept of an 'heir' in modern constitutional law compared with Kiganda tradition. In modern law, an heir is a person designated in a will or, if there is no will, then it is the next of kin, who is commonly the spouse or the deceased's children. In traditional Buganda, an heir (*omusika*) is a person of the same gender and clan as the deceased. It is the duty of an *omusika* to take on the social responsibilities of the deceased. For example, if the deceased has young children, the *omusika* is responsible for their upbringing and, if the deceased was a chief, then the *omusika* takes over his responsibilities in the clan. *Abasika* ('heirs') are chosen by the people before their deaths, based on the candidates' suitability of character, or otherwise by their clan relatives.

Figure 4.2 illustrates graphically the good alignment that is now common between the concept of a family, as understood by the Anglican Church of Uganda, and the concept of a family, as understood traditionally. It is a photograph of A.H.'s wife and her sister in a graveyard (*ekiggya*) belonging to sub-branch (*essiga*) of the Grasshopper Clan (*Ekika ky'Enseenene*). They are attending the graves of their relatives. Most of the people who have been interred in this graveyard during recent years have been Christians, but the graveyard predates the date of arrival of Christianity in Uganda. Nowadays, the ceremonies performed when newly deceased people are buried at this site include rituals associated both with Christianity and Kiganda tradition. The forest seen behind the graveyard is a traditional sacred grove (see Section 8.1).

In some ways, the degree of cultural change in Buganda/Uganda since the date of its first contact with Britain/the West has been extraordinary. Even by 1900, some of the old ways in Buganda were beginning to fade from memory (Roscoe, 1921). By the 1950s, Buganda/Uganda had become similar to Britain/the West in that the majority of the people considered themselves to be Christians and had acquired much the same idea of what a proper Christian family and a proper Christian marriage should be. However, since then, a rift has developed between the two sides on the concept of the family and the definition of a marriage. The rift has been caused by the success of gay-rights activists in the West in their campaigns to allow people of the same sex to marry and raise children. In Britain in 2013, they achieved their ambition when the

Fig. 4.2. Clan graveyard and associated sacred forest, Uganda. Photo: A.H. (2019).

British Parliament passed the Marriage (Same Sex Couples) Act, which legalised homosexual marriage. The Church of England has tended to prevaricate on the issue, although, in 2005, it did allow priests to enter into same-sex civil partnerships. (A civil partnership is a weakened form of marriage, in that no vows are required.)

The Church of Uganda, with backing from the Ugandan Government, has stood behind the concept of the proper Christian family, as was taught by the early Western missionaries. The dispute has not been confined to Britain and Uganda. It has engulfed the whole Anglican community, including in the United States where it resulted in the Episcopalian Church becoming split into two. Other Anglican churches in sub-Saharan Africa have tended to be on the Ugandan side. A religious cum political tit-for-tat battle is now taking place between gay-rights activists in the West pushing for Africans to change their views on family and marriage, and politicians in Africa passing ever more draconian anti-gay laws (Kretz, 2013). There has been so much divergence on the issue that it has become a textbook example of how the worldview of one social group can be virtually incomprehensible to another.

A new British government elected in 1997 declared that it was adopting an 'ethical' foreign policy and that, from then on, the objective of British development aid was to be the elimination of world poverty. The morality of Britain's 'ethical' foreign policy in the way that it has actually been pursued in Africa is debatable. In Section 5.7, we describe how Britain's imposition of its neoliberal thinking on forestry in Uganda has contributed to the country's rapid rate of forest loss. In Section 7.5.3, we mention how Britain has been depriving Uganda of doctors and nurses to compensate for its homegrown shortage. With regard to enforcing its recently discovered ethical approval of same-sex marriage, Britain, along with other Western countries, has resorted to threatening to cut off, and sometimes actually cutting off, its development aid to African countries unless they withdraw their anti-homosexual laws (Heuler, 2013; Onapajo and Isike, 2016). In 2013, the World Bank aligned itself with the West, when it announced that no new funding would be made available for countries that refused to withdraw their anti-homosexual laws (Dasandi and Lior, 2023).

In Section 4.4, we mention that the collectivism of traditional Buganda, such as still exists today, provides a measure of social security to people when they become ill, old or financially broke. People's obligations in these respects are related to their biological gender, age and position with families. Children in societies of subsistence agriculturalists – a way of life still strongly represented in Buganda – acquire their knowledge of how to behave in society, from an early age, through learning for their elders and joining in the everyday activities and some of the rituals of their societies (see Section 7.1). So far as plants are concerned, the learning of growing children has a strong connection with biological gender, as we describe in relation to the plant-resource system of the Baganda in Section 7.2. Therefore, the efforts of the West to change the Baganda's traditional view of the family carries the risk of even further reducing the social security of the Baganda. The strong reaction in Buganda to the West's efforts to alter its traditional worldview has been exacerbated by the open displays of their sexuality by gay-rights activists in the West. According to the prevailing worldview of the Baganda, this is not a way that a civilised person (*omuntubulamu*) should behave.

5 Plants in Human History Since 1500 CE

Abstract

In about 1500, European explorers set sail and initiated a process of European, later Western, geographical and economic expansionism. Portugal and Spain were the first to establish overseas colonies, followed in about 1600 by the Netherlands, Britain and France, and, in the 19th century, by Belgium, Italy and Germany. Russian expansionism was terrestrial, eastwards across Asia. Botanical commodities lay at the economic hearts of the European empires, among them spices, sugarcane, cotton, tea, rubber and timber. In the mid-19th century, Britain created an imperial system of botanical gardens and research stations designed to maximise the economic benefits that it derived from plants. Other European powers, such as Germany, developed similar systems. The industrialisation of agriculture and forestry began in Europe in the 19th century. Uganda was introduced to the global marketplace at the turn of the 20th century, when it was accompanied by the redistribution of the ownership of land. New forestry systems, influenced by scientific thinking, were introduced in Germany and France in about 1800, in British India in the mid-19th century and in Britain itself after the First World War. A wave of decolonisation after the Second World War saw the emergence of a new type of relationship between richer and poorer countries, based on the concepts of development aid and a Global North/Global South divide. The chapter concludes with a discussion of the influence of development aid on the histories of botany and forestry in Uganda.

5.1 European Geographical and Economic Expansionism

The world has seen great changes during the last 500 years. In 1500, it had not yet been established that the world is a planet; today, it is known that it is one among many planets in a vast universe. Economies were much more localised in 1500 than they are now; today, the whole world has been drawn together into one giant economic web. Biologically and culturally, the world is

© Alan Hamilton and Pei Shengji 2024. *History and Future of Plants, Planet and People*
(Alan Hamilton and Pei Shengji)
DOI: 10.1079/9781789248944.0005

becoming increasingly globalised, with consequent loss of local biological and cultural diversity. As a measure of the extent of cultural loss, it is estimated that, of the $c.1175$ indigenous languages that were spoken in Brazil in 1500, fewer than 200 still survive (Rodrigues, 1993; Crystal, 2000).

It was certain countries in Western Europe that spearheaded the human journey along this globalising path. Later, these were joined by others elsewhere that had become dominated politically by people of European extraction. Later again, during about the last 60 years, several non-European countries, mainly East Asian, have risen to powerful positions on the global economic stage. The concept of 'the West' is one often used in discussions about modernity. We use the term to refer to those countries, mostly in Europe or dominated by people of European extraction, that, together, are widely seen as forming one of the world's most powerful political, economic and cultural blocs.

We mention in Section 4.3.2 that a major cultural revolution began in Europe in the mid-2nd millennium. An early indication of what was to come was the Renaissance in northern Italy ($c.1350$–1600), a period of remarkable cultural flourishing. For the most part, Western Europe was not a particularly distinguished place in 1500. Comprised of small rival states competing for influence, it compared unfavourably in sophistication with several large, politically well-organised, empires in other parts of the world. Among them were the empires of the Ming dynasty in China (1368–1644), the Khmer in Southeast Asia (802–1566), the Mughals in India (1526–1858), the Safalids in Persia (1504–1722), the Ottomans in Asia Minor (1281–1922), the Aztecs in Mexico (1300–1521) and the Inca in Peru (1400–1533) (Prescott, 1843, 1847; Masselos, 2010). In China, the Song dynasty (960–1279) was comparable to the Italian Renaissance in its cultural achievements. The inventions made during this dynasty included the world's first movable-type printing press (invented in 1040 by Bi Sheng), a water-driven spinning machine, the coke-fired blast furnace, steel smelting, gunpowder and paper money. Other achievements of the Song included the world's first national library and university, a high level of school enrolment and reform of the civil service.

One of the forces that propelled the geographic expansionism of Western Europe in the mid-2nd millennium was the desire of some of its rulers to gain economic advantage over their rivals through acquiring new sources of wealth abroad (Kocka, 2016; French, 2021). In particular, the fabled gold fields of tropical Africa and the riches of the East beckoned. At the time, the routes to the East via the eastern Mediterranean were blocked by the Ottoman Empire, effectively sealing off direct connections between Christian Europe and the terrestrial Ancient Silk Roads. This is why Christopher Columbus in 1492 and Vasco da Gama and John Cabot in 1497, sailing respectively for the Crowns of Spain, Portugal and England, tried to find other ways to reach the East. Both Columbus and Cabot, who sailed west, were Genoese – natives of Genoa in Italy, the long-time rival of Venice for commercial domination of the eastern Mediterranean. Da Gama, who travelled south to try and work his way around Africa, was Portuguese.

The Spanish and Portuguese started to explore the islands in the East Atlantic to the west of Africa during the 14th century (Map 5.1) (French, 2021).

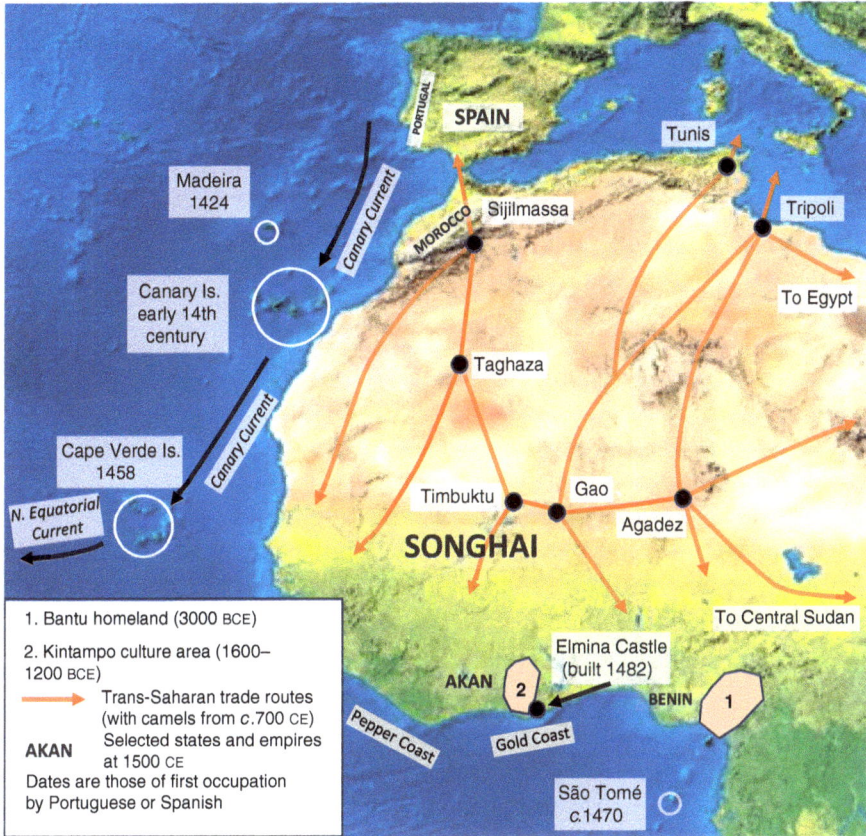

Map 5.1. Pre-1500 Northwest Africa and the East Atlantic islands.

In 1424, the Portuguese discovered Madeira, which had never before been inhabited, and followed this up by establishing plantations of sugarcane. They built aqueducts (*levadas*) in the hills to transfer water from the wetter western and northwestern aspects of the island to the drier southeast, which was better for cane. By the mid-15th century, Madeira was exporting 70 tonnes of sugar a year to Portugal. The labour needed for building the *levadas* and working in the cane fields was provided by enslaved Guanches, a hunter-gatherer people who were living on the Canary Islands. The Guanches fiercely resisted Portuguese and Spanish occupation, but finally succumbed to the Spanish in 1496. Meanwhile, the Portuguese were pushing on further around Africa. They discovered the island of São Tomé in the Gulf of Guinea in about 1470 and proceeded to turn it into a major sugar-production area.

After Christopher Columbus had discovered America in 1492, the East Atlantic Islands assumed a new strategic importance for the Portuguese and Spanish. Because of the way that the prevailing winds and surface currents revolve around the North Atlantic, the easiest way to travel by sail from the Iberian Peninsula to the Americas was first by heading southwest to about the latitude of Senegal and then turning westwards towards the Americas (Map 5.1). The best leg back was further north.

There were three commodities available in West tropical Africa that especially interested the Spanish and Portuguese. A Lisbon merchant, Fernão Gomes, who had been granted the monopoly to trade in the Gulf of Guinea by King Alfonso V of Portugal in 1469, discovered that one of them, malegueta pepper, was available along the so-called Pepper Coast (where Liberia is situated today) and a second, gold, along the Gold Coast (where Ghana is today) (French, 2021). The third commodity was slaves. The first overseas fortified European trading station built by Europeans during their age of geographic and economic expansionism was Elmina Castle on the Gold Coast.

Malegueta pepper consists of the seeds of a spicy, ginger-like, herb, *Aframomum melegueta*, which grows in swampy places in West African rainforests. Other species of *Aframomum* grow in rainforests elsewhere in Africa (see Section 7.3.2). From the 7th to 8th centuries CE, before the maritime trade route around Africa to the Indian Ocean was opened, malegueta pepper and gold had been exported to North Africa and Europe from West Africa by means of camel train across the Sahara Desert. Pre-1500 trans-Saharan trade routes are shown on Map 5.1. Marketed in Europe as 'grains of paradise', malegueta pepper was a popular spice in 14th and 15th century Europe but became increasingly substituted by chili pepper (*Capsicum frutescens*) after samples had been brought back from the New World in 1492 by Christopher Columbus and, from the 16th century, by black pepper (*Piper nigrum*) from Asia.

Competition between Spain and Portugal for the West African trade led to an incident in 1478 when Portuguese ships ambushed a Spanish convoy returning from Elmina Castle and managed to capture some of the vessels and seize their gold (French, 2021). This was the first intra-European colonial war to be fought at sea. Responding to this crisis, the Catholic Church, headed by Pope Sixtus IV, prompted Portugal and Spain to sit down together and agree on a deal. The resulting Treaty of Alcáçovas (1479) stipulated that Portugal was to enjoy rights to all islands discovered or to be discovered from the Canary Islands to beyond Guinea, except for the Canary Islands, which were to remain with the Kingdom of Castile (Spain). Later, another pope, Alexander VI, issued a papal bull, confirmed by the Treaty of Tordesillas (1494), that divided the world outside Europe between Portugal and Spain along a line of longitude about midway between the Cape Verde Islands off West Africa and the island of Hispaniola in the Caribbean. Spain was granted the lands to the west of this line and Portugal those to the east. Once it had been confirmed through its circumnavigation that the world was a sphere, a new agreement, the Treaty of Zaragoza (1529), was mediated by the Catholic Church to establish the line of demarcation between Portugal and Spain on the far side of the world.

Unknown to Columbus, the captain of the first European ship to visit the Americas since the Vikings (who were there in *c*.1000 CE), the American continent blocked the direct path from Europe to the East. Columbus' first sighting was of Cuba, which he thought was China, and his first landing was on Hispaniola, which he thought might be Japan. It was not known at the time of Columbus' visits to the New World that South America protruded far out eastwards into the Atlantic, which is why, when this was discovered fortuitously by the Portuguese navigator, Pedro Álvares Cabral, in 1500, Portugal acquired Brazil.

The first circumnavigation of the world was made in 1519–1522 by a small fleet of ships that sailed for Spain under the command of a Portuguese navigator, Ferdinand Magellan (who died during the voyage). In 1521, Magellan landed on the coast of the Philippines, which he claimed for Spain. Later, Spain used its base of Manila in the Philippines as the focal point of a maritime trading empire that it developed in Asia. It was the port of embarkation for fleets of treasure ships which periodically set sail bound for Acapulco in Mexico. Spanish trans-Pacific trade from Manila to Acapulco continued for 300 years, until disrupted by the Mexican War of Independence (1808–1821).

One arm of Spain's maritime empire in Southeast Asia became involved in a long history of conflicts with the sultanates of Brunei and Sulu, situated respectively on the north coast of the island of Borneo and in the Western Philippines (Map 3.6). From the mid-18th century onwards, other Western colonial interests, including the British East India Company, came into conflict with Spain's Asian maritime trading empire. A dispute between Spain and Britain about which of them owned North Borneo was not settled diplomatically until 1885 (Wright, 1966). In 1898, Spain ceded the Philippines to the United States after its defeat in the Spanish–American War.

For their part, the Portuguese were lucky to secure the Spice Islands (Moluccas) in Eastern Indonesia, which had been a particular target (Map 3.6). They also managed to establish trading stations at Macau in China and Nagasaki in Japan. From Macau, Western influences spread further into China, with Matteo Ricci (1552–1610), an Italian Jesuit, becoming a respected figure in the Chinese imperial court. The Jesuits engaged in cultural and philosophical discussions with Chinese scholars, who were particularly eager to learn about the discoveries in mathematics, astronomy and the visual arts that were being made in Europe. The influence of the Jesuit delegation at the imperial court continued throughout the 17th and 18th centuries, its members becoming among the most valued and trusted of the emperor's advisors.

Meanwhile, in the Americas, the conquest of the sophisticated Aztec and Inca empires proved relatively easy for the Spanish, who possessed certain crucial technological advantages, notably guns, steel swords, steel armour, the horse and the wheel. However, what really proved disastrous for the indigenous Americans, not just in Mexico and Peru, but throughout the Americas, was their lack of resistance to the infectious diseases which were introduced from the Old World. It is estimated that the population of indigenous people in the Americas fell by 90% during the hundred or so years that followed the arrival of the Spanish. One consequence of this was that, in places in Amazonia where urban-centred and well-organised societies had developed, there was a reversion back to small-scale communities of subsistence agriculturalists and hunter-gatherers (Heckenberger et al., 2008). It has been calculated that the regrowth of vegetation that resulted from the reduction of human pressure on the forests of the Americas was so great that sufficient carbon dioxide was sucked out of the atmosphere to cause worldwide climatic cooling from the 16th to 19th centuries (known in Europe as the Little Ice Age) (Koch *et al.*, 2019). It was these earlier civilisations in Amazonia that were responsible for forming its fertile *terra preta* soils (see Section 3.2).

In the event, the Catholic Church's effort to separate the expansionist ambitions of the Spanish and Portuguese proved quite successful, but this did not deter other European powers from trying to seize the lands that these two countries had acquired or take other lands for themselves. There was so much rivalry between the European powers, in various combinations of allies and enemies, that there was almost continuous conflict between the European powers from the beginning of the 16th century right up to, and including, the Second World War (1939–1945). Divisions in religion that had developed within European Christianity sharpened tensions, because now the competition was not just about acquiring wealth and prestige, but also about saving souls. The initial schism in post-1500 Christianity was a break between Catholicism and Protestantism (in 1529). Other schisms within Protestantism followed, backed up by disagreements over theology.

After Portugal and Spain, the next batch of European powers to acquire colonial territories were the Netherlands, Britain and France, all of which started to build up their overseas possessions from about 1600. Much later, Belgium, Italy and Germany joined in, once they had been created (Belgium in 1830) or become unified states (Italy in 1861, Germany in 1871). The British defeat of the combined fleets of Spain and France at the Battle of Trafalgar (1805) and the British and Prussian defeat of the French army at the Battle of Waterloo (1815) greatly boosted Britain's imperial success. Additionally strengthened by being the first country in the world to become industrialised (from the late 18th century), Britain was able to greatly expand its empire during the Victorian age (1837–1901) and, at the date of its maximum extent (1922), possessed the largest empire that the world has ever known.

There was another country in Europe that created a large empire for itself during the age of European expansionism, but, unlike the others, this was through expansion of its land territory, rather than by first sailing out to sea. This was Russia. The first step was taken in 1480, when Ivan the Great stopped paying tribute to the Mongols and took full control of the Duchy of Moscow, a large territory to the west of the Urals. Russia completed the exploration and conquest of Siberia during the 17th century, then moved east to Alaska and established trading posts down the west coast of North America as far as California. Later, in 1867, Russia sold Alaska to the United States and abandoned or sold its outposts in California. As mentioned in Section 19.2.2, Britain in the 19th century became anxious about Russian expansion southwards into India, leading it to adopt a forward policy of periodically sending punitive expeditions into Afghanistan to ensure that its foreign policy conformed to British, not Russian, interests.

5.2 How Quinine Assisted Europeans to Acquire Tropical Africa

The penetration of Europeans into tropical Africa came late, relative to the Americas and Asia, because of their lack of genetic immunity to falciparum malaria. The key that unlocked the door for their entry in force during the closing decades of the 19th century was their ability to secure access to good-quality supplies of the antimalarial drug, quinine (Brockway, 2002, 2020). It was only later that it was discovered that falciparum malaria is caused by a protozoan (*Plasmodium falciparum*) that lives in red blood cells and that it is transmitted between people by a secondary host (the blood-sucking females of *Anopheles* mosquitoes). The indigenous people who lived in the African interior possessed a measure of genetic resistance to infection by *P. falciparum* because of their possession of the sickle-cell mutation (see Section 2.3). Moreover, they were aware of a number of plants for treating fevers (Ssegawa and Kasenene, 2007).

The first Western explorer to reach Uganda (in 1862) was an officer in the British Indian Army named John Hanning Speke (1827–1864). Speke's success in making the journey, which began in Zanzibar on the East African coast, may have been because, by chance, the quinine that he was carrying in his baggage happened to be of good quality. Speke's published description of his journey informs us that, apart from quinine, the medicines which he carried included calomel and jalap (Speke, 1863). Calomel, which is mercuric chloride, was a popular medicine in 19th century Britain that probably did more harm than good. Jalap is a purgative drug made from a Mexican plant, *Ipomoea purga*. After he had introduced himself to Kabaka Muteesa I, the King of Buganda, at his court in Mengo, Speke was frequently requested by the queen-mother (*Nnamasole*) to visit her court and treat her with quinine, presumably because she had become convinced of its efficacy.

Quinine was commonly used in 19th century Europe to treat fever, but its quality could be highly variable. It is derived from the bark of *Cinchona*, a genus of trees found in montane forests on the eastern slopes of the Andes. *Cinchona* bark was once commonly known in Europe as Jesuits' bark, because, in the 17th century, Jesuit missionaries living in the Andes had learnt of its medicinal powers from the indigenous people and had begun dispensing it in their pharmacies and exporting it to Europe. The drug, quinine, was first isolated from the bark in 1820 (see Section 7.4.2).

The variability in efficacy of quinine was a well-known problem in 19th century Europe and, in 1857, the British Government decided to take the matter in hand and ensure that reliable supplies of good-quality quinine were available from plantations situated within the British Empire. The decision to do so was precipitated by the Sepoy Revolt in India (see Section 5.3.3), which had made the British Government realise that there was an urgency about improving the health of British people living in India (Brockway, 2002, 2020). The government commissioned the Royal Botanic Gardens, Kew to dispatch plant collectors to the Andes to bring back seeds and seedlings of the quinine tree. The intention was, after germinating the seeds at Kew and growing on the seedlings, to dispatch planting stock out to suitable places in the British Empire to establish plantations.

Knowing that the quality of quinine was variable, Kew decided to dispatch collectors to different parts of the Andes to increase the chance of obtaining good-quality material. Disregarding the inconvenient fact that the removal of quinine seeds and seedlings was illegal under the laws of the Andean countries, by 1859–1860 the collectors had succeeded in their missions and the seedlings were being grown in heated greenhouses at Kew. When sufficiently established, the seedlings were shipped out to botanical gardens in India and Ceylon (now Sri Lanka), which then became the sources of seedlings for plantations in India, Sri Lanka and the Malay Peninsula. The Netherlands carried out a similar operation and established its own plantations on the island of Java in the Dutch East Indies (now Indonesia). As it turned out, the types of *Cinchona* grown in the Dutch plantations proved to be of superior quality to those grown in the British and it was Dutch quinine that came to dominate the market.

Under the terms of the Treaty of Versailles (1919), which concluded the First World War, Germany lost its tropical colonies and thus the possibility of growing its own *Cinchona* trees. Responding to this challenge, German chemists started exploring the possibility of synthesising antimalarials, which resulted in the discovery of chloroquine in 1934. At first, chloroquine was considered too toxic to use routinely in medicine. However, subsequent research in the United States demonstrated that chloroquine was adequately safe, which was fortunate for its military in the Second World War when they were denied access to the *Cinchona* plantations of Southeast Asia.

However, for *P. falciparum*, the parasite that causes malaria, discovery of chloroquine only signalled a ceasefire. A new war began between its genetic capability of evolving new drug-resistant strains and the ability of scientists to discover new antimalarials. Chloroquine-resistant strains of *P. falciparum* were first detected in Colombia and at the Cambodia–Thailand border during the late 1950s, then spread steadily from these foci in the 1960s and 1970s through South America, Southeast Asia and India, arriving in Africa in the late 1970s (Wellems and Plowe, 2001). Malaria continues to be a scourge in Africa, although there is now some hope of a more permanent solution, thanks to the recent discovery of effective vaccines. On the other hand, global warming is expanding malaria's reach (see Section 16.3).

5.3   Plant Trade and European Imperialism

Trade in biological produce, grown or wild-harvested, lay at the economic hearts of the European empires (Brockway, 2002; Collingham, 2017). Initially, trade was often conducted by means of trading stations established along coasts, as the Portuguese did on the West African coast from 1482 and around the Indian Ocean from 1498. However, sooner or later, the Europeans tended to penetrate inland, settle down, establish plantations and take control of the harvesting of wild biological produce that could be sold profitably on the world market.

5.3.1   Sugarcane, slavery, indentured labour, Mauritius and Uganda

The arrival of Europeans in the New World led quickly to the establishment of plantations. Christopher Columbus, after returning to Spain from his first visit to the West Indies in 1492, sailed back west again in 1493, this time taking with him sugarcane stock from plantations on the East Atlantic islands. The usefulness of sugarcane as a plantation crop sprang from the high market value of the crystalline sugar that could be extracted from its canes (a technology discovered in India in the 4th century CE). By the time of Columbus' journeys, sugarcane, a crop that had originally been domesticated in New Guinea, had become widely distributed in the Old World and on islands across the Pacific. Sugarcane was known in Zanzibar in 1100 and was introduced into Spain following its conquest by the Moors in the 8th century (McMartin, 1961; Warner, 1962).

The indigenous inhabitants of the Americas were reluctant to work in the plantations of sugarcane and other crops that the Europeans established in the Americas, preferring to follow their established ways of life instead. The response of the plantation owners was to turn to the use of slaves, most of whom were purchased in Africa. A total of 12.5 million Africans were transported across the Atlantic during the days of the Atlantic slave trade, which, in total, extended from 1519 to 1875 (Franses and van den Heuvel, 2019).

Slave rebellions, an increased knowledge in Britain of the horrors of slavery and the rise of the humanitarian movement resulted in 1807 in slave trading being banned throughout the British Empire. Owning slaves was made illegal in 1833. A consequence of this was that the owners of plantations in America and elsewhere turned instead to the use of indentured labour, a way of employing people that has been likened to semi-slavery (Leopold, 2021). Indentured labour systems operate by signing people up to work away from their homelands for specified periods of time (often 5 or 10 years) in return for meagre wages, free accommodation and, on completion of contract, either small plots of land in the places where they were then living or free passage back to their homelands. The Indian subcontinent was the main place where the British recruited indentured workers, a total of 1.6 million being transported overseas from there between 1834 and 1920.

The place that received the largest number of indentured workers from the Indian subcontinent was Mauritius, a total of 450,000 being transported there between 1830 and 1920, the majority for work in sugarcane plantations. Mauritius is a textbook example of how the arrival of the human species in a place, from which it had previously been absent, can be devastating to the indigenous flora and fauna (see Chapter 1). It demonstrates the vulnerability of the endemic plants and animals of long-isolated oceanic islands to invasive species (see Section 2.3) (Strahm, 1994). Mauritius was apparently missed by the Austronesian people who travelled across the Indian Ocean to settle in Madagascar during the early centuries CE and, with the possible exception of a landing by Arab sailors, remained unvisited by people until the European colonial era. The Dutch discovered the island in 1598 and later settled on it and established sugarcane estates. Slaves were imported from Madagascar to work on the estates. The Dutch abandoned the island in 1710, which soon passed to the French. In 1810, the British took it over from the French, their purpose being to use it as a base to guard the sea passage from Britain to India that, at that time, passed around the south of Africa.

Loss of habitat and the ravages of invasive species are the principal causes of loss of indigenous species on Mauritius. Today, 85% of the island is covered by estates growing sugarcane, the few remaining fragments of indigenous eco-systems being mostly on rocky ground and infested by invasive plants. Figure 5.1 is a photograph of Mauritius showing a field of sugarcane. The forest on the hill in the background is Mondrian Nature Reserve, which was declared a protected area by an estate in 1979. Figure 5.2 is a photograph taken inside Macchabée Forest Reserve. It shows the fence on the boundary of an exclosure (to the right of the fence), which has been weeded of invasive plants. All the rather few remaining plants in the exclosure belong to indigenous species, graphically showing how abundant invasive plants had been in the unweeded forest. The fence is intended to exclude at least some of the invasive animal species that abound on Mauritius, preventing them from hampering the regeneration of the native flora. The most aggressive invasive plant in this forest is the guava tree (*Psidium guajava*), a native of South America.

Some species of plants and animals on Mauritius were directly targeted and driven to extinction. There were no indigenous terrestrial mammals on Mauritius; but the dodo, a type of giant flightless pigeon, proved an easy prey and was gone by 1662, and the two species of giant tortoises that lived on the island became extinct a few decades later. The ebony tree, *Diospyros tessellaria*, was a particular target for the Dutch, its black heartwood being valued for the manufacture of musical instruments and for use in marquetry. There were 12 species of indigenous *Diospyros* on Mauritius when people first arrived, but

Fig. 5.1. Sugarcane on Mauritius and forest remnant. Photo: A.H. (1990).

Fig. 5.2. Macchabée Forest Reserve with weeded exclosure, Mauritius. Photo: A.H. (1990).

one of them is now extinct and several of the others are endangered. Although *D. tessellaria* survives, its regeneration is poor, possibly because its seeds are adapted to germinate after passing through the guts of the dodo or a giant tortoise, now no longer available to fulfil this role.

Not all indentured workers from the Indian subcontinent worked in plantations. Some laboured on the Uganda Railway, a narrow-gauge line constructed by the British between 1896 and 1901 to connect the port of Mombasa on the African coast to Kisumu, a port in the northeastern corner of Lake Victoria (Hill, 1961). Onward connection to Uganda was initially by steamboat. According to Winston Churchill (the future British prime minister), the main reason for building the railway was political, specifically to secure British preponderance in the upper catchment of the Nile (Churchill, 1908). The railway was not expected to make an early profit. However, since it made the movement of British people and goods to and from Uganda much easier, it actually started to turn a profit in 1908. In 1916, Uganda, as a British territory, became financially self-supporting (Hansen, 1984).

Indentured labour was used to construct the railway for the same reason that slaves from Africa were used in the early plantations in the Americas – the reluctance of the indigenous people to forego their traditional ways of life and work all day long in labour gangs under the blazing sun. It was also because some of the indentured workers were literate, knew how to speak and write in English, and could serve as foremen and clerks. After fulfilling their contracts, most of the indentured workers returned to India, but others stayed on and helped to establish communities of South Asians in East Africa. There, they were joined by other Indians, such as Nanji Kalidas Mehta and Muljibhai Madhvani, who were both from Gujarat and who, at the ages of 13 and 14, respectively, made their own ways to East Africa to seek their fortunes (see Section 5.4.3).

The construction of the Uganda Railway had the by-product of drawing attention to the abundance of large wild animals then living in East Africa and the potential for 'big game' hunting. A book, *The Man-eaters of Tsavo and other East African Adventures*, published in 1907 by John Henry Patterson (1867–1947), one of the British supervisors (of Irish extraction) employed to oversee the gangs of indentured Indian labourers, became extremely popular (Patterson, 1928). One person who is known to have acquired the book was Theodore Roosevelt, the US president, who, once he had completed his second term as president in 1909, took the opportunity to travel to East Africa on an extended hunting tour (see Section 8.2.2). Patterson believed, as later did the British civil servants who worked in Uganda's Game and Fisheries Department, that game tourism could make a significant contribution towards maintaining the financial solvency of Britain's East African colonies (see Section 2.3.2). President Roosevelt's hunting party certainly encountered a lot of wild animals. Altogether, it killed more than 11,000 animals, including 5000 mammals, of which 13 were rhinos and nine lions (Thompson, 2010). Their preserved skins and other parts were shipped back to the United States for research and display in Roosevelt's home and in the Smithsonian Museum in Washington, DC and the American Museum of Natural History in New York.

5.3.2 Joseph Banks, Kew Gardens and economic botany

A pivotal figure in the botanical history of the British Empire was Joseph Banks (1743–1820), an aristocrat who developed an interest in botany early in life and, as the long-standing president of the Royal Society (1778–1820), assumed a pre-eminent position in British science. One of Banks' international scientific contacts was Alexander von Humboldt, who visited him in London in 1791 as part of his preparation for his travels in Central and South America (see Section 6.5.1).

As a young naturalist, Banks had accompanied the British explorer, Captain Cook, on a voyage of discovery to the Pacific (1768–1771) and had used the opportunity to collect large numbers of plants, many new to science. On Tahiti, he encountered the breadfruit tree (*Artocarpus altilis*) and, believing that its large nutritious fruit might be useful for feeding slaves in the West Indies, organised two follow-up expeditions to collect seed and seedlings, the second of which (in 1791) was successful. Cook's Pacific voyage included the first known landfall of Europeans on the eastern seaboard of Australia. Later, after he had become an advisor to the British Government on matters relating to Australia, Banks recommended Southeast Australia as a good place for a British settlement and Sydney Cove in Botany Bay as suitable for a penal colony. (Botany Bay lies within the modern city of Sydney.)

Banks arranged for many types of crops and trees to be sent to Australia to be tested for their suitability for Australian conditions and, in return, received a large number of Australian plants to try and grow in Britain. The latter included members of the genera *Acacia* and *Eucalyptus*, species of which have subsequently become widely used in forestry plantations in warmer parts of the world. Another of Banks' accomplishments was his founding, along with others, of the Horticultural Society of London in 1804 (renamed, the Royal Horticultural Society in 1861). During its time of existence, the Horticultural Society of London encouraged plant hunters to look for new types of plants to grow in Britain (see Sections 7.5.1 and 8.2.2).

In 1772, two royal estates near London, at Richmond and Kew, were merged and developed into an outstanding private botanical garden, containing many recently introduced species. The three people who were closely involved in the garden's early development were the king, George III, who was passionate about agriculture, his consort, Queen Charlotte, a lover of gardening, and Banks, a friend and key advisor to the king. Both George III and Banks died in 1820, after which Kew Gardens fell into a state of disrepair and, in 1840, was handed over to the state. The government considered closure, but a report written by a treasury commission recommended that, instead, the garden should remain open and be reorganised as a coordinating centre for the many botanical gardens that had been founded around the British Empire (Brockway, 2002, 2020). Kew was instructed to give these gardens a purpose and sense of direction. The motive was economic. By that time, the industrial revolution in Britain was in full swing and a considerable proportion of its industrial capacity was devoted to the processing of raw materials of botanical origin received from Britain's overseas possessions, for example cotton (see Section 5.3.3).

The system of organisation developed by Kew involved the enlistment of a network of corresponding individuals and institutions around the world to collect plants of potential economic value for dispatch to Kew. Map 5.2 shows the locations of botanical gardens involved in this network according to a note in the *Kew Bulletin of Miscellaneous Information* (Kew, 1889). Also on the map are Entebbe Botanical Garden (established 1898) and Limbe and Amani botanical gardens, both of which were acquired from Germany as spoils after the First World War. Germany had developed a similar system to Britain when it started acquiring its own colonies after 1871.

On receipt at Kew, the plants sent from correspondents were evaluated for their economic potential and, if promising, forwarded on to a network of botanical gardens, agricultural research stations and forestry research stations established at places in the British Empire. There, they were tested for their suitability for local conditions. To back up its mission, several centres of scientific expertise were created at Kew – an Economic Botany Collection (in 1847, initially called the General Museum), a Wood Museum (1847) and an herbarium (1852). A new scientific discipline, economic botany, emerged, involving the classification of plants according to the types of products that they yielded, rather than their taxonomic affinities (Cobley, 1956). Sometimes, Kew's enthusiasm for collecting plants strayed beyond the bounds of the British

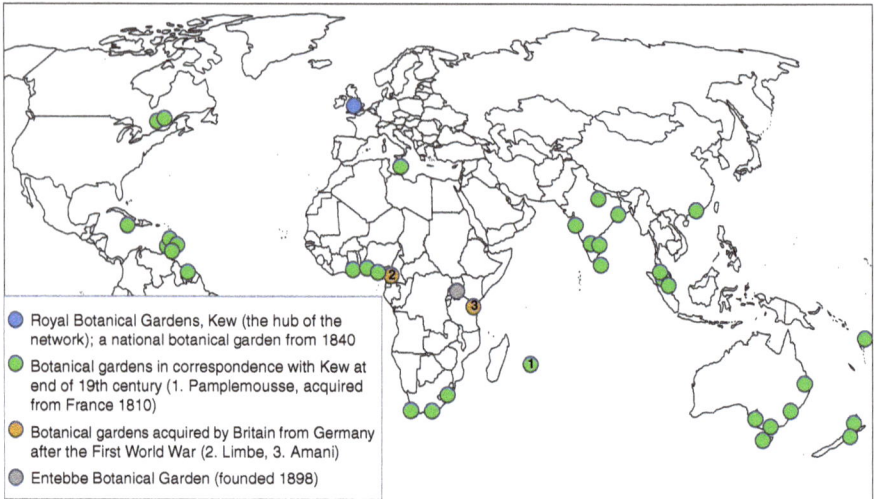

Royal Botanical Gardens, Kew (the hub of the network); a national botanical garden from 1840

Botanical gardens in correspondence with Kew at end of 19th century (1. Pamplemousse, acquired from France 1810)

Botanical gardens acquired by Britain from Germany after the First World War (2. Limbe, 3. Amani)

Entebbe Botanical Garden (founded 1898)

Map 5.2. Botanical gardens in correspondence with Kew in 1889. Note regarding the positions of country boundaries: see Disclaimer.

Empire, examples being the collection of seeds and saplings of the quinine tree from the Andean countries (see Section 5.2) and those of the *Hevea* rubber tree from Brazil (see Section 5.3.5).

For many of Britain's imperial possessions, the success of this Kew-centred system for economic development was integral to their overall economic success, for instance in terms of being able to build up their networks of roads and railways, educational systems and healthcare facilities. This was because the British Government in London expected the governors of its imperial possessions to balance their financial books and not repeatedly request subventions (funds) from the national treasury. Since the cores of the economies of many British possessions lay in their agricultural, forestry and pasture sectors, the key to their economic prosperity was often contingent on identifying biological produce that could be sold at a profit on the world market. This externally orientated approach to economic development fed back into the ways that the management of forests, farms and pastures were organised in the British Empire.

In forestry, systems of forest management were developed aimed primarily at producing timber of high value in export markets, with relatively little attention given to producing products that were only used locally (see Section 5.6). In agriculture, two different policies were followed, depending on the colony. One was the production of commodities for export through plantation agriculture (usually with British planters). The other was to find ways to encourage indigenous farmers to grow 'cash crops' on their farms.

5.3.3   The British East India Company and the story of cotton

It was common practice in medieval Europe for the Crown to grant charters to favoured people, giving them exclusive rights to benefit from the revenues from certain areas of land or to trade in specified commodities. The expansion of the European empires saw this system modified and extended, now with joint stock companies owned by companies of private investors being granted charters to trade in designated areas (Kocka, 2016). Two prime examples were the Dutch East India Company (VOC, 1602–1799) and the British East India Company (1600–1874), each of which was granted by its respective government the exclusive right to trade in the East Indies (a vague term at the time, referring roughly to the Indian subcontinent and those parts of Asia that lie to its east). The stock in these companies was initially traded in coffee shops, but later became traded in a more regulated way in authorised stock exchanges. The first base established by the British East India Company was in 1604 at Bantam in Java, where it negotiated with the Muslim sultan the right to trade in pepper. Later, hostility from Dutch traders led the company to move its base to the Indian subcontinent, where, subsequently, most of its activities were concentrated.

In practice, the chartered companies were run by a handful of Europe-based directors and were monopolies within their allocated territories. They enjoyed extraordinary measures of quasi-governmental power, being able to wage war, conclude treaties and take possession of land (Kocka, 2016). In India, the British East India Company steadily expanded its rule over the Indian subcontinent, overpowering its European competitors and assuming increasing dominance over the native rulers.

The most powerful political power in 18th century India was the Mughal Empire, which had been founded by Babur, an Uzbek chieftain, who had descended with his army on India via Afghanistan and the Khyber Pass in 1536. In 1749, the British East India Company was alerted to the fragility of the Mughal Empire, when Nader Shah, the Emperor of Persia, used the Khyber route to enter India, capture Delhi (the Mughal capital) and slaughter a large number of its inhabitants (Dalrymple, 2019). (Nader Shah then returned to Persia by the way he had come, laden with loot.) Critical events in the journey of the British East India Company to becoming the effective ruler of all India included the defeat of the Nawab of Bengal and his French ally at the Battle of Plessey (1757), the assumption of control over tax collection in Bengal (1765) and the capture of Delhi and deposition of the Mughal emperor (1803).

Intercontinental communications were very slow by modern standards prior to the mid-19th century and the representatives of the European chartered companies stationed abroad possessed considerable freedom to act on their own initiative. What this meant operationally was that the directors and owners of the companies in Europe sometimes knew little about what their companies were actually doing. The rule of the East India Company in India was sometimes criticised in Britain for its unaccountability, callous behaviour and the ruinous levels of taxation that it imposed on the Indian peasantry. Questions were asked in the British parliament about how it was possible for some company officials to secure vast fortunes, considering the relatively modest sizes of their salaries. However, such questioning tended to go nowhere, because many parliamentarians owned shares in the company or were happy to receive backhanders (Dalrymple, 2019). Matters came to a head in 1857, when a revolt erupted among the sepoys (Indian soldiers) in the company's army. The immediate cause was a suspicion among the sepoys that the new gunpowder cartridges with which they had been issued were greased with the fat of cows and pigs. The cartridges had to be torn open with the teeth prior to being loaded into muskets, but this was anathema both to Hindus (who regard the cow as sacred) and Muslims (who are required not to eat pork).

After supressing the Sepoy Revolt, a task in which the British were aided by the railway and the telegraph systems – both of which were beginning to be installed, the British Government decided to make some adjustments to the way that India was run. These included the replacement of the rule of the East India Company by direct rule, some land reforms (which included the establishment of a state forestry service – see Section 5.5), the admission of Indians into the junior ranks of the civil service and efforts to secure supplies of good-quality quinine to administer to the British people who lived in India (see Section 5.2).

Another influential development was the success of the French diplomat, Ferdinand de Lesseps (1805–1894), in securing an agreement with the Ottoman governor of Egypt in 1854 to build a canal across the Isthmus of Suez (opened 1869). Together with the introduction of steamships, this greatly reduced the travel time from Britain and India. Instead of the passage taking 5–8 months, it now took only 30–45 days (Frost, 2004). One consequence was a greatly increased number of British women travelling to live in India. Another was a tendency for the British in India to disengage from living in close contact with ordinary Indian society by retreating into guarded compounds – a development influenced by reports of atrocities committed against British women and children during the Sepoy Revolt.

The first three Western-style universities in India available for entry by Indians were opened in 1857 and, thereafter, the number of Indian graduates grew rapidly. These educated Indians were well aware of the regard with which educated Britons held the European enlightenment and its liberal values (see Section 6.1), which led to some of them questioning why they were not being treated by the educated British in India in the same way that they treated one another. Another cause of mounting resentment was disgust at the way that the British had reduced the once mighty Indian textile manufacturing industry to a shadow of its former self, while, simultaneously, building up its own domestic textile industry.

Mahatma Gandhi (1869–1948), an Indian who had trained as a lawyer in Britain, became a key figure in a subsequent non-violent political campaign that eventually led to India throwing off its colonial yoke and achieving independence (1947). As a symbol of his determination, Gandhi decided to stop wearing Western-style clothing in 1915, when he moved to India from South Africa (where he had been practising law), and then, in 1921, to change his attire once again from that of a better-off Gujarati gentleman to the simple clothing of the Indian poor. This clothing consisted of a short white *dhoti* (cotton loincloth) for wearing in summer and a *khadi* (shawl) to wear when it was chilly. His decision to regularly spend time spinning thread on a traditional Indian spinning wheel was a signal to his fellow Indians to take pride in their cultural heritage and to inspire them to join his campaign to boycott British goods (Brown and Fee, 2008).

Cotton is the world's most important natural fibre. The cotton genus, *Gossypium*, which contains about 50 species, is distributed naturally in the arid to semi-arid tropics of both the Old and New Worlds, and also Australia (Wendel, 2009). Four species of *Gossypium* became domesticated in antiquity, two in the New World (*Gossypium hirsutum* in Central America and *Gossypium barbadense* in South America) and two in the Old World (*Gossypium arboreum* in the Indian subcontinent and *Gossypium herbaceum* in Africa). The earliest archaeological records of cotton textiles date back to 5500 BCE in India and 4200 BCE in Peru (Fuller, 2008). Today, 90% of the world's cotton output is from *G. hirsutum* (which is favoured for its high yields), 8% from *G. barbadense* (which produces exceptionally high-quality fibres) and less than 2% from each of the Afro-Asiatic species (Hu Guanjing *et al.*, 2021).

Mughal India was the most important global manufacturing centre of textiles in the world in the 17th century, its subdivision of Bengal lying at the heart of a huge textile manufacturing and exporting industry. The British East India Company, once it had become established in India, started to export textiles to Britain, some produced by Indian manufacturers and some in its own factories. The quantities of textiles received from India in Britain became so great that British woollen and linen manufacturers, fearing that they might be put out of business, lobbied the government for protection. The result was the Calico Acts of 1720 and 1721, which banned the import of textiles from India, except for fustian (a heavy, hard-wearing, type of cloth favoured by manual labourers, country people and, later, left-wing political agitators). However, the importation of raw cotton continued to be permitted.

The response in Britain to the Calico Acts was the development of an indigenous cotton textile manufacturing industry, which led, at some time during the 19th century, to Britain replacing India as the world's leading manufacturer of cotton textiles. Britain's competitive advantage owed much to the shifting of textile manufacturing from home to factory settings and the adoption of new technologies in the cotton mills and weaving sheds, speeding up production and reducing unit costs. The new technologies included the flying shuttle (invented by John Kay in 1733), the spinning jenny (invented by James Hargreaves in 1764) and the spinning mule (invented by Samuel Crompton in 1779). From the early 1800s, James Watt's steam engine became widely used to drive factory machinery, its advantages (compared with using waterpower) being greater reliability and power output and allowing cotton mills to be established anywhere, not just by rivers.

The heyday of the British Empire was the 19th to early 20th centuries, at which time one pillar of Britain's economic strength was its domination of the global textile business. Cotton-growing was encouraged in suitable places around the empire, including Uganda. One of Britain's most ambitious cotton-growing projects was the Gezira Scheme in the Sudan, which was based on an irrigation system that included construction of a dam on the Blue Nile (the Sennar Dam, completed 1925). On the manufacturing side, other national players started to come into the global cotton and textile business from around the beginning of the 20th century, one of the earliest being the United States. After the Second World War, an attempt was made in Britain to reinvigorate its flagging textile industry by recruiting foreign labour to work in its mills. A major source of this labour was Pakistan, prominent among them being many of those who were displaced when the Mangla Dam was constructed on the Jhelum River in the upper reaches of the Indus (see Section 19.2.1).

5.3.4 Tea from China

Tea became the British national drink during the 18th century, becoming so popular by mid-century that John Wesley, the founder of Methodism, complained that the poor were spending too much of the little money that they had on tea (Wesley, 1748). Long a commodity widely traded in China and neighbourhood (see Section 3.5.2), the first direct shipment of tea by the British East India Company to arrive in London came in 1669, following the placing of an order for a small amount with the company's agent in Bantam in Java. In 1757, the Chinese Government decided to regularise the way that Western merchants traded with China by restricting their operations to a single neighbourhood (known as the Thirteen Factories) within the port city of Guangzhou (Canton). Here, each of the Western countries was assigned a 'factory', an area of land on which the Western merchants established their offices, trading posts and warehouses.

A number of events in the 18th century resulted in the British East India Company becoming faced in 1772 with a financial crisis. The Calico Acts had severely restricted the company's revenue (see Section 5.3.3), the imposition of crippling levels of taxation on the rural Indian population had contributed to a great famine (1769–1773) and the company had become over-dependent on paying for goods through issuing credit notes (Dalrymple, 2019). In desperation, the company appealed to the British Government for assistance and the government duly responded by easing the tax burden on the company though passing a Tea Act (1773). This enabled the company to sell its tea at prices that undercut its rivals. However, an unfortunate side-effect (from the British point of view) was that it ignited the American War of Independence (1775–1783), which resulted in Britain losing its colonies in the land that is now the United States.

Additional to this, the British East India Company responded to is financial woes by turning to selling opium to China, where it had a particular financial deficit. This problem arose because, while tea, porcelain and silk were popular in Britain, there was little made in Britain that interested the Chinese. The opium destined for sale in China was grown for the company by Indian peasant farmers in Bengal. The company did not wish to be seen to be directly involved in shipping opium to China, because the Chinese Government, worried about the numbers of its citizens who were becoming opium-addicts, had made its importation illegal. So, instead of selling directly, the company imposed an obligation on the independent merchants who bought the opium at auctions in Calcutta, that they had to sell it in China. The independent merchants resorted to smuggling.

In 1838, with the amount of smuggled opium approaching 1400 tons a year and the number of Chinese opium-addicts having grown to between 4 million and 12 million, the Chinese Government decided to impose a death

penalty for people involved in opium smuggling. A special imperial commis-
sioner, Lin Zexu, was dispatched to enforce the decision. Matters came to a
head in 1839–1842, when the First Opium War was fought between China
and Britain. This failed to completely resolve the matter and it was followed
by a Second Opium War in 1856–1860, this time with France joining in with
Britain. It turned out that the Chinese military was no match technologically for
the militaries of the British and French, which resulted in China being forced
to accept humiliating terms in the treaties that concluded the two wars. The
Treaty of Nanking (1842) ceded Hong Kong Island to Britain and stipulated
that, in addition to Guangzhou, four more ports were to be opened to foreign
trade (the treaty ports). The Treaty of Tientsin (1858) opened yet more ports
to foreign trade and permitted foreign Christian missionaries to move freely
within interior China (see Section 6.9).

In the 1830s, the British East India Company started to plant tea in the
Himalayan mid-hills of Assam, a part of India that it had just acquired (Mair and
Hoh, 2009). The crop was not performing well and, in 1848, the company
commissioned Robert Fortune to collect seeds and seedlings of tea plants from
China and ship them to India (Brockway, 1979, 2020). Fortune, who was
Director of Chelsea Physic Garden in London, had worked earlier (in 1843–1845)
as a plant hunter in China (see Section 7.5.1). Exporting tea plants and tea
seeds was illegal under Chinese law. Nevertheless, in 1851 Fortune succeeded
in shipping 2000 tea plants and 17,000 seeds of tea to India.

Chelsea Physic Garden, which had been founded in 1673 by the Worshipful
Societies of Apothecaries (pharmacists), has many connections with the story
of post-1500 British geographic and economic expansionism (Minter, 2000).
The garden was instrumental in the invention of the Wardian case, a sealed
greenhouse used for carrying plants on the decks of sailing ships, protecting
them from salt spray and self-sufficient in terms of nutrients and fresh water.
Robert Fortune's transportation of tea plants and seeds from China to India
was the first practical application of the Wardian case. Thereafter, it was used
by Kew, in 1857–1860 and 1873–1876, respectively, to transport seedlings
of quinine and rubber plants from South America to Britain, and then, later on,
plants grown on at Kew to various tropical destinations in the British Empire
(see Sections 5.2 and 5.3.5). Figure 5.3 is a photograph of Chelsea Physic
Garden, showing members of Plantlife International touring the garden during an
Annual General Meeting. The bed in the left foreground is dedicated to plants
used for anaesthesia and analgesia (see Section 7.5.2).

However, in the end, Fortune's shipment of tea plants and seeds to China
did little to further the development of the tea industry in India, because the
British preference was for a strong dark brew, which was found to be best
made from a variety of the tea plant that is native to Assam (*Camellia sinensis*
var. *assamica*). In 1853–1856, Fortune made yet another trip to China, the
main purpose this time being to recruit skilled manufacturers of tea in China
to take to India to help develop the Indian tea industry, a mission in which he
was successful. Earlier, Fortune had learnt on his travels in China something
that was previously unknown in the West – that all types of tea, whether black,
green or any other, come from the same species of plant, but are processed

Fig. 5.3. Chelsea Physic Garden in London, Britain. Photo: A.H. (2007).

differently. Therefore, the big influence of Fortune on the economic development of India and the other parts of the British Empire where tea was planted came not through his expertise in smuggling, but rather from the spreading of Chinese knowledge of tea manufacturing to other parts of the world.

5.3.5   The saga of rubber, Part 1 – Amazonia and Africa

Rubber, like quinine, sugar, cotton and tea, is one of those botanical commodities that have influenced the course of world history. Made from a latex that oozes out of the trunks and stems of certain types of trees and lianas when cut or damaged, all of the plants that have been exploited commercially for rubber during the last 200 years have belonged to one or another of three plant families – the *Apocynaceae*, *Euphorbiaceae* and *Moraceae*. However, nowadays, only one species is used, the pará rubber tree, *Hevea brasiliensis* (*Euphorbiaceae*), which is native to the Amazonian rainforest. Historically, rubber played a significant role in some pre-Columbian American civilisations. From as far back as the 2nd millennium BCE, balls made from rubber obtained from the tree *Castilla elastica* (*Moraceae*) were used to play a ritualised game with religious overtones – the world's first-known organised sport.

The modern rubber story started in 1839, when Charles Goodyear (1800–1860), a self-taught American chemist and manufacturing engineer, discovered a chemical process (vulcanisation) that stops rubber getting sticky in the heat and hard in the cold, so making it useful for industrial applications (Brockway, 1979). A boom in the price of rubber starting in 1850 led to Brazilian and foreign (especially British) investors making agreements with businessmen active in Amazonia to organise the indigenous Amerindians to collect rubber from wild *Hevea* trees on their behalf. The harsh ways in which this collection was organised proved disastrous for the indigenous people, contributing to the extinction of several ethnic groups.

Following up on its new mission of supporting Britain's imperial economy (see Section 5.3.2), Kew Gardens, under its second director, Joseph Hooker, was keen to see if *Hevea* plantations could be established somewhere within the British Empire. In 1873, Hooker requested Henry Wickham, who was a Kew correspondent living in Santarém in Amazonia, to send seeds of *Hevea* to Kew. In 1876, Wickham managed to do so, in the process slipping the seeds through the Brazilian customs. At Kew, the seeds proved difficult to germinate, but, eventually, its horticultural experts succeeded in germinating about 10% and, when sufficiently well established, the resulting seedlings were shipped out to Peradeniya Botanic Gardens in Ceylon (Sri Lanka). Later, plants from Peradeniya were dispatched to Singapore Botanic Gardens, whose researchers managed to find a way to ensure that the latex was released from the trees in quantity. This final obstacle overcome, the first commercial rubber plantations in the Malay States were planted in 1895. Once rubber from plantations came on stream, collecting wild rubber in the Amazonian rainforest lost its profitability and, in 1913, the commercial collection of wild rubber ceased.

Meanwhile in Belgium, a country created in 1830 as a buffer to keep the periodically warring states of France and Germany apart, King Leopold II (reigned, 1865–1909) developed an interest in acquiring land in Africa. In 1879, he contracted the Welsh-American journalist and explorer, Henry Stanley (1841–1904), to undertake a mission to the Congo and report back on the natural resources that it had to offer. Leopold's choice of investment advisor was due to Stanley's reputation as a successful explorer of Africa. In 1874–1877, he had traversed the whole width of tropical Africa – the first person known to have done so. His journey started in Zanzibar on the shore of the Indian Ocean, passed through Buganda, down the Lualaba and Congo rivers and ended on the Atlantic shore.

Leopold's interest in the resources of the Congo was part of a surge of interest in Europe in the natural resources of Africa that developed in the second half of the 19th century. Aware of the endless wars between the European powers that had broken out from the 16th century onwards as they fought for the land and natural resources of the Americas and Asia, those European powers that were interested in having a stake in Africa met at a series of congresses and conferences during 1876–1890 to decide how to divide up the continent between them (Pakenham, 1991). Apart from promoting commerce, another declared motive was to put an end to the slave trade. As mentioned in Section 5.3.1, the rise of humanitarianism in Europe had finally put an end to the European trans-Atlantic slave trade in 1875 and the slave trade that the European leaders had in mind was the Arab slave trade operating from bases along the Indian Ocean coast. Zanzibar, an island situated just off the coast of the modern country of Tanzania, was at the hub of the slave trade at the time. Said bin Sultan, the ruler of Muscat and Oman, had moved his capital there from Arabia in 1840, so as to better control the production of cloves on the island and the trade in slaves.

There were two other new members of the European colonial club in the 1880s, in addition to Belgium. They were Germany and Italy, which had only become unified countries in 1871 and 1861, respectively. At a conference in 1884–1885 of European ambassadors to Berlin called by Otto von Bismarck (the first chancellor of unified Germany), Germany succeeded in being granted the territories that today are the countries of Cameroon, Namibia, Tanzania, Rwanda and Burundi. Italy was allocated Libya, Eritrea and a part of Somalia, which was an insufficient amount in its opinion, leading to its decision to invade the Empire of Ethiopia in 1895–1896 to try and seize its territory (see Section 8.2.3). Leopold requested the Congo, which the other European powers were reluctant to grant him, wary of his character and intentions. However, eventually they agreed to do so after Leopold had assured them that he would commit himself to improving the welfare of its people.

At first, Leopold's hope in making money in the Congo lay in killing elephants to sell their tusks, but this turned out not to be particularly lucrative. Then, in the 1890s, a surge in the demand for rubber on the world market transformed the economic prospects of the Congo. The surge came about as a result of the invention and popularisation of the bicycle and motor vehicle, both of which had rubber tyres mounted on their wheels. There were two plants in the Congolese forest that were known to produce rubber of commercial quality, one a tree (*Funtumia elastica*) and the other a liana (*Landolphia owariensis*), both in the botanical family *Apocynaceae*. King Leopold, who never personally visited the Congo, granted the rights to collect the rubber to private companies, in the process imposing virtually no conditions on how they were to operate. The companies in question were not necessarily Belgian, because one of the agreements that had been reached by the European leaders at their meetings to discuss the future of Africa was that the Congo and Niger basins were to be free-trade areas, not territories assigned to single national companies, as had been the custom earlier.

What transpired in the Congo was an almost unbelievable level of cruelty inflicted on the Congolese people to persuade them to collect wild rubber. Between 1885 and 1908, millions of Congolese died directly as a result of harsh treatment and many more perished through the disruption caused to the pre-existing indigenous economic and cultural arrangements (Iliffe, 2017). When it became known in Europe what was happening, which was partly through reports received from Christian missionaries, an outcry ensued and, in 1908, the Belgian government took over the running of the Congo from the Belgian Crown.

On the 7th of December 1941, the Japanese launched a surprise attack on a US naval base at Pearl Harbour in Hawaii, which precipitated the United States into becoming a combatant in the Second World War. Very soon thereafter, the Japanese launched attacks on the Malaya Peninsula and the Dutch East Indies (Indonesia), both of which contained extensive plantations of rubber.

The Japanese succeeded in occupying Malaya and the Dutch East Indies very quickly, precipitating a crisis in rubber supply for the United States and its allies.

Richard Schultes, an American professor of biology at Harvard University and a pioneer ethnobotanist, played a key role in what happened next (see Chapter 10). Schultes was deep in the Colombian rainforest when the United States entered the Second World War. On learning of the attack on Pearl Harbour, Schultes, who was an American patriot, travelled post-haste to report for military duty at the US embassy in Bogotá. However, instead, he was given orders to return to Amazonia, organise the indigenous people to collect rubber from wild *Hevea* trees and thereby create an alternative source of rubber. Additionally, Schultes was ordered to collect seed from different stands of *Hevea* trees, the purpose of which was to see whether populations of *Hevea* could be found that were genetically resistant to leaf blight, a fungal disease that had hitherto prevented rubber plantations from being established in the New World (in the event, no such populations were found).

5.4 Agricultural Revolutions Since 1500

Agricultural changes since 1500 have profoundly influenced the course of history, just as did the inventions of farming and irrigation earlier. Various labels have been given to stages along this agricultural journey, among them the agricultural and green revolutions (terms that, in this context, have particular meanings), and industrial agriculture. All can be seen as both efforts to increase the production of food to feed an expanding population and as root causes of that expansion.

The agricultural revolution is a term used for a period in British agricultural history, dating to approximately 1500–1850, marked by land reform and technological innovation. The term, green revolution, is used for a phase in the agricultural history of certain countries in the Global South, dating from approximately 1950 to the 1980s, when some aspects of industrial agriculture became adopted. The term, industrial agriculture, is not specific to any particular part of the world or time in the same way and several dates have been proposed for when it began, from 1850 on. Much depends on the place under consideration and which aspects of industrialisation are regarded as significant. Industrial agriculture has since permeated to many parts of the world, although there are still substantial areas where traditional forms of agriculture are dominant.

5.4.1 The agricultural revolution in Britain

Farming in Britain prior to the agricultural revolution was organised along feudal lines. The land was held by the Crown and, under it, the aristocracy and the Church; at village level, land lay in the hands of the lords of the manor, the abbots of monasteries and the local clergy. The people who actually worked the land, the peasants and serfs, did have certain customary rights to natural resources. Notably, they had rights to cultivate strips of land in open fields on payment of dues to landlords, to harvest certain products from woodlands and to pasture their livestock on communal land (known as 'common land'). Outside this settled manorial order, there were extensive areas of land where specified rights were reserved for the Crown. Known as 'forests' (from Latin *foris*, 'outside' – meaning 'outside the common law'), chief among these rights was the hunting of 'game', referring especially to deer. Forests were not necessarily areas of wild, untouched land. They could include villages and even towns.

The agricultural revolution affected many aspects of the feudal order – the ownership of land, the rights of access to natural resources, the look of the countryside and farming practices. The most drastic legal change was the transfer of land from multiple to sole owners. The new landowners were often members of the aristocracy or lords of the manor, who were now freer from their feudal obligations to ordinary villagers. Because these changes were so drastic, with details varying from place to place, the introduction of this land reform often required changes to the law. In consequence, over 5200 individual acts of parliament authorising the reform were made between 1604 and 1914. For the most part, the acts were agreed in parliament without demur, a process smoothed along by the fact that, at the time, many parliamentarians were members of the aristocracy or lords of the manor themselves. The losers were the peasants, many of whom lost their means of livelihood and had little choice but to leave the countryside. From the late 18th century on, many went to live under very poor conditions in the rapidly expanding towns and cities of the industrial revolution. Others emigrated abroad.

The acts of parliament authorising the reorganisation of the countryside were known as Enclosure Acts, a name that refers to one of their major consequences – the dividing up of the open fields (in which cultivation was often formerly practised) into enclosed fields surrounded by hedges. Each of the enclosed fields was then subjected to rotations of annual crops, a practice that, ecologically, had two advantages – it helped to rejuvenate the soil and to reduce the likelihood of outbreaks of pests and diseases. Various sequences of crop rotation became recommended. For example, in one known as the four-year Norfolk System, the succession of crops was wheat, turnips (cultivars of the cabbage genus, *Brassica*), barley, a mixture of clover (*Trifolium*) and rye-grass (*Lolium*), and then back to wheat. The turnips were used for fodder, and the mixture of clover and ryegrass was cut, dried, stored and used for winter feed. Other technological innovations that came with the agricultural revolution included increased mechanisation (the horse providing the power) and the introduction of more productive varieties of crops and breeds of farm animals.

5.4.2 Industrial agriculture

In Britain, a country early to industrialise, the industrialisation of agriculture began in the mid-19th century. Today, industrial agriculture can be found all over the world. Typical features (compared with pre-industrial agriculture) are extensive use of fossil fuels (to drive farm machinery, for transport and to manufacture agrichemicals), the use of many types of specialist machinery, a reduced use of labour and an increased size of farms and fields. Plant and animal breeders involved in agriculture rapidly developed fresh techniques for producing new varieties of crops and livestock, including through making precise changes to their DNA. A wide variety of agrichemicals (synthetic chemicals, often made from petroleum) may be used, these including fertilisers, herbicides, arboricides, pesticides, fungicides and antibiotics. Nowadays, the use of information technology in agriculture is being continually updated. One of the more recent introductions is the use of a combination of big data, remote sensing and high-tech tractors to deliver precise doses of agrichemicals to particular parts of fields (Ullo and Sinha, 2021).

Backed up by the rise in information technology and with underpinning from the financial sector, the growth of industrial agriculture has been accompanied by the increasing globalisation of agrifood businesses (see Section 6.3.3). Countries that have adopted free-market, neoliberal, policies have seen increasing amounts of their farmland coming under the control of foreign investors (Scoppola, 2021). For example, in Uganda by the early 2010s, an estimated 4–8% of the land had been acquired by foreign investors (FOE, 2012; Lyons et al., 2014) and, since the government has been eager to attract yet more investors, the percentage is likely to be higher now.

People in rich countries and the rich in poor ones have become habituated to the expectation that they will be able to purchase, as they please, a huge variety of food products in their local supermarkets, these products containing ingredients from plants sourced all over the world. The effects of this economic pull have included increased inequality and levels of food insecurity, both between the Global South and the Global North and, within each of them, between the wealthy and the poor (Otero et al., 2017). At the same time, people have become increasingly distanced, physically and psychologically, from the realities of the natural world. Although it is common for supermarkets to make some efforts to be environmentally and socially responsible (Jayawardhena et al., 2016), in free-market economies the financial bottom line means that, to stay in business, grocery retailers are tempted to cut environmental and humanitarian standards (Fornari et al., 2020).

The end result of this chain of interconnected processes is that industrial agriculture, considered as a conglomerate of human activities, is one of the principal causes of environmental deterioration, including loss of plant species, global warming, and eutrophication of freshwater and marine habitats (Withers et al., 2014; Angelo, 2017; Nic Lughadha, 2020). To take just one example, the Haber–Bosch process, which is used for manufacturing nitrogen fertilisers, emits more carbon dioxide globally than any other single, chemical-making, industrial reaction (see Section 7.4.2) (Boerner, 2019).

As a generalisation, it is countries in the Global South that produce primary commodities, which includes produce from farms, pastures and forests, and countries in the Global North that manufacture consumer products and which house the headquarters of large transnational businesses. In a low regulatory environment, the key driver determining the prices paid to producers of primary commodities are those which consumers in the Global North are prepared to pay for products, because it is in the Global North that most profits are to be made and where competition to secure slices of the lucrative consumer cake is most intense (Otero *et al.*, 2017). Political leaders in the Global South can be tempted to offer favourable deals to large transnational companies to persuade them to operate in their countries, rather than others.

One consequence of these competitive pressures is that, at field level, it is tempting for managers of farms, pastures and forests to maximise yields of produce at the expense of environmental and social costs. This can lead (for example) to the use of excessive amounts of irrigation water and agrichemicals on farms, which, in effect, results in the Global North being subsidised by the Global South in terms of water usage and chemical pollution (White *et al.*, 2018).

Similarly, human health has been another casualty of industrial agriculture. In the absence of effective regulation, agrifood businesses are tempted to produce foods with high levels of salt, fat and sugar – substances that are attractive to the human palate, but contribute nothing to nutrition and, if taken in excess, are major underlying causes of ill-health (Downer *et al.*, 2020). There is a global pandemic in tooth decay, thanks to the high levels of availability of sugar-rich and highly processed foods and drinks (Ungar, 2020). High levels of nitrate in drinking water, caused by the excessive use of nitrogen fertilisers, are damaging human health. The common practice of adding antibiotics to livestock feed has contributed to a rise of antibiotic resistance in pathogenic bacteria, which is now a major public health concern (Yi Zhou *et al.*, 2021).

The fate of the Aral Sea illustrates how environmentally damaging the unrestrained adoption of industrial agriculture can be (Manschadi *et al.*, 2010). Prior to the mid-20th century, the Aral Sea was a large freshwater lake situated in an intercontinental basin in semi-arid Central Asia. Fed by the inflows of water from the Amu Darya (Oxus) and Syr Darya (Jaxartes) rivers, rising in the Pamir and Tian Shan mountains, respectively, it was an oasis of productive human life that had once supported the sophisticated Bronze Age Oxus Civilisation (c.2400–1900 BCE). After the takeover of Russia by the communists in 1917 and the creation of the Soviet Union (in 1922), control over agricultural production became the preserve of central planners. Casting around for likely prospects, the planners spotted an opportunity to generate foreign exchange by growing cotton in the basin of the Aral Sea and selling it on the world market. Growing the cotton required irrigation and, as the area of irrigated agriculture in the basin increased from 1924 to 1991, so more and more water became extracted from the rivers that flowed into the lake.

For a time, the plan succeeded, and Uzbekistan (at that time part of the Soviet Union) became the world's leading exporter of cotton (or 'white gold', as it came to be called). The peak was reached in 1988. However, the extraction of water for irrigation from the rivers had resulted in such a great reduction in inflows into the lake that it began to shrink. From being the fourth largest lake in the world in 1960, by 2005 it had become only 20% of its former size, both in area and volume, and, from being a single body of water, it had become split into smaller parts. A once-thriving fishing industry was lost and a combination of salinisation of the irrigated land and excessive applications of agrichemicals had turned huge areas of once-productive farmland into polluted deserts.

Farmers are at the sharp end of industrial agriculture, because they work at the interface between the world of big business, which demands financial efficiency, and the world of nature, with its long-term requirement for ecological sustainability. James Rebanks is an example of a British farmer who, recognising the environmental challenges posed by industrial agriculture, decided that he would do his best to farm his land in a way that is both financially viable and environmentally friendly. Rebanks' farm, which lies in the hilly Lake District area of England, has been in his family's hands for several generations. Having gained practical experience over a number of years, he decided to publish his conclusions so that others could benefit from what he has learnt (Rebanks, 2020). He makes three suggestions about what other farmers in similar circumstances can do. One is to learn from the past about how organic or relatively low-input farming used to be done. The second is to undertake research on their own farms on how to increase the productivity of organic or low-input agriculture. The third is to find ways to reconnect urban people with the countryside, so that they can learn something about how their food is produced and the realities of farming. He hoped that, as a consequence, urban people would become more aware of the implications of their food-purchasing decisions and not just take for granted the ready availability of food on supermarket shelves.

Jared Diamond, a biogeographer, has wondered whether there is anything to be learnt about what to do about modern agriculture by studying the fates of human societies that, in the past, have run into the environmental buffers (Diamond, 2005). The sample of societies that he selected for his study, some modern and some in the past, included a number on isolated islands in the Pacific Ocean, some situated in isolated valleys in the highlands of New Guinea (largely separated from one another by the deeply incised mountainous topography), the Greenland Norse, the Maya, and several other ancient civilisations in the Americas. Some of these societies thrived and others collapsed, and the question asked by Diamond was, 'Why?' Diamond defined a societal collapse as a 'drastic decrease in human population size and/or political/economic/ social complexity, over a considerable area, for a considerable time'.

From his analysis, Diamond found that all the past societies that collapsed had followed a common path. First, growth in the sizes of their populations had forced the people to adopt intensified methods of agricultural production and expand agriculture on to land of only marginal suitability. This then led to environmental damage, food shortages and outbreaks of war between rival factions competing for the remaining, increasingly scarce, resources. After this, the governing elites were overthrown by the disillusioned masses – which, in turn, caused even further social chaos. This eventually resulted in reductions in the size of the populations through starvation, war and/or disease and the loss of some of the political, economic and social complexities that the societies had achieved at their peaks.

Easter Island, a small, isolated island in the Pacific Ocean separated by vast distances from other habitable islands, provides a textbook example of how a society collapses according to the findings of Diamond's research. The first people to occupy Easter Island were Austronesian-speaking Polynesians (see Section 3.3.3), who, arriving during the 1st millennium CE, found a bountiful, well-forested land. The soils were excellent for growing many of the food crops that they brought with them in their canoes. However, eventually, the soils became impoverished through their overexploitation and all the trees were gone. When the next people from outside arrived, which was in 1722, the inhabitants of the island were found to have been reduced to scratching out a precarious living by growing crops on degraded soils and involved in incessant warfare between rival factions. They had become trapped on their island, unable to leave. Their predecessors had destroyed all the trees, so that they could no longer fashion hulls for boats, and they had been unable to grow *Pandanus*, the tropical plant whose leaves their ancestors had woven into sails (McCoy, 1973; Thomson *et al.*, 2006).

Diamond was wary of being accused by historians of being ignorant about the details of their pet past civilisations. In consequence, he decided to extend his analysis a stage further by exploring other factors that might have contributed to whether the societies had thrived or died. What this extended analysis revealed was that there was one set of additional factors that had always been influential and that there were three other sets of factors that in some, but not all, cases had played a hand. The set found to have always contributed consisted of the basic cultural values of the societies and their types of political, economic and social institutions. The sets that only sometimes contributed were climate change, the existence of hostile neighbours and the existence of friendly trading partners. Diamond's finding that basic cultural values are necessary for societies to stand a chance of achieving ecological sustainability is resonant with Adam Smith's conclusion that permitting the unregulated operation of markets will only work as a driver of general prosperity if practised in well-governed societies that uphold high standards of public morality (see Section 6.2.1).

In Britain, the coming of industrial agriculture continued the process of expelling peasants and agricultural workers from the countryside. Although conditions for those displaced could be terrible, there were two safety valves that gave them a chance of economic survival. One was the relatively high possibility of being able to emigrate, since Britain at the time was expanding her empire. The other was to find work in the new factories that were springing up in the early industrial revolution. Neither of these solutions was (and is) so readily available to the rural people in the Global South who have found (and are finding) themselves displaced by the arrival of industrial agriculture. Nowadays, few, if any, countries in the Global North welcome large numbers of immigrants from the Global South. Opposition to mass immigration from the Global South has hardened up in Britain over the decades. Earlier, as mentioned in Section 19.2.1, Britain was welcoming to the around 110,000 people who were displaced in Pakistan when the Mangla Dam was built during the 1960s.

Another reason why the prospects of the poor in the Global South have declined is because the automation of industry has reduced the need for unspecialised workers. In partial compensation, some new industries connected to the international market have emerged in the Global South to take advantage of the ready supply of cheap labour, for example to manufacture textiles. However, because many countries in the Global South are competing with one another to secure shares in these businesses and there can be several businesses within each of the countries competing with one another, wages can be minimal, working conditions dire and environmental standards non-existent.

5.4.3   The agricultural revolution in Uganda from 1900

By the end of 1890, the European powers had agreed on how to divide up the African continent between them (see Section 5.3.5). Among other allocations, nearly a quarter of the continent went to France (mostly in its drier northern part), Belgium received the Congo, and Britain and Germany were handed the northern and southern parts of East Africa, respectively. The British portion of East Africa essentially comprised the land now contained within the modern countries of Uganda and Kenya, and the German portion that contained within the modern countries of Tanzania, Rwanda and Burundi. Britain was especially pleased to receive Uganda, because of its strategic position in the headwaters of the river Nile and its agricultural fertility (Churchill, 1908).

Kabaka Muteesa I, the king who had greeted the first European explorer to reach Buganda, died in 1884 (see Section 5.2). His successor, Mwanga II, inherited a kingdom in which four, religiously connected, factions were competing for power – tensions that erupted during 1888–1898 into a series of civil wars (see Section 6.8.2). Mwanga, who was only 16 when he assumed office, was faced with the almost superhuman task of maintaining control over his kingdom in the face of this internal disorder, plus being faced with the demands of forceful representatives of European businesses. The first of these representatives to introduce himself at Mwanga's court was Carl Peters of the German East African Company. Mwanga and Peters signed a 'treaty of friendship' in February 1890, but, later in the same year, this became null and void in European legal eyes due to the decision made by the European governments that Uganda should go to Britain.

Queen Victoria granted a royal charter to the British East African Company (BEAC) in 1888 and, in December 1890, Captain Lugard, its representative, arrived in Buganda and quickly managed to get Mwanga and his senior chiefs to sign a treaty with the BEAC. On paper, this treaty was less favourable to Buganda than that which had been offered by Peters, because it transferred to the BEAC certain powers over revenue, trade and the administration of justice (Pakenham, 1991). The British followed up this coup by building a railway from the port of Mombasa to Lake Victoria in 1896–1901, establishing a Botanical, Forestry and Scientific Department (BFSD) in 1898 and signing an agreement (the Uganda Agreement) with the Kingdom of Buganda in 1900 that redistributed the ownership of land within the kingdom (Table 5.1). Agreements similar to the Uganda Agreement, although without including the *mailo* element, were made with the kingdoms of Toro, Ankole and Bunyoro-Kitara in 1900, 1901 and 1933, respectively.

As can be seen in Table 5.1, the two main beneficiaries of the Uganda Agreement were the British colonial administration and 'one thousand chiefs

Table 5.1. Allocation of land in Buganda under the Uganda Agreement of 1900. Source of information: West (1965).

Beneficiaries	Total area (km²)	Subcategories	Area (km²)
Protectorate Government (British colonial administration)	27,324	Forests Waste and uncultivated land Government stations	3,885 23,310 129
One thousand chiefs and private landowners (*mailo* land)	20,720		
The Kabaka and other dignitaries	2,481	Personal private property Land attached to the offices	1,942 539
Mission societies	238	Subdivided between Anglicans and Roman Catholics	
Estimated total area of Buganda	50,764		

and private landlords'. The initial allocation of land to each of the chiefs and landlords was 1 square mile, which resulted in the land coming to be called *mailo* land. The choice of those to receive *mailo* land was left to the *lukiiko* (the Buganda parliament), so that they tended to be members of the social elite. As time went by, *mailo* land has become increasingly subdivided and, today, there are many more than 1000 people who hold freehold titles to land in Buganda. They are not necessarily Baganda. The absence of a *mailo* element in the agreements signed by the British with the other three large kingdoms has meant that subsequent demands for private land in Uganda have fallen heavily on Buganda.

Nowadays, the principal law governing land issues is the Land Act (1998), which recognises four forms of land tenure, namely customary, leasehold, freehold and *mailo* (the latter applying only to Buganda). In addition, some land is held in trust on behalf of the people (including Forest Reserves and National Parks). Provisions in the Land Act allow customary and leasehold land to be converted into freehold, which gives farmers greater security against being evicted from their landholdings, but, in reality, little customary land has been so converted because the procedures to obtain the required Certificates of Customary Ownership are extremely cumbersome (FOE, 2012). Farmers farming customary land are vulnerable to losing their holdings when confronted by individuals and companies that have come into the possession of land titles.

The increasing authority of the British in Buganda and Uganda during the last decade of the 19th century was accompanied by the introduction of the Indian rupee as an official currency. The colonial administration started to pay salaries to chiefs in rupees and expect to receive tax revenues from them in the same currency (Pallaver, 2016). However, for some time, transactions between ordinary people at local level continued to be in the traditional currency of cowrie shells (see Section 3.5.3). In 1900, the colonial government introduced a hut tax – in effect, a tax on rural homesteads, one intention being to force farmers to grow cash crops, since this was the easiest way for them to raise the necessary cash. An Agricultural Department was created in 1907 and this was followed by the establishment of agricultural research stations, such as those that housed collections of banana varieties listed in Table 2.2. Later, an agricultural extension service was established to convey information from the research stations to farmers on how to grow crops, the emphasis being on crops of high export potential. Chiefs were given the job of distributing seed and conveying information to the indigenous farmers (Mukembo and Edwards, 2015).

The original intention of the colonial administration was to develop a plantation economy based on arabica coffee, cocoa and rubber, but, in 1909, this policy was superseded by one favouring the production of cotton and coffee by indigenous farmers (Dunbar, 1965; Hansen, 1984). The result was the creation of a dual system of cash-crop agriculture, involving both planters and indigenous farmers. Following experimentation, the main plantation crops turned out to be sugarcane and tea. The tea estates were mainly created after 1945 by European planters in the higher-altitude districts of Western Uganda. The principal owners of the sugarcane plantations were Indians, one of them being Indar Singh Gill, who was also involved in sawmilling, including on the East Usambara Mountains of Tanzania (see Section 16.2). Two of the others were Nanji Kalidas Mehta and Muljibhai Madhvani, who arrived in Uganda in 1900 and 1912, respectively (see Section 5.3.1), and proved to be highly successful businessmen. Madhvani's fortune eventually amounted to 10% of Uganda's gross domestic product. In 1972, along with other Asians, the Mehta and Madhvani families were expelled by President Idi Amin (see Section 5.7), but, following a change in government policy in the 1980s, the Asians who owned properties in Uganda were allowed to return and repossess them. Members of both the Mehta and Madhvani families came back and re-established control over their holdings.

The Uganda Agreement was a hastily drawn-up document which has led to many subsequent complexities and disputes (West, 1965; Lunyiigo, 2011b). One of its inadequacies was its apparent assumption that what is referred to as 'waste and uncultivated land' was free for the taking. However, as mentioned in Section 7.2, the land being referred to was actually an integral part of the plant-resource system of the Baganda. One historian, Professor Samwiri Lwanga-Lunyiigo, has put it this way (Lunyiigo, 2011b):

> In a hunting and shifting cultivation economy it is not possible to describe land as either wasteland or unoccupied. These so-called waste and unoccupied lands were either hunting grounds or areas preserved to move into when land is exhausted elsewhere.

There is no mention of ordinary people in the Uganda Agreement, more or less all of whom were small-scale farmers. Traditionally, they were provided with land by their clan chiefs, who also receive no mention in the agreement. Inevitably, the result has been conflicts breaking out between the customary landowners (the chiefs and their clansmen) and the new legal landowners – conflicts that are still very much in evidence today. Nowadays, ordinary villagers, living on land that their ancestors have farmed for generations, can find themselves summarily evicted by people who have come into the possession of land titles. The colonial government eventually realised its mistake and, in 1927, made an attempt to try and correct it by passing a law (the Busuulu and Envujjo Law) which gave rights to people who were occupying land through traditional understandings the right to pass it on to their heirs. (*Busuulu* is a labour obligation to a chief and *envujjo* is a tribute given to a chief, typically in kind.) However, misunderstandings about land rights continued, complicated by the failure of the colonial authorities to recognise that the giving of tribute by a peasant to a chief in return for protection (under the traditional feudal system) cannot be equated, in terms of the perceptions of the people, with the paying of rent by a renter to a landowner (as in a monetary economy) (Hansen, 1984).

Particularly painful consequences of the Uganda Agreement and the way that it has unfolded are the obstacles that it has placed on the people being able to visit and attend the graves of their ancestors (see Section 4.5). The pain has been described as follows (Roscoe, 1921):

> The only freehold lands [in traditional Buganda] were the clan burial grounds and they remained such until, through ignorance of the importance of the customs, the British Government abolished these grounds. Events have proved that it would have been beneficial for the country, had these freehold burial grounds been preserved, and the British Government would have been saved both trouble and expense by allowing the rights of the clans.

He adds:

> It was the universal belief that the clan burial ground was the home of the ghosts of the departed members, and that they remained there until they were reincarnated in the form of some child of the clan.

5.5   Forestry in Britain Since 1919

The First World War (1914–1918) shocked the British Government into abandoning its earlier laissez-faire policy of leaving the supply of timber to the mercy of the market (Tsouvalis, 2000). Large quantities of timber were required during the war to construct trenches on the frontline and, on the home front, to prop up the working faces of coal mines. However, it soon became apparent that Britain's sources of timber were not secure. Before the war, Britain had been importing 93% of its timber requirements, but the cargo ships on which the timber was carried proved vulnerable to being sunk by torpedo. In response to the challenge, the government established a Forestry Commission in 1919, charging it with the remit of creating a strategic reserve of timber. This was the first time that Britain had possessed a national forestry service, possibly because it had been lulled into a false sense of security through its possession of an empire. After all, so it may have been imagined, 'British' timber did not necessarily have to come from Britain.

The Forestry Commission set about its task by encouraging private landowners to plant trees and purchasing land for tree-planting on the open market. It turned out that much of the land that it was able to purchase was either low-grade agricultural land on hills or sandy, infertile, soils in the lowlands. More or less all of the species of trees planted in quantity by the Forestry Commission during its earlier decades were conifers, rather than broad-leaved trees. The conifers were preferred because of their straighter forms and potentially faster growth rates on the low-grade land that the Commission had acquired. Only one of the conifers, Scots pine (*Pinus sylvestris*), was a native. Most of the rest were North American, notably Sitka spruce (*Picea sitchensis*), lodgepole pine (*Pinus contorta*) and Douglas fir (*Pseudotsuga menziesii*). Hybrid larch (*Larix* × *eurolepis*), which is a cross between the European and Japanese species, was frequently planted on relatively fertile soils. One of the foresters responsible for promoting the foreign conifers was Augustine Henry, who, at one time, had been a plant hunter in China (see Section 7.5.1) (Durand, 1992).

Decisions by the Forestry Commission about which species to plant and how to manage the plantations were based on trial plots and experiments designed to determine, through statistical analyses, the ideal ways to raise seedlings, plant them out, thin out the stands and harvest the crops. Plantings were made in single-species stands, the intention being to have entire coups (areas of trees) ready for harvest at the same time. In that way, gangs of men, armed with tools and machinery, could arrive at work sites and, in efficient operations, clear the areas of standing trees and detritus, plant up the land with fresh batches of seedlings and launch fresh crops of trees on their way.

Although new at scale to Britain, forestry of this type had been practised in Germany and France from about the beginning of the 19th century. In Germany, a key figure in its development was George Ludwig Hartig (1764–1837), who initially worked as a forest manager for the princely county of Solms-Braunfels. Hartwig went on to found a number of forestry schools and ended his career as an honorary professor at the University of Berlin. In France, the state became strongly involved in forestry following the confiscation of large areas of forest from the aristocracy during the French Revolution of 1789. A national forestry school was opened at Nancy in 1824 (Oosthoek, 2010). The British aristocracy developed an interest in forestry in continental Europe during the 19th century and it became fashionable for them to employ German foresters on their country estates.

The 'scientific', experimentally based, German approach to forestry became influential at national level in British India about 50 years earlier than it did in Britain itself. What happened was that, in 1837, the British East India Company, which was running India on the British Government's behalf, began to construct a railway system, which required large quantities of timber. Concern was mounting about potential timber shortages and there were worries, too, that deforestation was causing soil erosion, climate change and shortages of water (Oosthoek, 2010). The British Government, which took over direct control of India after the Sepoy Revolt of 1857, decided in 1864 to tackle this problem by creating a National Forestry Service with the aim of improving the management of the country's forests. There being no pool of professionally trained British foresters available, the first three appointed heads of the department were all German, the first being Dietrich Brandis (1824–1907), who had trained at Nancy, and the second Brandis' student, William Schlich (1840–1935).

After completing his stint in India, Schlich moved to Britain, where, in 1885, he founded a forestry section at the Royal Indian Engineering College at Windsor and, subsequently in 1905, a school of forestry at the University of Oxford (Burley et al., 2009). Unlike Germany, which has traditionally placed a high value on technical education, Oxford had earlier been resistant to establishing a department to teach forestry, because it deemed forestry to be a technical subject unworthy of being a subject taught at university. In 1924, a separate institute, the Imperial Forestry Institute (later, the Oxford Forestry Institute), was established at Oxford, aimed specifically at supporting forestry in the British Empire. In 1901, William Schlich was elected a fellow of the Royal Society and, in 1909, received a knighthood from the king for his services in India.

From the start, the form of 'scientific' forestry adopted by the Forestry Commission was more narrowly reductionist than the ways that it was applied in France and Germany. For instance, French forestry left more room for the incorporation of traditional forestry practices (Oosthoek, 2010). Concentrating narrowly on its remit of producing a strategic supply of timber, a common attitude in the Forestry Commission was to consider its critics as irrational and emotional (Tsouvalis, 2000). They imagined that, given time, the nation would come round to appreciating what it was trying to do. Impressed by its own logic, a view developed within the Commission that the ways in which the woodlands had been traditionally managed in Britain were primitive and haphazard. The result was that patches of native woodland that were being managed in traditional ways could be marked as 'scrub' or 'derelict' on forestry planning maps, destined for clear-felling and replacement with conifers. The result was an historically unprecedented loss of native woodland between 1945 and 1975 (Rackham, 1976). The concept of land being 'derelict' or 'wasteland' was one that Britain employed also in its empire (see Table 5.1) (Whitehead, 2010).

German forestry and environmental thinking also penetrated China (Liu Liang, 2014; Wang Xi-qun, 2017). The person responsible was Gottlieb Fenzel, who came to the country twice. The first of his visits, which was in 1927–1929, was at the invitation of the invitation of Tao Xi-sheng, the president of Sun Yatsen University in Guangzhou in Southeast China and at the time minister of education. While Fenzel was there, he created the first modern forestry course in China and established the first pine plantation according to the principles of 'scientific forestry'. (The pine in question was *Pinus thunbergii*, a native of East Asia.) Frenzel's second visit, which was in 1932–1936, was to Northwest Agricultural and Forestry Academy in Shaanxi, where he succeeded in raising the awareness of the importance of forests for the conservation of soil and water. Apart from his duties at the universities, Frenzel campaigned for the strengthening of the laws protecting the forests. He died in China in 1937 and is buried in Lianhu Park in Xi'an.

Dissatisfaction with what the Forestry Commission was doing became vocal during the 1980s. A feeling had developed among the public that the geometrically shaped patches of dark green conifers on the hills did not belong in Britain (Schama, 1995). They jarred with the ideals spread by the 19th century nature romantics that the English countryside should consist of rolling farmland with enclosed fields, interrupted by patches of native woodland and rustic villages, and that Scotland should have purple heather-covered hills surmounted by craggy summits (see Section 8.2.1). Germany had had its critics of 'scientific' forestry too, but earlier, during the 19th century. However, rather than just sticking stubbornly to its 'scientific principles', forestry in Germany had proved to be more flexible than forestry in Britain. This was partly because, according to the traditional German worldview, untamed forest is close to the soul of the nation (Schama, 1995). It must also owe something to the environmental legacy of Humboldt (see Section 6.5.1).

A turnaround in thinking in professional forestry in Britain came in about 1990, when the Forestry Commission replaced its laser-like focus on producing a strategic supply of timber with a multipurpose policy that included, among its management objectives, the conservation of biodiversity, carbon capture, public education and making provision for public recreation – all while retaining the initial reason for its existence, namely to supply forest products (Tsouvalis, 2000). The conversion of the Commission and the public was facilitated by successful campaigns organised by environmentalists that rebranded heathland (formerly considered to be wasteland) as 'wildlife habitat', native woodland as 'ancient woodland' and large gnarled, old, partially dead, trees as 'heritage trees' (Schama, 1995).

The term 'ancient woodland' was an invention of two people – one an academic, Oliver Rackham (1939–2015), an expert on the history of the British countryside (Rackham, 1976), and the other a forester, George Peterken, who was familiar with the jargon of the Forestry Commission (Peterken and Mountford, 2017). The word 'ancient' was well chosen. It resonated with the British interest in old-fashioned things, including the idea of a once-verdant Britain existing before the imagined traditional freedoms of its people had become trampled underfoot by foreign invaders (Schama, 1995). Popularisation of the 'ancient woodland' concept was achieved by defining it in terms that 'ordinary' foresters and amateur naturalists could appreciate and apply for themselves (Tsouvalis, 2000). The criteria used for its definition included that a woodland should have been present on a site for a specified number of years (a matter that, in Britain, can be investigated by studying old maps) and by the presence of certain 'indicator' species of plants, mostly species of attractive appearance familiar to natural historians.

The Forestry Commission was dissolved in 2013 as part of a general devolution of power in the United Kingdom that was essentially unrelated to considerations of forestry or, indeed, to consideration of any other aspect of the natural environment. The Forestry Commission in England became Forestry England. In Wales, it was merged with other agencies to become Natural Resources Wales. In Scotland, the Forestry Commission was replaced in 2019 by two new bodies, Land Scotland and Forestry Scotland.

The story of forestry in Northern Ireland, the other contemporary part of the United Kingdom, has been intertwined with the stories of both the island of Ireland and the island of Great Britain. These histories include the experience of the Irish potato famine (see Section 2.3), the separation of Ireland into two self-governing parts in 1921 and the uncertainties and conflicts that have continued to plague its politics. The roaming life of Augustine Henry (1857–1930), who was one of the early plant hunters in China, epitomises these complexities.

Henry's father was an Irish Catholic businessman involved in the linen trade. His mother was accompanying her husband on a business trip to Dundee in Scotland when she came to full term and gave birth to Henry. Later, Henry attended school in Cookstown in the north of Ireland and then studied medicine in Galway in the south of Ireland, Belfast in the north and Edinburgh in Scotland. Next, Henry worked as a doctor and plant hunter in China (see Section 7.5.1). Retiring from this work in 1900, he retrained as a forester at Nancy in France, co-wrote an exhaustive seven-volume monograph on trees in cultivation in Britain and Ireland (Elwes and Henry, 1906–1913), became an advocate for planting foreign conifers in the British Isles and became the first person to receive a professorship in forestry in Ireland. Henry's institute, the Royal College of Science in Dublin, was the scene of a violent conflict during the Irish Civil War (1922–1923).

Evaluating the achievements of the Forestry Commission at the close of the 20th century, it can claim to have been quite successful in meeting its original remit (Aldhous, 1997). The forest cover of Great Britain had increased from 4% of its land area in 1919 to 13% in 2013. After its conversion to a more environmentally friendly approach in about 1990, it gained a reputation in international forestry circles for being one of the most progressive national forestry departments in the world. This said, there was (and is) a long way to go. Britain is one of the most biodiversity-depleted countries in the world (Stafford et al., 2021). About 80% of the timber and wood pulp used in Britain is imported from other countries, in some cases at the expense of high environmental and social costs in the source countries.

A.H. has some connections with Britain's forestry story. He was an undergraduate in the same college at Cambridge University as Oliver Rackham and occasionally accompanied him to local woodlands to assist with coppicing. More recently, he co-founded a community group, Greening Godalming (2009–2015), in his home town of Godalming, its purposes being to raise public awareness about the dangers of climate change and encourage local people to adopt more environmentally friendly lifestyles. One of the group's campaigns was the promotion of the use of local wood fuel harvested from sustainably managed sources (Greening Godalming, 2015). It so happens that Godalming lies in the part of Britain which has the highest cover of native woodland in the country and where both Forestry England and the local county natural history society (the Surrey Wildlife Trust) agree that greater extractive use of woodlands would be beneficial to both the economy and conservation of biodiversity. Forestry England was particularly keen to make more use of the woodlands for the production of wood fuel.

Many of the patches of native woodland in the Godalming area are of a type known as coppice-with-standards, consisting of an upper canopy of large, well-spaced, trees, known as 'standards' (typically oaks – Quercus petraea and Quercus robur), and an understorey of coppiced trees (often with abundant hazel – Corylus avellana). These woodlands are a product of a traditional management system designed to produce a combination of large trees with widely spreading branches for use as timber and smaller trees to produce products for local village use. Traditionally, the widely spreading branches of the oak trees were needed to provide timber for making the curved side-timbers of the hulls of wooden sailing ships. Prior to the introduction of iron-hulled vessels towards the end of the 19th century, the British merchant navy relied on such ships to transport products, such as sugar, cotton and tea, from their places of production overseas to factories in Britain; then, subsequently, to carry manufactured products out of Britain to sell abroad. To ensure safe passage for its merchant ships, Britain maintained a formidable navy of oak-constructed warships.

Three things were learnt by the committee of Greening Godalming during its wood fuel campaign. One was that many of the woodlands have been essentially unused and unmanaged for decades. Figure 5.4 is a photograph of such a 'neglected' woodland. The large tree on the right is an oak (*Q. robur*) and the small, multi-stemmed trees are hazel (*C. avellana*). Hazel is an understorey tree that was traditionally coppiced to provide many products for local village use, but the spindly shapes of the hazel trees in this picture show that they have not been properly managed for years. A second discovery was that knowledge of traditional woodland management had largely died out. The third was that a substantial number of people object to felling local trees as a matter of principle, especially if the trees are of indigenous species. However, a few local people have taken an interest in traditional woodland management and trying to produce forest products in sustainable ways. Figure 5.5 is a photograph of an iron ring charcoal kiln belonging to one of the Godalming woodmen.

Several historical factors have contributed towards creating this state of affairs – the decreased access to the woodlands by ordinary villagers during the agricultural revolution, the increased availability in Britain of wood imported from the British Empire, the replacement of wooden-hulled sailing ships by iron-hulled steamships (reducing the demand for large pieces of oak) and the availability of cheap coal (replacing local wood fuel). One hundred years ago, every village would have had its expert woodsmen, but, as time went by, their ancient craft has withered away. An apparent dissociation in people's minds has developed between the large quantities of products that they buy containing products from trees and the need to fell trees to provide the raw materials for the products.

Neoliberalism began to influence forestry in Britain during the 1970s. (See Section 6.3.1 for an explanation of neoliberalism.) Several attempts have been made to privatise the Forestry Commission (or its successor organisations), but, so far, these have been resisted successfully by environmentalists. Another avenue of neoliberal intervention was launched when hedge-fund managers discovered that there is money to be made through acquiring companies involved in tree-planting (Garside and Wyn, 2021). They had spotted an opportunity to profit from a combination of being able to purchase low-grade agricultural land at cheap prices, receive government grants for tree-planting and exploit changes in the tax laws (Tsouvalis, 2000). The public began to pay attention to what was happening when a number of celebrities, who had grown rich and famous through being in the media, were revealed to be the owners of tree plantations on land of value for wildlife. Some of this land was unsuitable for tree plantations anyway, because the soils were of such low quality that it is doubtful whether there would ever be a crop of timber. Some was covered by deep peat, requiring the digging of deep drains prior to planting the seedlings, which was causing the organic matter to oxidise, the peat bodies to shrink, streams to become acidified and increased risks of downstream flooding.

Fig. 5.4. Neglected woodland near Godalming, Britain. Photo: A.H. (2013).

5.6 Forestry in Uganda Since 1898

As mentioned in Section 5.4.3, the British administration in Uganda began a concerted search for botanical commodities with export potential with the founding of a BFSD in 1898. As in the Belgian Congo, a botanical commodity of particular interest was rubber. Morley Dawe (1880–1943), who had been a student at Kew, was appointed the first Director of the BFSD in 1902 and

Fig. 5.5. Charcoal-making near Godalming, Britain. Photo: A.H. (2013).

began assessing the territory's forests for valuable natural products, including the availability of the rubber-producing species, *F. africana*, *L. owariensis* and *Clitandra* (a genus of lianas in the *Apocynaceae*) (Colonial Report, 1905; Dawe, 1906). The first tapping concession for wild rubber was granted in 1902 and the quantity of rubber exported climbed to a peak in 1910, after which Ugandan rubber came under fierce competition on the world market. By 1919, exports had virtually ceased (Dunbar, 1965). There was a brief resurgence during the Second World War (Osmaston, 1959).

It was not until 1917 that a Forest Department was created to manage the land designated as 'forests' under the Uganda Agreement (Table 5.1) and, not until 1929, that a forest policy was officially adopted by the colonial administration. The formulation of the policy was achieved with the assistance of advice received from two British foresters employed in India, who came on successive visits (Troup, 1922; Nicholson, 1929). Because German environmental thinking had permeated Indian forestry, emphasis was placed in the reports of these consultants on the value of the forests for environmental protection, the production of forest products being only a secondary consideration. Nicholson stressed the importance of protecting the forests to moderate the local agricultural climate and for the delivery of water supplies. In production forestry, Nicholson recommended that emphasis should be placed on high-grade timber because of its value in the international market. When management plans were drawn up for individual forests, little attention was given to the management and production of non-timber forest products (NTFPs), such as medicinal plants and wild foods.

The advice received from the foresters in the Indian Forest Service was to establish a system of Forest Reserves, which the colonial administration in Uganda duly set about doing. Larger bodies of forest were declared Central Forest

Reserves (CFRs) and placed under the direct control of the Forest Department, while smaller forest patches were classified as Local Forest Reserves (LFRs) and made the responsibility of the local governments. At the time when the first LFRs were gazetted, the local governments were organised largely along ethnic lines, which provided them with a measure of received authority. The purposes designated to the two types of Forest Reserves were different. The CFRs were to be primarily for environmental protection and, in some cases, timber production, while LFRs were to be managed primarily to meet local demands for forest products, notably wood fuel and poles. A special effort was made to establish CFRs at altitudes above 2150 m, because of the value of a forest cover for protecting upland catchments. The first LFRs were small patches of indigenous forest, but soon these became supplemented by a large number of plantations mainly of *Eucalyptus*, a fast-growing Australian tree.

The forestry situation in Uganda has changed drastically during the last 50 years, prior to which there were few cases of illegality within the Forest Reserves according to the Annual Reports of the Forest Department (Hamilton, 1984). Following the political independence of Uganda in 1962, there was a 4-year period when the local governments were still organised largely along ethnic lines. During this period, the Kingdom of Buganda (the local government for its area) showed considerable enthusiasm for establishing LFRs, declaring 1373 km² of new ones between 1964 and 1966 alone.

Today, the rate of deforestation in Uganda, at ~2.7% per annum, is one of the highest in the world. Many of the remaining Forest Reserves have been degraded by virtually unrestricted collection of forest produce (Hamilton *et al.*, 2016). Extensive areas of Forest Reserves have been illegally cleared for agriculture. The LFRs have almost completely disappeared, declining from a total area of 3028 km² in 1965 (of which 1036 km² was rainforest) to 2.43 km² in 2011 (Webster and Osmaston, 2003; Kabogozza, 2011). Some Forest Reserves have been handed over to agrifood businesses for conversion to plantations of sugarcane and oil palm, a development that has attracted internal and external criticism (BirdLife International, 2008; Nakkazi, 2011; Tenywa, 2013). In 1993, in a move to try and save some of the larger forests from further loss and degradation, five of the largest CFRs were transferred from the Forest Department to the more disciplined Uganda National Parks (now the Uganda Wildlife Authority) and were turned into National Parks. We describe how this happened in the case of Bwindi Impenetrable Forest in Chapter 18.

Some of the woes of forests and forestry since the 1960s can be attributed to the increased pressure on the forests resulting from a massive rise in the human population, which grew from 7 million in 1962 to 45 million in 2020. Another influential demographic trend has been a burgeoning in the size of the urban population. The capital, Kampala, is today one of the fastest-growing cities in the world. The principal fuel used in Kampala is charcoal, supplies of which nowadays are being trucked from trees felled as far away as the South Sudanese border, a distance of 300 km.

As can be seen in Table 5.2, the institutional structure of forestry in Uganda has been unstable, especially since 1967. Some of these institutional changes were introduced for political reasons that, essentially, had nothing to do with

Table 5.2. Institutional history of forestry in Uganda, 1898–2016. Source: Hamilton *et al.* (2016).

1898	First director of a new Scientific and Forestry Department appointed
1917	Forest Department created
1929	First formulation of forest policy, concentrating on forest reservation for environmental protection and timber production
1929–1951	Large forests made Central Forest Reserves (CFRs) under the central government and small forests made Local Forest Reserves (LFRs) under local governments. (However, a survey in 1956–1960 found that a considerable area of forest in Buganda had fallen into private hands)
1967	CFRs and LFRs merged into the unitary category of Forest Reserves under the central government
1971–1973	Forest policy adjusted favouring enlargement of conifer plantations for volume wood production. Little emphasis given to natural forest, either for its protective functions or for productive purposes
1993	All Forest Reserves decentralised to local government, except for five of the larger forests (Bwindi Impenetrable, Elgon, Kibale, Mgahinga, Rwenzori) which were made National Parks and were transferred to Uganda National Parks
1995	All Forest Reserves recentralised
1997	All Forest Reserves decentralised
1998	Forest Reserves over 100 ha recentralised (and labelled CFRs); Forest Reserves under 100 ha remaining with local authorities (and labelled LFRs)
2001	A new forest policy adopted, emphasising a greater role for the private sector in forestry operations
2003	Forest Department replaced in principle by: (i) National Forestry Authority (NFA), responsible for CFRs; (ii) District Forest Services (DFS), responsible for LFRs and advice to private forest owners; (iii) Forest Sector Support Department (FSSD), responsible for coordination and regulation. The latter two organisations never materialised
2006	Board members of NFA resign or summarily dismissed, relating to conflict with the government over allocating parts of Mabira CFR and forests on the Ssese Islands to investors to establish plantations. Replacement board members appointed

forestry. Globally, this is not unusual. The same thing happened in the United Kingdom in 2013 (see Section 5.5). The political event that precipitated the centralisation of all forestry services in Uganda in 1967 was a military coup in 1966 mounted against the president, the Kabaka of Buganda, by the prime minister, Milton Obote (see Section 6.8.5). In 1968, the government belatedly produced a document entitled 'The Common Man's Charter', in which it explained why a *centralised* system of government was needed to accelerate the country's development (Aasland, 1974). Interestingly, exactly the opposite conclusion – that a *decentralised* system of government was needed to accelerate development – was the conclusion reached in 1972 in Tanzania, following the Arusha Declaration of Julius Nyerere in 1967 (see Section 16.2).

For plant conservationists interested in forestry, two points are emphasised from these experiences. One is that forestry is rarely of much interest to the political classes. The other is the desirability of plant conservationists taking time to consider the ideal structures of governance of natural resources from a sustainability viewpoint. Having some ready-formed ideas about this could serve them well in the event of opportunities arising that give them a chance to influence how the governments of their countries are institutionally structured.

5.7   Development Aid – The Cases of Botany and Forestry in Uganda

A wave of decolonisation swept through the world after the Second World War. Among the first countries to acquire their independence were the Asian countries of Burma (Myanmar), Ceylon (Sri Lanka), Indonesia, India and Pakistan. Many African countries followed in the 1960s. Together with the former colonies of Spain and Portugal in the Americas, most of which had gained their independence about 150 years earlier, it became customary to refer to this bloc of newly independent states as 'developing', 'underdeveloped', 'the Third World', 'the Global South' or just 'the South'. The last two of these terms, which have become increasingly fashionable, draw a contrast with another newly named bloc of countries, 'the Global North' (or just 'the North'). The Global North is essentially equivalent to the longer-established concept of 'the West' (see Section 5.1). Generally, countries in the Global South are financially poorer and lie in the tropics or subtropics and those in the Global North are richer and lie in the temperate climatic zones (especially in the northern hemisphere – hence the use of 'North' in the name). Countries in the Global South tend to be recipients of development aid and those in the Global North aid providers (Di Nicola, 2020). It is widely acknowledged that the Global South/Global North split is crude, but it can be useful as a generalisation.

For the first ten years after Uganda became an independent country in 1962, the transition from British rule to fully self-governing was smoothed along by many Britons who were working at senior levels in the civil service, education, the Church and business staying on at their posts and gradually becoming replaced by Ugandans who had been trained up for their jobs as counterparts. As this was happening, development aid – a new phenomenon – started to flow from the Global North to Uganda. Apart from Britain with its imperial hangover, the first country to supply development aid to botany and forestry to Uganda was Norway, which, from the late 1960s, supported a few Norwegian botanists teaching in the Department of Botany at Makerere University, helped to establish a new academic Department of Forestry, also at Makerere, and provided assistance to the government's Forest Department to assist in the management of its forests.

Norway's expectations of Uganda as an aid partner received a jolt in 1972, when Idi Amin, the President of Uganda, decided to expel all the 'Asians' from the country. (By 'Asians' was meant those residents of Uganda whose heritage lay in the Indian subcontinent.) Norway saw this as a gross abuse of human rights and responded by abruptly closing down its aid operations and quitting. A.H. was affected by this, because it came at the time when he had just delivered to Makerere Printery the manuscript of a Ugandan forest flora, which had been supported in its preparation by NORAD (Norwegian aid). It delayed the printing of the book by 10 years (see Preface).

Not all Asians were unwelcome in Idi Amin's Uganda. A number of Pakistanis, a predominantly Muslim country, were recruited to work at Makerere University. One of them, Professor Salim, an expert on algae, was a colleague of A.H. when he taught in its Department of Botany in 1981–1982 (see Preface). Amin's willingness to allow Pakistanis to be recruited was likely related to his intention of turning Uganda into an Islamic state (see Section 6.8.3).

In the 1980s, after Idi Amin had been overthrown (1979) and a semblance of political order restored in Uganda, Norway returned to support the development of professional capacity in botany and forestry. It provided many scholarships for postgraduate training between 1988 and 2008 and supported a biomass survey of the country to provide information for planning the management of its wood resources (Drichi, 2002). Another post-Amin development at Makerere was the building of a new herbarium, funded by the Global Environment Facility (GEF) through the Food and Agricultural Organization (FAO) and the United Nations Development Programme (UNDP). The intention was to create a national herbarium attached to the university's Department of Botany, to incorporate the herbarium of the Forest Department, which was then in Entebbe, and an herbarium at Kawanda Agricultural Research Station (as of 2024, its transfer was still pending).

A.H. has first-hand knowledge of the fate of the Forest Department's herbarium during the 1970s. He had known it well during 1966–1971, when he frequently consulted it during his doctoral research and when writing his forest flora. Later, when he taught at Makerere University in 1981–1982, he visited the herbarium on a number of occasions, often with students. At that time, the herbarium specimens were no longer in cupboards, but were tied up in bundles piled in a heap on the floor. The forestry staff told him that, during the invasion of Uganda by the Tanzanian army in 1978–1979 (which led to the ousting of Idi Amin), looters had entered the forestry building and stolen the cupboards. When the specimens were transferred to the new herbarium at Makerere after it had opened in 1996, it was found that some had deteriorated to such an extent that they had to be discarded (J. Kalema, Kampala, 2020, personal communication).

In 1973 the Canadian International Development Agency (CIDA) supported the preparation of a report about Ugandan forestry written by a Canadian firm of forestry consultants (Lockwood Consultants Ltd, 1973). This report, which was influential in how forestry subsequently developed, argued for a big expansion of conifer planting to meet bulk demands for wood, while, at the same time, taking a remarkably casual approach to the values and management of the indigenous forests. To quote from the report:

> there are secondary objectives such as the protection of water catchments, soils, wildlife and amenity of land. These however cannot be measured and are dependent on responsible behaviour by (Forest) Department officials in their provision.

At a stroke, environmental protection, which had been a top priority when the British colonial government had started to take a serious interest in forestry in 1929, had been relegated to being a matter of personal, rather than institutional, responsibility. There seems to have been an assumption in the minds of the consultants that all forestry staff in whatever country and whatever rank have the same worldview as their own, but this is unlikely (see Chapter 4). It is also suspected that the consultants who compiled the report for CIDA were unfamiliar with the differences between managing species-poor temperate forests on recently formed Quaternary soils (such as characterise Canada) from managing species-rich tropical rainforest on impoverished lateritic soils (as in Uganda). A similar question mark about worldviews surrounds the development aid that was supplied by Finnish development aid (FINNIDA) to forestry in Tanzania and Mexico during the 1980s (see Chapter 16).

Another big blow to the fate of Uganda's indigenous forests came in the 1990s. Under the terms of loans made by the International Monetary Fund (IMF) to bail out Uganda's debt-ridden economy, the IMF insisted that Uganda restructure its economy. What this meant for forestry was a decreased level of support for activities at field level. For example, in Mpigi District between 1993 and 1995, funding for forestry was cut by 89% and the staffing level reduced by 68% (Banana et al., 2007).

Another significant development in development aid was the election of a new British government in 1997 which decided to adopt a new foreign policy that was to be 'ethical' and driven by 'global values' as defined by an 'international community' (Blair, 2006). From now on, it was declared, the aim of Britain's foreign policy would be to eliminate world poverty. A two-step process was to be followed in its delivery – first the use of 'soft power', which meant working through inducement and persuasion, and, if this failed, to apply 'hard power', which meant resorting to the military. The three elements of 'soft power' under Britain's new 'ethical' foreign policy were the British Council, the BBC World Service and a new development aid organisation called the Department for International Development (DFID). DFID was established in 1997 and soon superseded the Overseas Development Agency (ODA), which had been created in 1964.

Countries which early experienced Britain's 'hard power' were Kosovo (1999), Sierra Leone (2000), Afghanistan (2001) and Iraq (2003). The invasion of Iraq made it clearer what the British government meant by the 'international community'. It was not the United Nations, as might be supposed, but those states that considered themselves to be liberal democracies and belonging to the 'free world' (see Section 6.1) (Buchan, 2007; Shannon and Keller, 2007). The leading member of the 'international community' was seen as the United States, the argument being that, to maintain world order, whichever liberal democracy had the greatest current military ability should take the lead in international interventions (Keohane and Nye, 1977) – an aspect of the 'West knows best' way of thinking associated with the Western worldview (see Section 4.3.2).

At the tail end of the 20th century, DFID took the funding lead in a programme of forestry reform called the Uganda Forest Sector Umbrella Programme (1999–2003) (Harrison et al., 2004; Hamilton et al., 2016). Influenced by neoliberal ideology, this involved a reorganisation of Uganda's forestry sector that included the abolition of the Forest Department, its replacement (in principle) by three new agencies and a much bigger role given to the private sector in operations in the forests. In practice, only one of the three intended agencies, the National Forestry Authority (NFA), was actually created, its job being to manage the country's CFRs. The other two intended agencies were one to manage the LFRs and the other to be an overall regulator. The failure to complete the whole restructuring programme must surely have contributed to the virtual disappearance of the LFRs, which, as mentioned in Section 5.6, declined in total area from 3028 km^2 in 1965 to 2.43 km^2 in 2011.

One aspect of forestry in Uganda in which private companies have been involved since 1997 is the planting of trees using carbon credits. The idea behind carbon credits is for those people whose activities increase the amounts of greenhouse gases in the atmosphere pay money to those whose activities decrease them, the net result being that the carbon emissions associated with the first group are 'offset' (counterbalanced) by carbon sequestration associated with the second. Uganda has been a recipient of carbon credits for tree-planting, the carbon credits mainly being available through a free-market orientated programme called Reducing Emissions from Deforestation and Forest Degradation (REDD and REDD+). Criticisms of the workability of REDD have been made both in general and for Kenya and Uganda, in particular (Brown, 2013; Entenmann et al., 2014; Twongyirwe et al., 2015).

Carbon credits schemes involving tree-planting in the Global South are sometimes lauded by politicians in the Global North as win/win ways of tackling climate change, but they are risky. They depend on clarity on the mathematics of carbon budgets (how much carbon dioxide is produced and how much sequestered), transparency about transfers of money, a stable forestry policy environment and the ability to carry out prescriptions for managing the forests that will endure for a long time (to ensure that the carbon captured by the trees remains permanently sequestered). Carbon offset schemes can be seen as ethically reprehensible, in that they avoid politicians in the Global North having to face the harder task of persuading their own citizens to reduce their use of fossil fuels (Caney and Hepburn, 2011). The high level of perceived corruption that has existed in recent years in Uganda, according to Transparency International, throws question marks over calculations of carbon budgets and about transfers of money (https://www.transparency.org/en/cpi/2023/index/uga, accessed 15 May 2024).

A.H. carried out research through literature scanning and interviews in the mid-2010s to try and find out what was really happening with carbon credit schemes in Uganda (Hamilton et al., 2016). What emerged was not reassuring. As evident from Table 5.2, forest policy in Uganda has not shown the long-term stability necessary for carbon credit schemes to work. All the businesses that had received carbon credits for tree-planting in Uganda were Europe-owned and all, except one, were planting pines, not indigenous trees. (The exception was a scheme on Mt Elgon to offset the building of a coal-fired power station in the Netherlands.) One of the disadvantages of planting pines, rather than using indigenous species, is that it deprives the local people who live near the new plantations of ecosystem services, from which they would otherwise have been able to benefit. These include better protection against flooding and landslides, more reliable water supplies and a greater availability of some minor forest products, such as certain species of medicinal plants.

Operationally, Britain's new 'ethical' foreign policy required the introduction of new administrative procedures, based on the assumption that Britain's development partners in the Global South would know better how to spend the money than the British. One of the new procedures was to move away from providing aid in relatively small, targeted, short-term packets with detailed accounting required (known as 'projects') to providing aid in larger packets (known as 'programmes') with larger budgets and multi-year time spans. Details of expenditure and accounting were left to the recipients of the 'block grants', typically government departments in the Global South and the British branches of international non-governmental organisations (NGOs) (whose job it was to pass DFID's aid on in large amounts to the branches of their organisations in the Global South). Technical tools, such as the logical framework, were introduced to assist the recipients of the development aid to plan their schedules of work and report back on the results (Wiggins and Shields, 1995). It was an approach to giving aid that depended strongly on trust and which left little room for technical contributions from British expertise.

A.H. had a grandstand view of how Britain's new 'ethical' foreign policy worked in practice. This was because WWF-UK was one of the British NGOs that received block grants from DFID, because he was stationed within the office of WWF-UK and because some of his projects, such as the People and Plants Initiative described in Chapter 10, were supported by ODA and later DFID. While ODA existed, the fact that the state of the natural environment contributed to people's well-being was not in question; also, that expertise in the natural sciences, as well as in the social sciences, had a role to play in development projects. Indeed, ODA maintained a small advisory team of foresters experienced in the technical aspects of tropical forestry in its headquarters in London. When DFID was created in 1997 with its aim of 'eliminating world poverty', it was no longer taken for granted that the state of the natural environment was relevant to people's well-being and ODA's in-house tropical forestry team was disbanded. The sidelining of the environment by DFID posed a challenge for WWF-UK, because it is an environmental organisation and by the end of the 20th century had become highly dependent on ODA/DFID for the funding of its overseas work.

The changeover from ODA to DFID as a funder was accompanied by several of those officers within WWF-UK who were most experienced in knowing about how to achieve conservation in the Global South losing their jobs. The obvious interpretation was that, now that the details of what was to be done were to be decided in the Global South, the assistance of British conservation expertise was no longer required. Without this input of technical expertise, WWF-UK was reduced in its overseas work to being a largely administrative organisation, its role being to write applications for block grants to submit to DFID, prepare contracts for WWF branches in the Global South, send money to them in large tranches, receive back their reports of what had been achieved and how the money had been spent, and then summarise these reports into a short document to report back to DFID. During the next four years, while A.H. remained in WWF, the staff of WWF-UK's international conservation unit were required to attend workshops, facilitated by outside consultants, on administrative subjects such as 'filling up logical frameworks', 'managing change' and 'the project management cycle'. We never discussed conservation, in the sense of having discussions about how to actually achieve conservation of biodiversity and the sustainable use of natural resources on the ground.

WWF-UK was not alone among British institutions in its expertise in natural resource management in the Global South becoming lost or diminished by the UK's new 'ethical' foreign policy. In 2002, the Oxford Forestry Institute (OFI), which had existed since 1924 (initially, as the Imperial Forestry Institute), was closed, possibly because it was imagined in government policy circles to be an outdated colonial relic. In fact, OFI had kept up to date with international developments in forestry, was engaged in many international exchange programmes and was one of the world's leading institutions involved in research and training in tropical forestry (Mills, 2006; Burley *et al.*, 2009).

Other British institutions with expertise in tropical natural resources that were disbanded or radically changed in nature when Britain adopted its new 'ethical' foreign policy included the Natural Resources Institute (NRI), the Commonwealth Development Corporation (CDC) and Wye College. At their times of closure or radical change, these were among the world's leading centres of excellence in subjects such as tropical soils, tropical agriculture, and control of tsetse and quelea (a tropical African bird that can be very damaging to crops). The Royal Botanic Gardens, Kew, which, in the 1990s, was probably the world's premier botanical institute, was repeatedly challenged by the government to justify its budget and suffered successive culls of its expert staff.

6 Cultural Influences on People/Plant Relationships, Part 2 – Post-1500 Developments in Europe and the West

Abstract

Certain cultural developments with roots in Europe and the West have been influential for determining the way that the whole world has developed since 1500. Two early developments were the rise of science from about 1600 and of the ideas and values associated with the cultural movement known as the enlightenment from about 1650. Adam Smith's contributions to theoretical economics in the 18th century and Charles Darwin's to biology in the 19th century have, subsequently, had major consequences for people's relationships with plants. So too, since the 1970s, have been the rises of neoliberalism, information technology, large transnational businesses and financialization. A growing concern about the harm that people were inflicting on one another and the natural world gave rise in 19th century Europe to the ideologies and social movements of humanitarianism and environmentalism. The chapter concludes with an account of the role of Christianity in spreading European/Western cultural developments to the rest of the world.

6.1 The Mid-Second Millennium European Cultural Revolution

In Section 4.3.2, we mention a cultural revolution in Europe that began in the middle of the 2nd millennium CE at about the same time as Europe was expanding its geographic and economic interests abroad (see Section 5.1). The forcefulness of European, later Western, expansionism has meant that this cultural revolution has influenced how the whole world has developed during the last 500 years.

Aspects of the European cultural revolution included the Italian Renaissance (c.1350–1600), the split of Protestantism from Roman Catholicism (1517), the rise of science (from the 16th century), the enlightenment (c.1650–1800) and the romantic movement (peaking 1800–1850 – see Section 8.2).

The advance guard in this cultural march were certain independently minded people who were prepared to oppose established authority by advancing radical ideas. According to their talents, the material expressions of their insights could be in the form of the written word, artistic works or practical inventions. The cultural norms of the societies in which this revolution unfolded were initially those associated with medieval feudalism – the claim of the Catholic Church to have exclusive access to religious truth and the claims of kings to have divine rights to rule. Inevitably, the authorities sometimes attempted to suppress the heretical upstarts. Fortunately for the writers, the moveable-type printing press had just been invented (for Europe) by Johannes Gutenberg (in 1440), providing them with a handy vehicle to disseminate their thoughts.

One aspect of the intellectual movement known as the enlightenment was a move in the universities away from teaching knowledge as established fact to paying more attention to the discovery and advancement of new knowledge (Russell, 1979). More weight came to be given to the evidence of the senses as a guide to what was significant and true. There was a holism within the academic environment of the time, in comparison with the numerous specialist fields of knowledge that characterise modern academia. Key figures of the enlightenment, such as Humboldt, were convinced of the unity of the sciences and the arts (see Section 6.5.1). Artists and philosophers held science in high esteem.

During the 18th century, the enlightenment gave rise to a school of philosophy known as philosophic liberalism, which emphasised the fundamental equality of all people and that each person is due certain fundamental rights (Russell, 1979). The most fundamental of the rights recognised were freedom of speech, freedom to practice the religion of one's choice, equality before the law and the right to have a say in the selection of one's ruler. The motto of the French Revolution (1789) – 'Liberté, Egalité, Fraternité' ('Liberty, Equality, Brotherhood') – sums up the concept succinctly. Philosophical liberalism has been influential in moulding the character of those modern states that refer to themselves as liberal democracies and consider that they belong to the 'free world'. It is an optimistic school of philosophy with respect to the abilities recognised of ordinary people and the possibilities of human progress.

At the beginning of the 19th century, a French philosopher, Antoine Destutt de Tracy (1754–1836), invented the word 'ideology' to describe what he considered to be the science of ideas. Ideologies have subsequently had a big influence on the course of history. The ideologies referred to in this chapter are capitalism, communism, conservatism, environmentalism, humanitarianism, liberalism, neoliberalism and socialism. Each of these refers to a set of attitudes, beliefs and values shared by groups of people about priorities in politics and about how societies should be organised. De Tracy's thinking was influenced by the French Revolution, an event that also gave rise to the use of the words 'left' and 'right' for those who adhere more towards socialism or more towards conservatism (and, more recently, capitalism), which has been the main ideological axis of variation in politics during the last 200 years. 'Left' and 'right' referred to the seating order adopted by those who attended the National Assembly of the first French Republic (1792–1804) (Johnston, 1996).

6.2 Two Powerful Ideas

Two ideas that originated in Europe/the West during its cultural revolution have had major impacts on the way that the whole world has subsequently developed, including in the relationships between people and plants. They are the virtue of the free market for serving the common good, as argued by Adam Smith (1723–1790), and the theory of evolution of Charles Darwin (1809–1882). Smith was a professor of theoretical economics at the University of Glasgow and a friend of David Hume (1711–1776), a fellow Scot and an influential member of the school of philosophic liberalism (Russell, 1979). Darwin was a privately wealthy Cambridge graduate in theology who had been inspired by reading Humboldt (see Section 6.5.1). In 1831, at the age of 22, he had managed to obtain the position of gentleman-companion to Captain Robert Fitzroy (1805–1865) on the Royal Navy ship, HMS *Beagle*, on a five-year surveying expedition around the world.

Darwin's ability to come up with his revolutionary theory of the origin of species by natural selection owes much to his having the unusual opportunity of comparing nature at different places around the world. The leisurely rate of travel of the *Beagle* was integral to its mission, which was to undertake detailed hydrographic surveys around the coast of South America and link the positions of the places surveyed back to the same start and end point in Britain. The voyage of the *Beagle* marked the first time that a linked chain of surveyed reference points circumscribing the world had been achieved. On completion of the voyage, it was found that the measurements of position made on the *Beagle* had an error (in time) of only 33 seconds (equivalent to 15.28 km). This was an impressive achievement for Fitzroy, who was an eminent scientist in his own right. He was one of the inventors of the weather forecast.

The voyage of the *Beagle* is an example of how important the invention of new devices has been for the advancement of science. It was a test of the accuracy of the marine chronometer – a clock designed to work at sea. A prototype for a marine chronometer had first been invented by John Harrison (1693–1776), a self-educated London carpenter and clockmaker. The marine chronometer was essential for the accurate calculation of longitude at sea, making transoceanic voyages much safer and delivering a strategic advantage to those maritime nations that carried it on their ships. At the time, it was normal for British naval ships to carry three marine chronometers (in case one malfunctioned). However, the *Beagle* was carrying 22, the purpose being to test the accuracy of their designs.

6.2.1 Adam Smith: the virtue of the free market

Adam Smith was the founding father of Western theoretical economics and a passionate believer in the rights of people to liberty, fair trial and freedom of speech. He set out in a book, *The Wealth of Nations* (Smith, 1776), why he believed that the common good of a nation is best served by allowing the largely uninhibited pursuit of self-interest. The aphorism, 'the market knows best', and the metaphor of 'the invisible hand' are sometimes used to describe the advantages that he claimed for his belief. They refer to the guiding power of the sum of all the individual bargains struck in a free market between the buyers and sellers of goods and services. This, so the theory goes, is the best way to determine that the prices of goods and services are optimal, given the prevailing levels of supply and demand. Because of his views, Adam Smith is commonly taken to be the founding father of capitalism, but actually capitalism as a term descriptive of an economic system only entered into widespread use about 100 years later.

Smith added a caveat to his espousal of the free market, which was that it would only work as a driver of general prosperity if practised in well-governed societies that uphold high standards of morality. He cautioned vigilance against businesspeople colluding to enrich themselves at the public's expense and was fearful of businesses obtaining sufficient influence over governments to cause them to pass legislation that was in their favour, but not in the public interest. As explained in Section 6.3.1, in the last few decades, an ideology known as neoliberalism has become influential, especially in the United States and Britain. This is an extremist version of Adam Smith's theory that sees a minimalist role for the state in public affairs. The argument is that 'interventions' by the state in the operation of the free market will introduce 'distortions' to the economy that will detract from freedom of the individual and, in the end, do more harm than good. However, Adam Smith himself did not take such a fundamentalist approach, recognising that governments serve certain essential functions, for instance building infrastructure for everyone's benefit and regulation of the banks (Oreskes and Conway, 2023).

One of the most significant theoretical economists for environmentalists after Adam Smith was Robert Malthus (1766–1834), a professor of history and political economy at the East India College at Haileybury, a training centre near London founded in 1806 for British administrators in India (see Section 5.3.3). Malthus was also a Protestant cleric. Influenced by what he observed was happening with the agricultural and industrial revolutions, Malthus observed in his book, *An Essay on the Principle of Population*, that trying to increase the prosperity of a nation by increasing its food supply merely results in its population expanding until the same ratio of food availability to population size is reached as before, the result being that there is no net gain in overall prosperity (Malthus, 1798). Malthus further observed that, in practice, it is particularly members of what he referred to as the 'lower classes' that are the principal victims when nations overrun their food supplies. This is because they are the first to be affected when there are episodes of disease and famine and are the foot soldiers when wars break out between nations competing for scarce resources. In later versions of his book, Malthus pointed out that it is possible to avoid cycling between economic plenty and economic want if measures are taken to limit the size of the population, which, he suggested, might be achieved by delaying the age of marriage.

The German economist and social historian Jürgen Kocka has laid stress on the importance of trading systems in human history and of the key roles played by people with entrepreneurial spirit in making long-distance trade happen (Kocka, 2016). He pointed out that, historically, entrepreneurs have been able to find ways of doing business under a variety of political and economic circumstances. For instance, during recent decades, entrepreneurs both in the Western countries of the United States and Britain, as well as those in the East Asian countries of China, Japan and South Korea, have found ways to develop highly profitable businesses, despite their governments differing in their political ideologies and approaches to economic policy. The US/British economic approach has been characterised by low levels of governmental controls over how businesses operate, while the East Asian approach is characterised by their governments singling out sectors of the economy to be given support. The former can be seen as being influenced by the individualistic thinking that has permeated the West and the latter by the more collectivistic thinking of East Asia (see Section 4.4).

China experienced tumultuous political and economic changes during the 20th century. Major events have included: the end of the last imperial dynasty (1911) and its replacement by a republican state (1912); a long civil war (1927–1949); the invasion and partial occupation of the country by Japan (1937–1945); the closure of the last remaining treaty ports (during the 1940s); the victory of Mao Zedong's communist forces over the Kuomintang (nationalist) forces of Chiang Kai-shek (1949); the founding of the People's Republic of China (1949); the Great Leap Forward (1958–1962); the Great Proletarian Cultural Revolution (1966–1976); and market liberalisation (from 1978).

The Great Leap Forward and the Cultural Revolution were inspirations of Mao Zedong (1893–1976), the founding father of the People's Republic of China. The purpose of the Great Leap Forward was to accelerate the industrialisation of the country through creating collectives and work brigades and the purpose of the Cultural Revolution was to 'Destroy the old and cultivate the new', referring to customs, culture, habits and ideas. The Great Leap Forward witnessed the abolition of private farms and estates and their replacement by cooperatives. What then happened was a plummeting in agricultural productivity and a massive degradation of the forests (Smil, 1999). The result was the largest famine known in history, an estimated 30 million to 45 million people perishing through starvation. The forest degradation was due to vast numbers of trees being felled to provide fuel for small, inefficient, blast furnaces (see Section 21.1). In the Cultural Revolution, disorderly gangs of Red Guards deliberately destroyed huge quantities of irreplaceable treasures from China's rich cultural past, including temples, books and works of art.

Mao died in 1976 and, in 1978, under the leadership of Deng Ziaoping, China adopted a new form of economic system, known as a 'socialist market economy' or 'socialism with Chinese characteristics'. According to the British historian Michael Wood, the reason why the change was made was because Deng concluded that China would never be a modern state without opening up its economy to market forces and making space for individual initiative (Wood, 2020). The new economic system, which remains in place to this day, gives considerable freedom to both Chinese entrepreneurs and foreign investors to operate within China, but only within a framework of strong government control and planning. Chinese entrepreneurs are encouraged to seek new markets abroad. Figure 6.1 illustrates the encouragement that is now given

Fig. 6.1. Statue of Steve Jobs in Huhhot, China. Photo: A.H. (2016).

to entrepreneurism in China. It is a photograph of a statue of Steve Jobs, the American founder of Apple Inc., outside a supermarket in Huhhot, the capital of Inner Mongolia. The adverts on the store include some for Adidas, a large transnational company with its headquarters in Germany.

In the agricultural sector, the switch to a socialist market economy has seen the abandonment of agricultural collectives. Farmers are now encouraged to choose for themselves the agricultural systems that they wish to follow (this is known as the Household Responsibility System). In 2008, the Household Responsibility System was extended to forestry (see Section 21.2.5).

Market liberalisation, Chinese-style, has proved immensely successful at growing the Chinese economy, in the process lifting hundreds of millions of people out of poverty. China is now the world's largest manufacturing country and the second largest global economy (after the United States). However, this economic success has come at the price of the country becoming a major polluter, including now being the world's largest emitter of the greenhouse gas, carbon dioxide. In recognition of this problem, the Communist Party of China incorporated at its 18th National Congress in 2012 the target of China becoming an eco-civilisation, a term resonant with the Daoist belief that a state of harmony should exist between people and nature (see Section 4.3.1). Time will tell how successful China will be in its eco-civilisation mission. There is a long way to go. Notable achievements to date include a rapid increase in the production of sustainable energy (China produces much more than any other country) and having the world's largest tree-planting programme. Recognition of the value of forests for guarding against flooding has led to the introduction of large-scale schemes in the upper catchment of the Yangtze River to protect forests and promote reafforestation, which has proved a boon for biodiversity (Yin Hongfu and Li Changan, 2001; Zhao and Shao, 2002). Some of the effects of measures taken to protect the Yangtze River catchment are described for two places in Chapter 21.

6.2.2   Charles Darwin: the theory of evolution

Published in 1859, Darwin's theory of evolution – presented in his book, *On the Origin of Species by Means of Natural Selection, or the Preservation of Favoured Races in the Struggle for Life* (Darwin, 1859) – today forms the foundation stone of the biological sciences. When first published, Darwin's theory proved highly disruptive of the way people in the West thought about the human species, which was as a special type of being possessing a fundamental essence absent from any other form of life. Most people were caught up in a Christian worldview that held that the first man and first woman were created as the final acts in God's creation of the universe, the date of which was calculated on biblical evidence to have been at a few thousand years BCE. Some of the early scientists, such as Isaac Newton (1643–1727), who began the process of dislodging the medieval view of the world, believed that they had been granted the privilege of being able to glimpse into the mind of God.

Following on from the publication of Darwin's theory, a number of theories, collectedly referred to as social Darwinism, were advanced to account for the variations between humans in different parts of the world. The genetic distinctiveness and evolutionary connections between the human races (as they were then conceived) were matters of special fascination. Europeans living in the colonies as missionaries, settlers and administrators sometimes used social Darwinian theories to justify the colonial social order, including to account for their own elevated social positions (Roome, 1927). Social Darwinian theories bear little resemblance to modern knowledge about how human populations in different parts of the world have originated.

Darwin himself was unsure of the mechanism of heredity but thought that the various parts of the bodies of organisms might continually emit small organic particles (gemmules), these then aggregating in the gonads, thus in the gametes, and so being passed on to future generations. Two rival theories about the mechanism of heredity have vied for scientific and political attention since the early 20th century, in the process often being at the forefront of debates about the importance of nature versus nurture and policies on education, immigration and eugenics. One of the theories was initially proposed by Gregor Mendel (1822–1884), a monk at St Thomas's Abbey in Brno (now in the Czech Republic). Mendel carried out hybridisation experiments (mostly on pea plants – *Pisum sativum*) in the monastery's garden and found that some of the characteristics of the plants, such as whether the seeds of peas were smooth or wrinkly, were inherited as distinct units, not as blended mixtures. Mendel published his findings in 1866 in an obscure scientific journal and they failed to reach a wider scientific audience until the early 20th century, when their far-reaching implications became appreciated. The modern discipline of genetics was launched.

The rival theory to Mendel's was initially proposed by the French zoologist Jean-Baptiste Lamarck (1744–1829). Its hypothesis, that individual organisms pass on to their offspring characteristics they have acquired during their life-times, has subsequently been forced to retreat in the face of the spectacular advances made in Mendelian genetics.

Each side of the principal divide in modern politics (socialist versus conservative, or left versus right) has claimed that Darwin's theory validates its ideological position. Ideologues on the left claim that the human is fundamentally a collaborative species and those on the right that it is fundamentally competitive. The former tend to favour direct interventions by the state to serve the common good, while the latter tend to favour a 'small state' to encourage individualism and release enterprise. The left/right division in political thinking became hardened up in the Cold War that followed the Second World War, a confrontation that was led by the Soviet Union on the socialist side and the United States on the capitalist Western side. It pulled in many other countries, sometimes resulting in them fighting proxy wars on behalf of their powerful patrons.

Karl Marx (1818–1883), the son of a prosperous German lawyer, was a key figure on the political left and the father of communism. He was an admirer of Darwin and, reportedly, wanted to dedicate to him the book that summed up his life's intellectual achievement (*Capital*, published 1867) (Fay, 1978). Also reportedly, Darwin did not reply. Friedrich Engels, who co-authored the Communist Manifesto of 1848 with Marx, argued that, just as Darwin had discovered the law of organic nature, so Marx had discovered the law of human history (see Section 6.6).

The life of the Russian agricultural scientist, Nicolai Vavilov (1887–1943), demonstrates how a scientist of integrity can fall foul of a political regime that, for ideological reasons, is committed to a particular scientific theory. Vavilov was ahead of his time in recognising the significance of Mendelian genetics to agricultural development. He was the first person to propose that there are a number of places (Vavilov Centres) where the domestication of food plants initially took place (see Section 3.2) and the first to compile global gene collections of crops (Fet and Golubovsky, 2008). Vavilov became convinced of the correctness of Mendelian genetics during a visit that he made in 1913–1914 to a genetic research group led by William Bateson at the University of Cambridge in Britain (Edwards, 2013). William Bateson (1861–1926), the person responsible for inventing the word 'genetics', had extended and developed Mendel's work by carrying out a series of breeding experiments on plants and animals.

For the rest of his career, Vavilov devoted his life to the improvement of agriculture through genetic research (Pringle, 2008; Witkowski, 2008). A humanitarian and a Russian patriot, Vavilov was motivated in his work by a desire to alleviate the sufferings of the poor during famines, especially when these happened in Russia. A devastating famine had struck Russia in 1891 and there were more between 1921 and 1947. The immediate cause of death for many in the 1891 famine was cholera, related to exceptionally bad weather that had created a shortage of food (Johnson, 2015). An underlying problem with the 1921–1947 famines was the inefficiency of agricultural production on the collective farms established after the Russian Revolution – the same problem that bedevilled China later in its Great Leap Forward (see Section 6.2.1). An additional reason for 1921–1947 famines was that, for ideological reasons, Lamarckism (rather than Mendelian genetics) was the mechanism of heredity favoured by the rulers of the Soviet Union. Lamarckism fed into their ideological commitment to the view that the reason why people occupy lower positions in socio-economic orders is not because they lack inherent ability, but because they have been repressed by the ruling classes.

The nemesis of Vavilov was Trofim Lysenko (1898–1976), an agricultural researcher, advocate of Lamarckism and skilled political manipulator. Lysenko managed to convince Joseph Stalin, the Soviet ruler between 1924 and 1952, and his successor Nikita Khrushchev (ruled, 1953–1964) of the scientific correctness of Lamarckism. A result of this was that Vavilov, along with two scientific colleagues, were arrested (1940), tried and found guilty of pursuing Western ('bourgeois') science, after which Vavilov's two colleagues were shot. Vavilov starved to death in a Soviet jail in 1943. Although Lysenko's domination over crop breeding was over by 1965, Lamarckism continued to be the official line in the Soviet Union when it came to human evolution (Medvedev, 1977). A consequence of this was that some eminent scientists, such as the geneticist Raissa Berg (1913–2006), moved abroad so that they could continue their research (Conner and Lande, 2014).

At the other end of the left/right political spectrum, the phrase 'survival of the fittest' (used by Darwin in the fifth edition of his book) has resonated with some adherents of neoliberalism, a radical form of right-wing political ideology supportive of low-regulation capitalism (see Section 6.3.1). Several popular books by Richard Dawkins, notably *The Selfish Gene* (Dawkins, 1976) with its evocative title, have given an impression to some neoliberal theorists that their ideology has been authenticated by science (Grafen and Ridley, 2007).

John D. Rockefeller (1839–1937), an American tycoon of German extraction and possibly the richest American of all time, was a social Darwinian (Hofstadter, 1944). His stated belief was that the growth of a large business is nothing more than survival of the fittest. Rockefeller made his money in the oil business, at first by selling kerosene (used for lighting) and later, once the internal combustion engine had been developed to the state of being useful (in 1886), by selling gasoline. At his peak, Rockefeller controlled 90% of all oil in the United States. Rockefeller's success in business was due to his ruthless business tactics. In 1911, concerned about the implications of this extreme interpretation of capitalist theory, the US Supreme Court ordered his company, Standard Oil, to be broken up into 34 independent parts.

The problem with which the Supreme Court was grappling is one inherent in capitalism if fought to its logical conclusion, which is that, eventually, one company – the most scheming and unscrupulous of all – ends up achieving monopoly control of a sector of the economy. The danger for capitalism – as a theoretical model of how to create a prosperous society – is that the logic upon which it depends as being a force for good ceases to exist. The extent to which the break-up of Standard Oil has been successful in returning the oil business to a position of trust is doubtful, given that some of its modern progeny, which include ExxonMobil (a recent merger of Exxon and Mobil), Esso (a trading name of ExxonMobil) and Chevron, are business behemoths.

6.3 Four Influential Developments Since the 1970s

The last 50 years have witnessed an unprecedented rate of global socio-economic and cultural change. Four interrelated developments during this period are briefly discussed here to provide a flavour of what has been going on. These are the rises of neoliberalism, information technology, transnational businesses and financialization. To give this account coherence, we first provide a description and critique of neoliberalism and then consider the other three developments as potential multipliers of its influences.

6.3.1 The rise of neoliberal ideology

Neoliberalism is an ideology supportive of free trade, free markets and strong private property rights (Harvey, 2005). The role of the state is seen as being restricted largely to guaranteeing the quality and integrity of money, ensuring the sanctity of commercial contracts and creating markets where they do not currently exist (Kotz, 2000). Examples of the types of practical policies pursued by neoliberal politicians include privatising publicly owned enterprises, the slashing of regulations limiting how businesses can operate, reducing welfare budgets, lowering corporate taxes and abolishing taxes on wealth.

Because neoliberalism is sometimes advocated by the rich and powerful, it is easy to interpret the motivations that lie behind it as being driven by self-interest. However, a moral case can be made for neoliberalism, namely that, if people are provided with too much largesse by the state, it detracts from them taking responsibility for their lives and degenerates their moral fibre (a point with which William Beveridge, the founder of the British welfare state, agreed – see Section 6.6). On the other hand, in Britain the attacks that have been made by British neoliberal politicians on the civil service since 1980 have eroded the morale of those dedicated civil servants who are motivated by more altruistic values (Diamond, 2021).

Neoliberalism rose to prominence in the West when Margaret Thatcher became the British prime minister in 1979, and Ronald Reagan, her ideological soulmate, was elected president of the United States in 1981. Thatcher's election resulted in the breakdown of a political consensus in Britain that had held since the end of the Second World War, during which time the two main political parties had broadly agreed on the desirable balance between encouraging private enterprise, regulating its excesses and using national income to provide social services. At about the same time, neoliberalism received an international boost when the World Bank and the International Monetary Fund (IMF) adopted neoliberal policies (Raffer, 2011; Carroll and Jarvis, 2015).

This has had a big impact on those countries in the Global South that have found themselves in financial distress. This was because, in return for providing relief for their indebtedness, the IMF required them to restructure their economies along neoliberal lines. In Section 5.7, we mention how this contributed to the loss of tropical forest in Uganda.

Neoliberalism carries forward the belief, associated with the Western world-view, that economic growth is both possible and good – an example of the hold that the concept of progress has acquired on the Western worldview (see Section 4.3.2). It is optimistic about the power of science to deliver techno-logical solutions to human problems (a belief known as scientific optimism) (Carl *et al.*, 2016). Among neoliberals who accept that anthropogenic climate change is happening (and many do not), a belief has developed that scientists will find ways to control it. For instance, it has been suggested that climate change can be mitigated by adding iron to iron-poor parts of the oceans to fertilise the phytoplankton, thus stimulating photosynthetic activity and seques-tering carbon (Monastersky, 1995). Another suggestion is to place a giant sun-shade in space between the sun and the planet to reduce the amount of solar radiation reaching the Earth (Hickman, 2018).

The evidence of history suggests that introducing such drastic solutions is extremely risky. This is because, in actuality, scientists know rather little about how ecosystems function and, even when they think they have a good idea, translating scientific advice into political reality can be hard. The agricultural scientists who introduced new types of bananas into Uganda after 1898 had the intention of boosting the economy (see Section 2.3.1). They did not predict the damage that they would cause to the country's indigenous banana varieties and the diets of the people. The self-confident officers of the colonial Uganda Game and Fisheries Department thought that they understood wildlife when they released the Nile perch into Lake Victoria. They did not predict the harm that this would cause to the lake's ecosystem and to the health and livelihoods of the Ugandan people (see Section 2.3.2).

Politicians who are both neoliberal and charismatic, such as Donald Trump in the United States (president, 2017–2021) and Jair Bolsonaro in Brazil (president, 2019–2022), are able to convince large numbers of people that they have the answers to their problems (Signer, 2009). The optimistic mes-sages that these politicians deliver can be appealing psychologically in a world that can be unsettling, with people feeling that they have lost control over the courses of their lives. Under these circumstances, it is tempting for them to accept the propositions that the charismatics offer, for example that any damage caused to the environment by pursuing economic growth, as of now, can be rectified later. In Brazil, Hungary and the United States, there is a con-nection between neoliberal politicians and evangelical or nationalised versions of Christianity (Bozoki and Ádám, 2016; Margolis, 2020; Lapper, 2021). In Section 8.1, we note how, in Ethiopia, traditional beliefs and practices associ-ated with Coptic Christianity have been under attack from a Christian evangeli-cal movement led by charismatics (Kelbessa, 2017).

In the United States, during the presidencies of Ronald Reagan (in office, 1981–1989) and Donald Trump, there was a reversal in the progress towards

environmentally friendly forestry that had been made during the earlier presidencies of Theodore Roosevelt (in office, 1901–1909) and Richard Nixon (in office,
1969–1974). All of these four politicians were, or are, Republicans. The anti-
environmentalism of the most recent of the four, compared with the earlier two,
demonstrates the grip that neoliberalism has acquired on the Republican Party
during the last 40 years. Roosevelt's contributions to the environmental movement
are discussed in Section 8.2.2. Nixon supported the Clean Air Act (1963) and
the National Environmental Policy Act (1970) and established the Environmental
Protection Agency (1970). In contrast, these laconic words of the laid-back Ronald
Reagan sum up his indifferent attitude towards the environment: 'If you've seen one
redwood, you've seen them all.' In Brazil, under the presidency of Jair Bolsonaro,
the regulations designed to protect the Amazonian rainforest were dismantled,
causing consternation to both environmentalists and humanitarians, concerned
with the accelerated effect that this would have on the linked issues of global warming, species' extinctions and the fate of the indigenous people.

Margaret Thatcher, exceptionally for a British politician, had a good grounding in science and played instrumental roles in raising global awareness of the
dangers of ozone depletion and global warming (see Section 6.10) (Thatcher,
1989; Vidal, 2013). This is refreshingly different from the ignorance of science,
willingness to make deals with businesspeople to further their own interests and
failure to introduce serious measures to deal with pressing environmental concerns shown by some neoliberal politicians (Dillon et al., 2018; Hatzisavvidou,
2020; Deutsch, 2021).

One of the consequences of a country adopting neoliberal policies is to
widen the gap between a relatively small socio-economic elite and the rest of
the people (Amouzou et al., 2014). This is as expected, since members of the
elite are in better positions to hire smart lawyers and accountants to further
their interests, to be investors (rather than debtors) and to send their children
to schools that turn them out well-equipped to assume their places in the elite
(Bullough, 2018). A similar gap has opened up between countries in the Global
North and Global South (UN, 2020). (For definitions of Global North and
Global South, see Section 5.7.)

A handful of international companies and individuals have managed to
seize control of key nodes in the global economic system, which has made
them extraordinarily wealthy. Prominent among them are traders dealing in
critical commodities, examples being grain, fuel oil, eating oil, aluminium and
copper. It was collusion with Western traders of this type that enabled a small
bunch of people (kleptocrats) in the disintegrating Soviet Union of 1989–1992
to gain control over its major economic assets (Blas and Farchy, 2021).

Britain is a member of the Global North and a country that, within the
European context, has been particularly keen on neoliberalism (Bullough,
2018; Blas and Farchy, 2021). A lax regulatory regime has made it easy for
wealthy foreigners to gain residency in Britain – with few questions asked about
how they made their fortunes. Thanks to its imperial legacy, Britain still possesses
a number of overseas dependencies and territories that never achieved full
independence. Their legal systems are modelled on British law and, frequently,
they have close ties with members of the British socio-economic elite. Another

British oddity is the possession of laws that allow its citizens to declare for tax purposes that their main residences are outside Britain. What this achieves for those wealthy Britons who know how to work the tax system for their personal advantage is to provide them with non-domiciliary (non-dom) status, which reduces the amounts of tax that they are required to pay.

With this combination of easy immigration for wealthy foreigners and tax avoidance by people in the know, Britain has become a mecca for dirty money. Its overseas territories and possessions, such as Bermuda and Jersey, are among the world's most notorious tax havens. One consequence of this is that large numbers of properties in Britain are now owned by companies registered offshore, their real owners being next to impossible to trace (McKenzie and Atkinson, 2020). There are more than 100,000 such properties in England and Wales, an estimated half of them barely used. A knock-on effect of this has been to push up the prices of properties, especially in London and neighbourhood, to levels beyond the reach of ordinary Britons.

As part of its pursuit of neoliberalism, Britain, under the premiership of Margaret Thatcher, started to divest itself of its state-owned assets, selling them to private investors. Among these divested assets were water, which was privatised in 1989, and the railways, which followed suit in 1993 (Keating, 2018). Both water and railways are, by their nature, essentially monopoly enterprises and, because no government can afford to let them fail, they are risk-free investments for those seeking safe long-term homes for their money. Ironically, 70% of UK rail routes are now owned by the state-owned rail companies of Belgium, France, Germany, Italy or the Netherlands. Obviously, none of these countries has accepted Britain's extreme version of neoliberalism – on the contrary, they have used it for their own commercial advantage. More than 70% of English water companies are in foreign ownership.

6.3.2 The rise of information technology

Significant developments in information technology (IT) include widespread computer use (from the 1980s), the invention of the World Wide Web (1989), the rise of the social media (from the 2000s) and the use of artificial intelligence (especially since 2000). The use of artificial intelligence has been facilitated by the combination of a greater availability of big data, cheaper and faster computers, and improved techniques of machine-learning.

Allied to technological advances in transportation, a mesh of electronic communications now engulfs the world, moving huge quantities of information, goods and people from place to place at incredible speeds by historical standards. In the face of stiff competition, transnational businesses have adopted just-in-time deliveries to maximise profitability along supply chains. Psychologically, time and space have become foreshortened. The world has become a rushed place, especially for those more centrally involved in global affairs. It is fragile, with little resilience to disruption. Solar flares, tsunamis, power outages, armed conflicts, acts of sabotage, ill-considered decisions by powerful leaders and shortages of key workers can easily precipitate disasters.

For many people, especially wealthier and younger ones, sending and receiving information through the Internet has become an addiction. The corollary of this is reduced attention being given to face-to-face contacts with people and nature, both of which have traditionally grounded people in reality (see Section 7.1). The detachment for some people may soon become even more total, now that several giant transnational companies, including Facebook, Google and Microsoft, are investing heavily to create metaverses – totally immersive virtual worlds, in which people will be able to live out much of their lives, whether at work, rest or play.

Truth can be a casualty in today's information economy. Nowadays, anyone with access to the Internet and a modicum of technical skill has the potential to deliver information on any subject almost instantaneously to millions of people. In practice, the attention given to ensuring the accuracy of the information can be minimal and, for those who wish to do so, information can be an easy tool with which to manipulate the minds and behaviour of other people. With the freeing up of regulatory controls under neoliberal regimes, businesses can be tempted to use the doleful arts of advertising, branding and subliminal messaging to create beguiling images of themselves and their products with little basis in reality. Greenwashing is common – the use of deceitful wording and imagery to convince the public that a product is environmentally friendly when it is not. Sustained campaigns have been run by businesses, for instance those involved in the tobacco and petrochemical sectors, to hide the truth about the damage that their products cause (Bates and Rowell, 2004; Banerjee *et al.*, 2015). Certification of products is an approach that has been widely adopted by environmentalists to assist consumers to make ethical choices, but such branding can be deceptive, even when third-party certifiers are involved (Cramer *et al.*, 2014).

6.3.3 The rise of large transnational businesses

During the last 50 years, large businesses, having substantial control over the transnational flows of goods, financial capital and skilled workers, have become increasingly prominent in the global economy (Kotz, 2000; Degain *et al.*, 2017). Their business structures can be complex. According to what suits, they may operate within countries through holding companies, wholly owned subsidiaries or joint business ventures. Some transnational businesses have close links with particular countries, as do the oil and gas giants Saudi Aramco with Saudi Arabia and Sinopec Group with China. Others have more tenuous links with the countries in which they originated, as do Amazon (information technology), JP Morgan Chase (financials) and Pfizer (pharmaceuticals), all of which originated in the United States.

Agriculture and food are natural fits for large transnational businesses (see Section 5.4.2) (Scoppola, 2021). This is because modern, industrially produced food products often contain ingredients sourced from many parts of the world. The best places to grow the plants providing the ingredients can be a long way away from those where the products are most profitably sold and the ideal locations of intermediate stages in the food-production-and-marketing system can be different again (these include primary and secondary processing plants, manufacturing factories, wholesalers and warehouses). It is tempting for transnational food businesses to produce ultra-processed foods. These are manufactured carefully to maintain a uniformity in quality (useful for branding and marketing purposes) and contain ingredients to increase shelf life and ensure that they are highly palatable to consumers. An advantage to the businesses in producing ultra-processed foods is that it increases profitability. A disadvantage for consumers is that the products can be damaging to their health (see Section 6.5.2) (Miclotte and Van de Wiele, 2019).

The histories of the confectionery brands, Cadbury's, Fry's and Rowntree's, provide examples of large transnational businesses that have become involved in producing plant-based food products that can be highly profitable, but damaging to human health. Two principal ingredients in their products are sugar and chocolate, the latter derived from the seeds (known as cocoa beans) of an understorey tree (*Theobroma cacao*) that is native to the rainforests on the eastern slopes of the Andes (Thomas *et al.*, 2012). Nowadays, the main producer of cocoa beans is Ghana, where half a million smallholder families are involved in their production in ways that have raised environmental and humanitarian concerns (Dompreh *et al.*, 2020). The three confectionery brands have their origins in businesses founded in 19th century Britain by Quaker families whose humanitarian principles included providing their long-serving employees with excellent housing and which launched charitable foundations that are still active in promoting social improvements. The founders of these businesses believed that chocolate-based beverages could serve as substitutions for alcoholic drinks, the drinking of which they thought was a road to ruin for the poor (Wilson, 2010). At the time, both cocoa and sugar were regarded as medicinal. Cadbury's merged with Fry's in 1923 and, beginning in 1969, both Cadbury's and Rowntree's experienced a spate of business mergers, demergers, acquisitions and sell-offs. Cadbury's ended up being acquired by the giant American corporation Kraft Foods, and Rowntree's fell into the hands of the giant Swiss-based conglomerate, Nestlé, which drove the last nails into the Quaker philanthropic coffin.

The pharmaceutical and botanical business sectors are also natural fits for large transnational businesses. As with the agrifood sector, the botanical ingredients in the products that these sectors produce can be sourced from many parts of the world, which may be distant from the best places to site processing and manufacturing plants, and these can be distant again from those places where the greatest profits can be made. In Chapter 13, we describe how businesses involved in these sectors have been approaching the Kunming Institute of Botany in China to seek its cooperation in sourcing new botanical ingredients to enhance their product ranges.

The same types of plants tend to be of interest to both the pharmaceutical and botanical sectors, and it is customary to refer to them in business and conservation circles as medicinal and aromatic plants (MAPs). Many species of MAPs are traded, the great majority harvested in the wild, rather than cultivated (Hamilton, 2004a). The harvesters tend to be among the poorest members of their societies, often herders, shepherds, landless people and women. Germany lies at the international heart of trade in MAPs. Research carried out in 1998 revealed that, at that time, there were at least 2000 species of MAPs, originating from over 120 countries, being traded in Germany (Lange, 1998; Schippmann, 1998). Elsewhere, China is a major importer and exporter of MAPs and, like Germany, is also a major manufacturer of pharmaceutical and botanical products (Hinsley et al., 2020). During recent decades, there has been an upsurge in demand for MAPs, which has placed many of those that are wild-collected in danger of extinction (Pei Shengji et al., 2009a; Roberson, 2009).

The Himalayan Range is one of the major global sources of MAPs. Plants collected in the mountains enter into the global MAP trade through a number of large wholesale markets situated in an arc around their base, notably in Rawalpindi, Lahore, Amritsar, Delhi, Kolkata, Mandalay, Kunming, Dali and Chengdu. Figure 6.2 is a photograph of Tibetan stallholders in the wholesale herbal market in Dali City in China. Research carried out in this market in the 1990s revealed that 517 species of medicinal plants were being traded, many (but not all) originating in the Himalayas (Pei Shengji et al., 1996). In Diqing Tibetan Autonomous Prefecture in Northwest Yunnan, which is a part of the Himalayas that is connected to the Dali wholesale market, 25–80% of household income is derived from the sale of non-timber forest products (NTFPs), a category which includes MAPs as well as some other wild-collected products (such as mushrooms) (Salick et al., 2006). The yew tree (Taxus) is an example

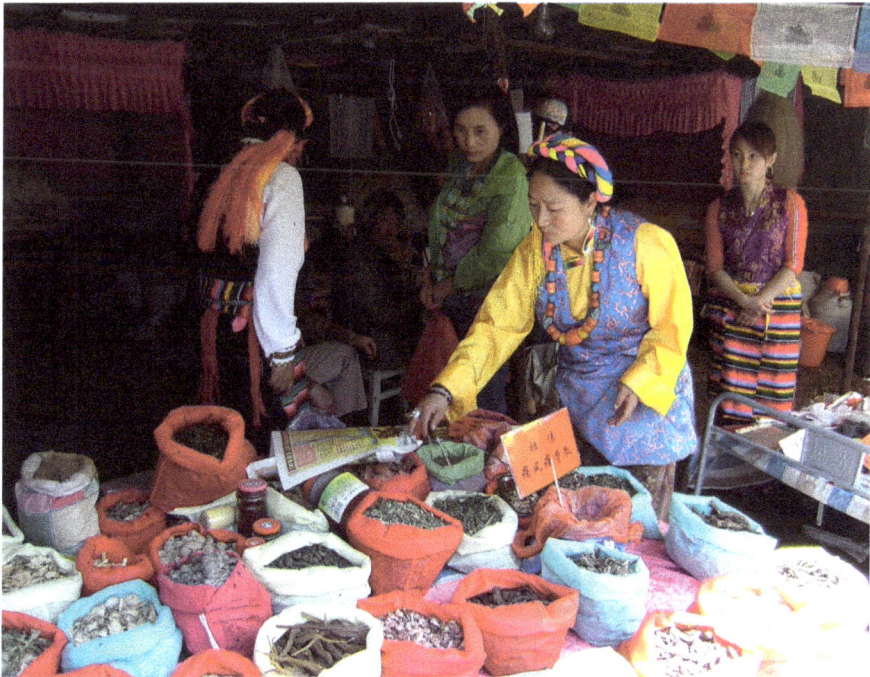

Fig. 6.2. Stallholders in the wholesale medicinal plant market in Dali, China. Photo: A.H. (2007).

of a medicinal plant whose populations in several places in the Himalayas have been depleted over recent years by commercial collection, in its case to sell the bark for use in the manufacture of anticancer drugs (McGuire *et al.*, 1989). All species of *Taxus* are affected, including *Taxus wallichiana* at Ayubia in Pakistan and *Taxus yunnanensis* at Ludian in China, two of the places where we have been involved in conservation projects (see Chapters 19 and 21).

Across the whole of Nepal, a Himalayan country with a huge altitudinal range (64–8848 m), an estimated 323,000–470,000 households (2.6 million people) are engaged in the collection of wild MAPs for sale (Olsen, 1997; Olsen and Larsen, 2003; Hamilton, 2008). The collectors in Nepal can be extremely poor. For instance, in Rasuwa District (north of Kathmandu), most of the herb collectors are continually in debit to roadhead traders, who are prepared to advance money to them to enable them to meet their immediate needs, but who hold them to account and require payment later. Not having much money, the collectors tend to repay the traders in kind through the delivery of yet more MAPs. The result is a relentless attack on the populations of MAPs, which are becoming increasingly scarce. Ignorant of the true values of MAPs to industry, the collectors at Rasuwa and elsewhere often sell MAPs at well below their true market value (see Section 21.2.6).

There are certain species of MAPs in international trade that have psycho-active effects and can be addictive. The exceptional properties of these plants have long been recognised by shamans, who have used them to assist in their communications with the spirits (see Section 7.3.2). European expansionism during recent centuries has resulted in some of these plants being introduced into international trade. One is opium, a commodity which the British East India Company traded from India to China (see Section 5.3.4). Awareness of the harm that opium can cause to people led to 12 of the leading international powers to agree in 1912 on an International Opium Convention, the world's first international drug control treaty (other countries became parties to the Convention from 1919). The primary objective of the convention was to impose restrictions on exports.

Since the introduction of the Opium Convention, two forces have vied with one another for control of the trade – one composed of criminal businesses, which have spotted an opportunity to make enormous amounts of money, and the other composed of the forces of law and order, which have taken increasingly stringent steps to restrict the growing, trading and use of opium and its derivatives. This is not a simple battle of pitting easily recognised good guys against easily recognised bad guys, because there are many pharmaceutical and other businesses operating in the grey zone in-between. For example, HSBC, one of the world's biggest transnational banks, was charged with showing 'stunning failures of oversight' when its Mexican branch was discovered to have laundered money for the Mexican Sinaloa and Colombian Norte del Valle drug cartels between 2005 and 2010 (Lewis, 2022). The case ended up with HSBC agreeing to pay US$1.9 billion in fines.

Today, some criminal businesses involved in the trade in narcotics have become so powerful that they have captured the governments of the countries in which they operate, turning them into narco-states (Ballvé, 2012). Apart from the violence, social disruption and the misery that this inflicts on society, the illegal cultivation of narcotic plants is directly responsible for considerable loss and degradation of natural habitats, including rainforest (Negret *et al.*, 2019). The three main centres of agricultural production of narcotic plants are the Golden Crescent (in Pakistan, Afghanistan and Iran), the Golden Triangle (mountainous Myanmar, Thailand and Lao), and Colombia and Mexico. A.H. has a little direct experience of the difficulties that this has caused to sustainable forest management in Mexico (see Section 16.6). Afghanistan, which is the principal source of opiates sold on the world market, is plagued by collusion between criminal businesses involved in narcotics and government officials, hampering the development of good governance and contributing towards keeping the bulk of the people in grinding poverty (Bullough, 2018).

6.3.4 The rise of financialization

There has long been an association between speculation and trade, with more daring, desperate or naïve investors being prepared to exchange high risk for high potential gain. The first example in post-1500 Europe of an investment bubble caused by people speculating on the future prices of goods was 'tulip mania' in the Dutch Republic (the Netherlands) between 1634 and 1637 (Mackay, 1841). Carried away by a speculating frenzy, some rich people risked and lost everything that they owned when the bubble burst. During recent years, speculating about the future price of money has led to a phenomenal growth in financialization – the aspect of finance in which finance itself is seen as a commodity to be bought and sold, just as if it was an actual commodity (e.g. a loaf of bread) or an actual service (e.g. the work of the shopkeeper who sells the bread) (Kocka, 2016). Combined with the rises in information technology and large transnational businesses, financialization has led in recent years to a massive proportion of capital investments globally being undertaken for speculative, rather than productive, reasons (Lui and Lamb, 2018).

A worldwide financial crash in 2008 revealed just how damaging to society strongly financialised capitalism had become. In Britain, the sequence of events that led up to this economic crisis included the introduction of legislation in the 1970s allowing short selling (which is a way to profit when the values of investments fall, rather than rise, in the markets) and the deregulation of the stock market in 1986 (Raffer, 2011). The 2008 crash revealed the vast size of the electronic and paper edifice that had become constructed on a foundation of financial derivatives (sold as 'financial products'), each consisting of a bundle of investments, each of which could be a derivative itself. Some of the financial products were so complex that their contents and values were extremely difficult to disentangle. Greed and fraud were deeply implicated. It was discovered in 2012 that the value of the London Interbank Offered Rate (LIBOR) had been fixed since 2003 by the employees of banks colluding illegally with one another to benefit their banks (Duffie and Stein, 2015). The LIBOR rate is supposed to be an objective monetary figure calculated at the end of each day's trading to establish the interest rates to be charged between banks in overnight loans.

Reforms were introduced into the banking system after the 2008 financial crash to reduce the risk of banks becoming insolvent. However, these failed to solve the underlying problem, because risky financial behaviour then became transferred from mainstream banking to the shadow banking sector, which lies beyond the reach of regulators and the law (Khan, 2023). The shadow banking sector consists of a plethora of businesses specialising in private equity, such as hedge funds, asset managers, insurers and pension funds. At the least ethical end of these businesses, near-criminal organisations are happy to relieve anyone of their money without shame. Periodic crises have drawn attention to this problem, but little progress has been achieved on its regulation. The collapse in 2022 of Greensill Capital, a company operating in the shadow banking sector in Britain, caused an investment panic when the financial incompetence of the governing party was revealed (Ruggiero, 2022).

The LIBOR scandal and the irresponsible behaviour of the shadow banking sector draw attention to Adam Smith's caveat to his espousal of the free market – that it would only work well as a driver of general public prosperity if practised in a well-governed society upheld by high standards of public morality. Smith was writing in an intellectual climate in which the basic tenets of the Christian religion were accepted with little question and which, in his case, was suffused with the values of philosophic liberalism (see Section 6.1). Historically, Christianity has opposed usury (the charging of excessive interest on monetary loans) and the same has been true of Buddhism, Hinduism, Islam and Judaism (Jain, 1929; Jafri and Margolis, 1999; Karim, 2010; Meeks, 2011). The Holy Quran and the Hadith have much to say about how societies should be organised and businesses run. The details of Islamic finance started to be formulated soon after the death of the Prophet Muhammad in 632 CE and were refined later by scholars, such as the Persian, Al-Ghazali (c.1058–1111). Islam stipulates that wealth should be used only for productive purposes, such as for manufacture and trade, and not to generate more wealth from itself, as when interest is paid on investments. In Islam, charitable giving is a religious obligation, not a choice left to the individual, as in the West.

6.4   Environmentalism and Humanitarianism

Environmentalism and humanitarianism are ideologies and social movements that have become prominent in human affairs during the last 200 years. They are both concerned with well-being, the former with that of the natural world and the latter with that of people. They resonate both with the concept of harmony – a feature of traditional worldviews – and also with that of progress, which is particularly associated with the West (see Section 4.3.2). The common ground between environmentalism and humanitarianism is especially apparent when people following a traditional way of life are confronted by politicians who are unsympathetic either to them or their habitats, as has been happening during recent years in Brazil (see Section 6.3.1) (Stewart *et al.*, 2020).

6.5   Environmentalism

6.5.1   The European/Western roots of environmentalism

Both an ideology and a social movement, the core belief in environmentalism is that the human has destabilised the Earth's ecosystem and it is now its duty to prevent it deteriorating further and, if possible, restore it to a more ecologically sustainable state. Its European/Western roots lie in the cultural movement known as the enlightenment, when some literate people became interested in taking a fresh look at nature and publishing their observations in books and journals. Some collected specimens of plants, animals, fossils and minerals, either for themselves or to pass on to those scientists who, at the time, were laying the foundations of the biological and earth sciences (see Section 7.4). Georges-Louis Leclerc, Comte de Buffon (1707–1788), was one of the most influential early natural historians, inspiring others with his series of books, *Histoire Naturelle*, containing accounts of quadrupeds (animals with four feet), fishes and minerals (Leclerc, 1749–1804). In Britain, the father of natural history was Gilbert White (1720–1793), a cleric who took a profound interest in the plants and animals found in the quiet English country parish of Selbourne where he lived (White, 1789).

Natural historians, like Buffon and Gilbert White, reinvigorated the spirit of personal enquiry into the natural world that that had lain largely dormant in Europe since the Greco-Romano Classical Age more than 1500 years earlier. Prior to the enlightenment, the Catholic Church in medieval Europe had decreed that, on matters relating to the natural world, it was enough for the laity to accept the findings of the Ancient Greeks and Romans, especially those of Aristotle (see Section 4.3.2). White was fascinated by what happens to swallows and other insect-catching birds in winter, a subject that had also intrigued Aristotle. Aristotle had hypothesised that swallows pass the winter in holes in the ground, an opinion with which White was inclined to concur, although he also conjectured that they might go to places where flying insects were available at that time of year. The truth about the matter was not actually established scientifically until the middle of the 19th century, when it was discovered that the European swallow passes the northern winter 12,000 km away in South Africa.

A key figure in the development of environmentalism was Alexander von Humboldt (1769–1859), a German polymath who personally knew other leaders of the German enlightenment (Wulf, 2015). Humboldt was a prolific writer, author of a number of influential books and in correspondence with the leading scientists of the day. Backed by personal wealth and good social connections, Humboldt was able to travel extensively in Central and South America (1799–1804) and Russia and Siberia (1829) – journeys in which he studied and noted how the people lived, as well as observing and recording nature. On his trip to the Americas, he was accompanied by the French botanist, Aimé Bonpland (1773–1858), who collected and classified about 6000 specimens of plants, most of them new to science. Humboldt was a humanitarian, as well as a pioneer environmentalist. A friend of Simón Bolivar (1783–1830), the liberator of Latin America from Spanish colonial rule, Humboldt campaigned against the slave trade in all parts of the Americas. He wanted to go to India to see what was happening there, but the directors of the British East India Company, fearful of what he might discover, managed to block his passage (see Section 5.3.3).

Humboldt carried meteorological instruments on his travels, which he used systematically to record climatic variables. In this, he was an example of the generality that long-term, disciplined, immersion in a scientific field is one of the prerequisites necessary for making major advances in science (see Section 7.4.1). He was the first scientist to note that vegetation of similar general appearance reoccurs in different parts of the world under similar climatic conditions, even though composed of more or less entirely different species. Through observing links between people and nature, he noted that different types of society influence the natural world differently. For example, he observed how the agricultural practices of the indigenous people of South America were more environmentally friendly than those of their colonial neighbours. On a visit in 1800 to Lake Valencia in modern-day Venezuela, he noted how the plantation agriculture of the colonial farmers had increased soil erosion, altered the lake level and caused the local climate to change. This was the first time that a scientist had noted a connection between human behaviour and climatic change.

In addition to Darwin, Humboldt influenced a number of people involved in the early development of environmentalism. Among them were Ernst Henry Thoreau, John Muir, Thomas Jefferson and Ernst Haeckel (for the first two, see Section 8.2.2). Jefferson, who became president of the United States (in office, 1801–1809), was, like Humboldt, a person of broad interests. As well as a politician, he was a farmer, botanist and member of the American Philosophical Society. Humboldt admired the United States for being a newly independent republic that had managed to throw off colonial rule and had been founded on the principles of individual liberty and the equality of people (principles not extended at the time to the indigenous people, Afro-Americans and women). Haeckel (1834–1919) was the person who invented the word 'ecology' for the science concerned with the study of relationships between organisms and their environments (Haeckel, 1866).

At the end of his trip to Latin America in 1804, Humboldt made a diversion to meet with Jefferson in the United States. As it happened, this was soon after Jefferson had dispatched the Lewis and Clark Expedition (1804–1806) to find a direct and practical route up the Missouri River from St Louis to the Pacific Ocean. Publicly declared to be a scouting trip to determine what natural resources were on offer, the hidden purposes of the expedition were to impress on the indigenous people that they were now subject to US sovereignty, establish a US presence in the Louisiana Purchase (which was bought from France in 1803) and combat the territorial ambitions of the Spanish, who were pushing up north from Mexico. The westwards expansion of the United States that followed the Lewis and Clark Expedition proved disastrous for the indigenous people, who became displaced from their lands.

6.5.2   The intimacy of relationships between the living and non-living worlds

Humboldt's realisation of the interconnectedness of the natural and human worlds is a truth well-known to people living in small-scale societies of hunter-gatherers, subsistence farmers and traditional pastoralists (see Section 4.2). Research in recent years into the connections between plants and the soil and between humans and the food in their guts has given some insight into just how intimate the connections between the living and the non-living worlds can be. The cells in root hairs, the tiny parts of roots that are in direct contact with the soil, are connected to the non-living components of the soil by organised complexes of bacteria and other microorganisms known as rhizospheres (Philippot *et al.*, 2013). Similar complexes of microorganisms (gut floras) connect the living cells lining the human gut to the gut's contents (Sender *et al.*, 2016) – complexes that have to cope with whatever humans choose to put into their guts, whether regarded as food, drink or medicine.

Recent research has woken up the medical profession to the importance of maintaining the right balances of microorganisms inside people's guts (Guarner and Malagelada, 2003; Mozaffarian *et al.*, 2022). The damage that can be caused to the gut flora by the consumption of ultra-processed foods is now better appreciated (see Section 6.3.3). The wisdom of maintaining a balance in life, as emphasised in traditional medical systems, is supported by modern dietary research (see Section 7.3.3). It has recently been discovered that the Hadza, who are a hunter-gathering people who live in Tanzania, have much higher diversities of microorganisms in their gut floras than is characteristic of people living in modern industrialised societies – no doubt, one reason for their excellent health (see Section 7.3.1) (Carter *et al.*, 2023).

The rhizospheres of most species of plants include types of fungi called mycorrhizal fungi. The plants and these fungi have a mutually beneficial relationship that involves the plants passing carbohydrates to the fungi and the fungi passing water and nutrients to the plants. Research has revealed that individual plants can be connected to one another through networks of roots and mycorrhizae that can be very extensive and through which messages, as well as water and nutrients, pass (Song Yuanyuan *et al.*, 2019; Nogia and Pati, 2021; Simard, 2021). One of the functions of the messages is for plants to inform others that they are being attacked by herbivores or pathogens, warning them to prepare their chemical defences (Sheldrake, 2020). The carriers of the messages are secondary metabolites – a term used for chemicals manufactured by plants that are not involved in the most basic physiological processes of plants (such as staying alive and growing), but rather serve specialist functions, many of them connected to transmitting information, such as attracting insects to flowers (Böttger *et al.*, 2018).

6.5.3   The Alliance of Religions and Conservation

Thinking positively about the connections between environmentalism and faith traditions led Prince Philip, the Duke of Edinburgh (President of WWF-International),

Martin Palmer (a British theologian) and Ivan Hattingh (Head of Education, WWF-UK) to found an organisation in the 1980s called the Alliance of Religions and Conservation (ARC). A key step in its establishment was a meeting in 1986 attended by leaders of five of the world's major faith traditions (Buddhism, Christianity, Hinduism, Islam and Judaism), the purpose of which was to provide an open forum in which they could discuss among themselves what their faiths could contribute towards saving the natural world. The meeting was held at Assisi, a city in Italy selected because of its connection with Francesco di Pietro di Bernardone (1181/2 to 1226 CE), better known as St Francis of Assisi. St Francis is regarded by many Christians as the patron saint of the natural world. Figure 6.3 is a photograph of Eremo delle Carceri, a hermitage near Assisi to which St Francis frequently retreated to contemplate and pray.

The culmination of ARC's efforts came in 2009, when, at a meeting at Windsor Castle (a royal residence in Britain), representatives of more than 30 faith traditions formally committed them to the environmental cause. This meeting, which was organised by the United Nations Development Programme (UNDP), together with ARC, was attended by UN Secretary-General, Ban Ki-Moon. It was described by Olav Kjørven, UNDP Assistant Secretary-General, as 'the biggest mobilisation of people and communities that we have ever seen on this issue [environmentalism]' (UNEP, 2021). The reason why Kjørven's reference to human numbers is significant is because the majority of people alive today are members of one or another faith tradition. For example, it is estimated that 71% of the human population is either Christian, Hindu or Muslim.

Fig. 6.3. Eremo delle Carceri near Assisi, Italy. Photo: A.H. (2017).

ARC was dissolved in 2019, but its legacy lives on through its founding of a new organisation called FaithInvest, which encourages faith-based groups to campaign for financial decision-making that is beneficial to both people and the planet (www.faithinvest.org, accessed 8 May 2024). In 2021, shortly before a meeting of members of the United Nations (UN) in Glasgow to try and achieve progress in controlling climate change (COP26), FaithInvest helped to organise a meeting in the Vatican attended by representatives of many faith traditions, including Buddhism, several Christian denominations, Confucianism, Daoism, Hinduism, both Sunni and Shi'a Islam, Jainism, Judaism and Zoroastrianism. At the conclusion of the meeting, a resolution was passed and forwarded to the world leaders who were about to assemble at COP26. It urged them to do three things: (i) find a way to achieve net-zero carbon emissions as soon as possible; (ii) for wealthier nations to take the lead; and (iii) for banks and investors to be environmentally responsible in their financial decision-making. For their part, they committed themselves to redoubling their efforts to educate and convince members of their own traditions.

6.6   The European/Western Roots of Humanitarianism

Humanitarianism, like environmentalism, is an ideology and social movement concerned with well-being, but differing in that it is the well-being of people, rather than the natural world, that is at the centre of attention. Humanitarianism became prominent in Europe earlier than environmentalism, a foundational event being the French Revolution of 1789 with its espousal of freedom, equality and brotherhood (see Section 6.1). One reason for humanitarianism's earlier appearance may be because the targets within society for replacement or reform can seem obvious. In the French Revolution's case, they were the monarchy, the aristocracy and the Catholic Church. In contrast, the connection between environmental decay and who, if anyone, is responsible for it happening can be opaque. In any case, much environmental decay happens little-by-little, year-by-year, and is hardly noticed. It is only when tipping points are reached and catastrophes strike that people stir themselves into taking action.

In Britain, humanitarianism began as a social movement towards the end of the 18th century. William Wilberforce (1759–1833), a member of parliament and an evangelical Christian, was a prominent early figure, best remembered for his campaign to abolish the Atlantic slave trade. He was a friend of William Pitt, the British prime minister (in office, 1783–1801 and 1804–1806), who was sympathetic to Wilberforce's campaign, partly because he was persuaded by an argument that had originally been made by Adam Smith, that slavery is an inefficient means of production, because slaves have no prospect of owning property and therefore no positive incentive to work (Haig, 2004). The early humanitarians had several concerns. One was the large number of impoverished rural people that had been created by the agricultural revolution through the replacement of farm labourers by machinery. Another was the squalid and unhealthy living conditions that had developed in the slums of the rapidly expanding towns and cities of the early industrial revolution. A third was the poorly paid and dangerous conditions experienced by those who laboured in factories and mines.

Gradually during the 19th century, thanks to actions taken by agricultural labourers and industrial workers, backed by humanitarian sympathisers, standards of social welfare improved, along with the political representation of the people. Achievements included: a series of Reform Acts that increased the parliamentary representation of the people (from 1832); the Factory Act (1833), which prohibited employers from making children aged 12 and under work for more than nine hours a day; the Mines and Collieries Bill, which prohibited all underground work by women, and also girls and boys under ten (1842); the repeal of the Corn Laws (1846), which had protected the interests of landowners and increased the price of bread for the poor; the restructuring of the corrupt civil service (1854 – see Section 4.3.1); the recognition of trade unions (1871); and the introduction of free elementary education (1891).

The economist William Beveridge drafted a report in 1942 proposing social reforms to reward the sacrifices that were being made by people during the Second World War and cure the social ills (as he saw it) of poverty, disease, ignorance, squalor and idleness. The last was included as one of the ills, because Beveridge believed that idleness destroys wealth and corrupts people. After the war, a new type of socio-economic system was created in Britain through the passage of the National Health Services Act (1946), which established a free national health service for all (some expensive treatments excluded), and a National Assistance Act (1948), which empowered local authorities to provide accommodation and care for those who could not provide them for themselves. Through creating an alternative to some of the social support that was earlier provided by the family, the creation of the welfare state has contributed to the turning of Britain into the highly individualist state that it has now become. In Section 4.5, in a comparison of changes in worldviews in Britain/the West and Buganda/Uganda since the 1950s, we note how Britain has been pressurising Uganda to abandon its conservative collectivistic attitudes towards the family and marriage, and how, if these were to succeed, it would reduce the family-based social security from which the people currently benefit. Uganda is far too poor to afford a state-supported social security system like the British. In any case, the weakening of the traditional family in Uganda carries the risk of generating an epidemic of loneliness, just as is now gripping Britain (Matthews *et al.*, 2019). One measure of the increasing fragmentation of British society is that, for the first time ever, the 2021 British census revealed that more births were taking place outside of marriage or civil partnership than within them.

In Section 6.3.3, we mentioned that the Quaker confectionery families of the Cadburys and Rowntrees launched charitable foundations to promote social improvements. These foundations (the Barrow Cadbury Trust and the Joseph Rowntree Foundation) are examples of philanthropic organisations which use a targeted approach to try and improve conditions in selected aspects of human affairs. The invention of the targeted approach is credited to John D. Rockefeller, whose ascent to the domination of the US oil industry by using ruthless business tactics is told in Section 6.2.2. Rockefeller was a devout Christian of Baptist persuasion and gave generously to charity throughout his life, even when he was relatively poor. The Rockefeller Foundation, one of Rockefeller's many charitable creations, was founded in 1913 with a focus on public health, medical training and the arts. One institution that received support from the Rockefeller Foundation was Peking Union Medical College in China, helping to place it on a firm financial footing (see Section 7.5.2). Rockefeller's involvement in China started during the 1890s, when his company, Standard Oil, began exporting kerosene to China – a business that proved extremely lucrative for Rockefeller. Prior to this time, the Chinese people had depended on substances obtained from plants to provide them with fuel for lighting. For over 2000 years in northern China, the resinous wood of pine trees has been burnt to provide light, while, in the centre and south, it has been oil extracted from the fruits of the tallow and tung oil trees and those of candlenut (*Aleurites moluccanus*) (see Preface).

6.7 Christianity and the Spread of Humanitarianism from Europe/the West

The humanitarian movement started to spread from Britain into Africa in the 19th century. The Church Missionary Society (CMS), an organisation founded in London in 1799 by evangelical clergy of the Church of England, sent its first missionaries to Africa in 1825 (to Egypt) and 1827 (to Ethiopia). The missionaries campaigned against the Arab slave trade, introduced literacy into places from which it had previously been absent (such as much of sub-Saharan Africa), studied the local cultures, translated the Bible into the local languages and opened Western-style schools and clinics. The European colonial administrators appreciated the religious, educational and medical work of the missions. It aligned with their own worldviews, removed some of the costs of colonial rule from their budgets, made the people easier to govern, produced literate people able to fill junior posts in the civil service and others, familiar with European ways, suitable for serving in the lower ranks of the army and police.

Prior to the emergence of humanitarianism in Europe, the focus of European missionary efforts in foreign lands had been to reach the heathen and baptise them in the Christian faith. The missionaries believed that the act of baptism opened up the possibility for people to dwell in the Kingdom of Heaven after their lives on Earth were over. In this way, Christianity acted as a third force, along with the drives for national enrichment and personal aggrandisement, that propelled the expansionism of Europe (Tutino, 2021). An illustration of how these drives could work together is provided by the fate that befell Atahualpa, the Inca emperor, after he had been captured in 1535 by the Spanish conquistadors who were invading Peru. After his capture, Atahualpa was ransomed for a huge amount of silver and gold, but, after it had been delivered, instead of releasing him, the Spanish forced him to accept baptism and had him killed (Prescott, 1847).

The perception that the Christian Church was aligned with the interests of an oppressive ruling class resulted, from the late 18th century, in it coming under attack by political revolutionaries, as in the French Revolution (1789), the Russian Revolution (1917), the Mexican Revolution (1910–1920) and the Spanish Civil War (1936–1939). In the Mexican Revolution, 90% of Catholic priests in Mexico were assassinated, expelled or emigrated. Not until the 1960s did the Catholic Church in Latin America, inspired by the doctrine of liberation theology, start to reorientate itself away from alignment with the ruling class towards the causes of the poor (Muller and Gutierrez, 2015).

Prior to the 19th century, slave-owning was not necessarily seen in the West as un-Christian. In Britain, the records of compensation paid under the Slave Compensation Act of 1837 reveal that large numbers of ordinary British people were slave-owners, among them many clergy (Draper, 2010; Olusoga, 2016). The Society for the Propagation of the Gospel in Foreign Parts, a Church of England missionary organisation, continued to run its sugar plantations on Barbados in the West Indies using slave labour until 1833, which was when the British government banned the owning of slaves.

Nineteenth century Protestant missionaries, inspired by humanitarian sentiments, ventured into all parts of the world that they could reach, regardless of whether or not they had become incorporated into one or another of the European empires. Here, we give two examples of countries outside Europe to illustrate their impacts. One is Uganda, which was part of the British Empire, and the other China, which, apart from Macau on its seaboard and the island of Hong Kong, escaped full-blown European colonialism.

6.8 Arrival of Christianity in Buganda/Uganda and Its Consequences

6.8.1 Some socio-cultural similarities and dissimilarities between 19th century Britain and Buganda

John Hanning Speke, the first Western explorer to visit Buganda/Uganda (in 1862) and Henry Stanley, who visited in 1875, made it known when they arrived back in Britain that they believed that the Baganda would appreciate a Christian mission. In response, in 1877, the CMS dispatched a small group of missionaries, who, as predicted, received a warm welcome and quickly made many converts. Several of the newly converted Baganda demonstrated their commitment to their new faith by dying for it during a period of religious persecution in 1885–1887 (Furry, 2008). Later, people in other parts of Uganda proved similarly receptive to the Christian message.

Many of the early conversions to Christianity in Uganda were not made directly by the European missionaries, but rather by Ugandans who had become converted and, on their own initiatives, taken Christianity to tribes that were other than their own. The following psychological explanation has been offered to account for this behaviour (Wild-Wood, 2021):

> They self-consciously allied themselves with a worldwide movement and they operated on a regional stage. They cast their identity beyond the local. They attempted to persuade others of the attraction of a transnational community

which they believed was offered to them in Christianity. Like their European counterparts, they were compelled to leave home and settle in new areas by a desire to preach and plant churches.

In other words, it can be hypothesised that the arrival of the Western missionaries with their strange ideas, possessions and ways of behaving had made the people realise that there was a bigger world out there, larger than that of their previous experience. This fits in with the psychological suggestion offered in Section 3.4 for why new belief-systems emerged on the transition from small-scale societies to states and empires.

The welcoming attitude of the Baganda contrasted with the wariness and sometimes outright hostility that the early missionaries experienced from the people whom they encountered on the first part of their treks inland from Zanzibar to Buganda (Mullins, 1904; Roscoe, 1921). Not having their own means of transport, the custom of the missionaries was to hire porters in Zanzibar to accompany them to the southern shore of Lake Victoria, from where they either walked around its western margin to reach Buganda or took canoe transport across the lake. The regular hirers of these porters were Arabs engaged in trading into the African interior. One of the commodities in which the Arabs were interested was African slaves, who, after capture, were taken for sale in the slave markets in Zanzibar. Arab slavery had intensified after Said bin Sultan, the ruler of Muscat and Oman, had taken up residence in Zanzibar in 1840 (see Section 16.2). It is not surprising that the Africans met by the missionaries on the initial legs of their journeys from Zanzibar to Buganda were wary of caravans.

The British and Baganda were mutually attracted to one another, as is evident from the success of the missionaries at making converts and the accounts of early British residents and visitors (Churchill, 1908; Roscoe, 1921). It can be hypothesised that one reason for this was the considerable similarity between the societies and cultures of the British and Baganda at the time. Both were hierarchical societies headed by monarchs. In both, the monarch (king = *kabaka* in Luganda) was advised by a parliament (*lukiiko* in Luganda) and there was a prime minister (*katikkiro*) in charge of administration. As in Britain, the system of administration in Buganda was organised hierarchically, its units of governance (and their British equivalents) being *ssaza* (county), *ggombolola* (sub-county) and *muluka* (parish). Both societies possessed roads, currencies and taxes, those in Buganda including a fee of 10% of the value of transactions in markets (see Sections 3.5.1 and 3.5.3) (Roscoe, 1911). Both frequently engaged in wars against their neighbours, and both were slave-owning societies (or, in the British case, had been until recently – see Section 5.3.1). Both had systems of higher education accessible only to a few (see Section 4.4). Healthcare in both was strongly rooted in local herbal traditions and, in both, it was generally believed that the world was suffused with spiritual influences.

Buganda in the 19th century was a tribe as well as a feudal kingdom, its tribal aspect being expressed in its clan structure and the totems attached to the clans (see Section 8.1). Every person had a position in a nested set of categories of family belonging, namely *ekika* ('clan'), *essiga* ('sub-clan'), *olunyiriri* ('lineage') and *oluggya* ('courtyard'). Members of each of these categories claimed descent along the male line from a common ancestor. As explained in Section 7.2, there were deep connections between membership of clans and particular areas of land.

As mentioned in Section 4.4, the monarchy in Buganda was chosen by a council of senior chiefs, which meant that the person chosen had a good chance of having the qualities required to meet the challenges of his day. The method of selecting the monarch in Britain was much less inclusive. The throne was inherited according to the rules of primogeniture, which meant that it passed to the next eligible person in a prescribed order of succession. There were no checks on fitness for the job. Another feature was that the marriages of royal heirs were often contracted between a small number of European ruling families, an inbreeding system that carried the danger of genetic disorders becoming inherited. Queen Victoria, the monarch who ruled Britain for most of the 19th century, provides an example. Her husband, Prince Albert of Saxe-Coburg and Gotha (a dutchy in Germany), was a first cousin and her children married into the ruling families of Denmark, Prussia, Russia and others. Queen Victoria was a carrier of haemophilia, a serious blood-clotting disease which she passed on to several of her children. The forcible introduction by the British into Buganda in 1897 of their own way of choosing monarchs (see Section 6.8.2) was a violation of the traditional customs of the Baganda and, from a genetic perspective, a step backwards in political development.

6.8.2   Complications and socio-cultural disruption in Buganda

The progress of the CMS mission in Buganda received a complication in 1879, when French Catholic missionaries arrived and, as the British Protestants had done, established a mission station in Mengo close to the court of the king and began to proselytise. The complication arose because the Protestant and Catholic missionaries disagreed openly about which of their versions of Christianity was correct. For his part, the king, Kabaka Muteesa I, could not understand why two white men (Alexander Mackay and Father Pierre Lourdel), both of whom claimed to belong to the same religion, were competing for his attention. Muteesa was anyway deeply involved in Kiganda traditional religion, as was inevitable for someone in his position. He may have been a Muslim, with Islam having been introduced into Buganda in about 1840 by Arab/Swahili caravan traders operating out of bases along the East African coast.

In the event, the French Catholics proved just as successful as the British Protestants at gaining converts, with the result that tension built up between two factions of the Baganda, the *Abangereza* ('English faction') and the *Abafalansa* ('French faction'). This spilt over in 1888 into a decade of civil wars, that also involved a Muslim faction (*Abasiraamu*), followers of Kiganda traditional religion and detachments of Nubian and other African troops under British command. According to Roscoe, who witnessed these events at first hand, these wars were not about religion, in the sense of not being disputes about theology, but rather were fights for power in a local political world that had become destabilised by foreign influences (Roscoe, 1921).

The figures given by the Christian churches for the number of Baganda who died for their faiths during the civil wars of 1888–1898 are 23 Anglicans and 22 Catholics. They are known by Christians as the Uganda Martyrs. Muslims too had their martyrs – an estimated 74 people who died for Islam in 1874–1876 during a period of anti-Muslim sentiment. However, the total number of people killed in Buganda as a result of religiously connected violence during the last three decades of the 19th century was much higher. Roscoe estimated that the wars were responsible for decreasing the population of Buganda from about 3 million in the mid-19th century to about 1.5 million when they ended in 1898 (Roscoe, 1911).

As mentioned in Section 5.4.3, in 1890, Kabaka Mwanga II (the King of Buganda in 1884–1888 and, again, in 1889–1897) signed a treaty with the British East African Company that transferred much of his political authority to the company, which hampered his ability to rule. Eventually, he became fed up with his fettered political position and the chaos that the foreigners had unleashed and, in 1897, made the fateful decision to stand up for his kingdom and fight (Lunyiigo, 2011a). However, his military forces proved no match for those of the British and, in 1899, he was captured and exiled to the Seychelles, where he died in 1903.

In 1897, with Mwanga rebelling and absent from his capital, the British authorities in Uganda decided to install a 1-year-old baby, Daudi Chwa II, in his place. This was a deliberate flouting of Kiganda traditions, since kings were traditionally selected by the senior chiefs according to their assessments of the suitability of character of the candidates (see Section 4.4). Daudi Chwa was far too young to be included in such an assessment. Having done the dead, the colonial authorities then provided Daudi Chwa with an upper-class British education and made sure that he was an Anglican, which is the version of Christianity that was (and is) the official state religion of England. Chwa's successor, Kabaka Muteesa II (reigned, 1942–1969), received a similar grounding in the culture of the British upper classes. He studied at the University of Cambridge and was commissioned as a captain in the Grenadier Guards, which is the premier infantry regiment in the British Army.

6.8.3   Influences on subsequent developments in education in Uganda

Alexander Mackay (1849–1890), the Protestant missionary who disputed publicly with Pierre Lourdel, was a Scot, an engineer and a Presbyterian, the version of Christianity that was (and is) the official state religion of Scotland. He introduced a printing press, printed passages of the Bible in Luganda and opened the first Western-style school. From this start, progress in establishing a Western-style system of education system in Buganda/Uganda was fast. The first Protestant elementary school was opened in 1886 (for boys only), the first Protestant secondary school in 1897 (again, for boys only) and the first Protestant boarding school for girls in 1906. The first Western-style hospital, which was in Mengo, was established in 1897 by Albert Cook (1870–1951), a medical missionary who benefited from receiving financial backing from Henry Wellcome (1853–1936). Wellcome, an American by birth but later a British subject, had made a fortune in pharmaceuticals (Roscoe, 1921). His generosity towards Mengo Hospital parallels that of the American tycoon, John D. Rockefeller, towards Peking Union Medical College in China (see Section 7.5.2). Wellcome's connection to Christianity began in his childhood. He was born in a log cabin in the American frontier state of Wisconsin, the son of an itinerant evangelical missionary who preached out of a covered wagon.

The British colonial administration in Uganda did not get involved in the educational sector until 1921, when it backed the founding of a technical school at Makerere in Kampala (expanded into a college in 1922). The purpose of the college was to train pupils received from the missions in practical subjects, such as carpentry, building, medical assistantship, agriculture, veterinary science and teaching. The government started to give grants to mission schools in 1924. The appointment of a progressive new British governor, Philip Mitchell, to Uganda in 1935 led to an acceleration in the process of promoting Ugandans to senior positions in education and the civil service (Apter, 1997). In 1945, Makerere College opened its doors to female students, in 1949 it became a university college affiliated to the University of London, in 1963 it became one of three colleges in a new University of East Africa (the other colleges were in Nairobi and Dar es Salaam) and, in 1970, it became an independent university offering degrees in its own right. Makerere was the only university in Uganda in 1970, but, since then, many more have been founded. There were 34 universities, as of 2013, five public and 29 private (Bisaso, 2017). Several of the private universities have religious affiliations (Protestantism, Catholicism, Seventh Day Adventism and Islam).

Scholarship in Uganda was aligned with a Western worldview throughout the period of British colonial rule. A leader in the introduction of African perspectives was John Mbiti (1931–2019), a Kenyan, Anglican priest and lecturer in religion and theology at Makerere from 1964 to 1974 (Peterson, 2019). Mbiti became a spokesman for African languages and cultures, demonstrating in his academic research that African philosophical and religious traditions are consistent and sensible and should not be dismissed as mere superstition (see Section 7.5.3) (Mbiti, 1969, 1970). He championed the opening up of dialogue between Christianity and African traditional religions, in the process pointing out that African traditions do not divide life into distinct religious and secular spheres (see Section 4.2).

Mbiti's pioneering work was responsible for introducing African knowledge and perspectives into academia at Makerere. The departments that have done so include those teaching botany, agriculture, forestry, languages, music and history. An undergraduate course in ethnobotany was introduced in the early 1990s, possibly the first such course in the world (Hamilton *et al.*, 2003). (Ethnobotany is normally taught at postgraduate level.) Teaching and research in agricultural and forestry have become more engaged with indigenous knowledge and practices, away from their former foci on cash crops, timber and protective forestry.

Mbiti witnessed two political coups while he was teaching at Makerere, the first when Milton Obote seized power in 1966 and the second when General

Idi Amin replaced him in 1971. Two weeks after his coup, Amin, who was a Muslim, sent a delegation to Mbiti to ask for his opinion about whether he should establish a new ministry to supervise religious affairs (Peterson, 2019). Mbiti replied in the positive, because he had become convinced of three things through his research: (i) that a social structure is essential for the well-being of a state; (ii) that secularism was spreading in the West; and (ii) that, if this were to happen in Africa, the resulting loss of social structure would undermine the state's ability to function. In the event, Mbiti need not have worried about the spread of secularism in Uganda (or the rest of Africa). The 2014 census in Uganda revealed that only 0.2% of the people had no religious affiliation. In contrast, the British Social Attitudes survey for 2018 revealed that 37% of Britons considered themselves as having 'no religion', a figure rising to over 50% among the under 40s. There is today a 'reverse mission' of African missionaries coming to Britain to help cater for the spiritual needs of the British people (Olofinjana, 2020). It parallels the 'reverse flow' of doctors and nurses from Africa to Britain to help plug deficiencies in the National Health Service.

Taking Mbiti's advice on board, Amin established a new Ministry of Religious Affairs in 1975. According to Peterson, Mbiti was unaware when Amin asked him for his opinion about establishing the new ministry that his real intention was to turn Uganda into an Islamic state. When Amin publicly announced his intention to do so, which was in 1973, Mbiti equally publicly opposed him and was forced to flee for his life. However, soon thereafter, he received some comfort in being offered and accepting the directorship of the Ecumenical Institute Bossey in Switzerland. Established in 1946 to support the World Council of Churches, its mission is to unite all branches of Christianity. Of the mainstream Christian churches, only Catholicism is not a full member.

6.8.4   Influences on subsequent developments in religion in Uganda

Amin's motivation in founding the Ministry of Religious Affairs was at least partly because he worked to heal the divides that had opened up within Ugandan society following the introductions of Islam and Christianity (Leopold, 2021). In 1972, he backed the formation of a Uganda Muslim Supreme Council to serve as an umbrella organisation to oversee the numerous branches of Islam that were, by then, present in the country. In the same year he made his first visit to the Arab world, which included a pilgrimage to Makkah and consultations with the rulers of Saudia Arabia and Libya, both of whom promised him support. Amin was then in a better position to begin the construction of the Uganda National Mosque in Kampala. Finally finished in 2007, it is now named the Gaddafi National Mosque after Libya's leader, Muammar Gaddafi (ruled, 1977–2011). Within Christianity, only Anglicanism, Catholicism and Orthodox Christianity were allowed to operate.

Ugandans have shown a lively interest in Christian theology since the early years of its introduction. Many varieties of Christianity have emerged. One of those responsible for this proliferation was a powerful Kiganda chief, Semei Kakungulu (1869–1928), who took it upon himself to bring Christianity and the Kiganda system of governance to a large part of Eastern Uganda

(Twaddle, 1993). Kakungulu was a careful reader of the Bible. Originally converted to Anglicanism by missionaries when he was serving as a page in the royal court during the 1880s, he later became convinced that the Anglican missionaries were not reading the Bible correctly. The Sabbath, he decided, should be observed on Saturday, not Sunday. Later, he became a Malakite, an anti-colonialist and anti-Western version of Christianity founded by a Muganda called Musajjakawa in 1914. Later again, he became convinced of the truth of Judaism and founded a religious community known in Luganda as the *Abayudaaya*, which means 'the people of Judah'.

Another Ugandan responsible for introducing a new version of Christianity to Uganda was Ruben Mukasa (1899–1982), who became convinced through his studies of the correctness of Orthodox Christianity and, on his own initiative, travelled to Egypt to consult with the leaders of the Greek Orthodox Patriarchate of Alexandria. He was ordained a bishop (becoming Bishop Christopher) and, together with other Ugandans who shared his religious persuasion, created in Uganda a community of Orthodox Christians (the *Abasokkookisi*). Today, an Orthodox cathedral crowns one of the hills of Kampala, just as other hills are crowned by the Protestant and Catholic cathedrals, the Gaddafi National Mosque and Kibuli Mosque (belonging to different versions of Islam) and the Bahá'í Mother Temple of Africa. Pentecostal Christianity has flourished in Uganda in recent years. Its adherents, the *Abalokole* (meaning 'the saved ones'), believe that they are guaranteed places in the Kingdom of Heaven. They have some connections with American right-wing evangelical Christians (see Section 6.3.1). The intense interest of Ugandans in Christianity has been accompanied by the emergence of unscrupulous businesspeople, who, through charismatic appeal, have exploited the vulnerability of the poor and fleeced them of their money (Lauterbach, 2020).

In a way, the eradication of traditional religion in Uganda is impossible, because of its closeness to the structure of society and the lack of a strict divide between the secular and religious spheres of life in the worldviews of many – a reality that has been accommodated by the Anglican Church (see Section 4.5). The large number of Baganda who visit the traditional shrines at Ttanda near Kampala exemplifies the continuing interest in traditional religion. Ttanda is a place famous in Kiganda folklore, because it was where Walumbe, the God of Death and Disease, was chased into the underworld by his brother, Kayikuuzi, who had been sent by Ggulu, the God of the Sky, to accompany Walumbe back to heaven. At Ttanda, there are more than 300 substantial pits, which, from an archaeological perspective, are actually the shafts of an ancient kaolinite mine (see Section 3.3.1). Several of the pits have associated shrines dedicated to Kiganda deities (*ba lubaale*). Figure 6.4 is a photograph of one dedicated to Wannema, the God of Cripples (*abalema*). Figure 6.5 shows the shrine of Muwanga, who was the leader of the gods and the one to be consulted in the event of sickness or disease (Roscoe, 1911; Welbourn, 1962).

There are parallels between Kiganda traditional religion and aspects of Christianity. The pile of crutches at the shrine to Wannema are votive objects. A similar pile was once present at the Catholic Shrine to Our Lady of Lourdes in France. Note that the spear in the hand of the man (Keefa Ssentoogo) in Fig. 6.4 is bent. This is symbolic of being a cripple. Note the hearth in Fig. 6.5.

Fig. 6.4. Votive objects at the shrine of Wannema at Ttanda, Uganda. Photo: A.H. (2018).

Fig. 6.5. The shrine of Wanga at Ttanda, Uganda. Photo: A.H. (2018).

This has a parallel with the candles lit by supplicants at the shrines of Christian saints. The deities at Ttanda have kept up with the times. Disused electrical equipment is being left at the shrine of Kiwanuka, the God of Thunder and Lightning, who has evidently expanded his realm to cover electrical devices. The whole of Ttanda is contained within a sizeable patch of native woodland, a good example of a sacred grove (see Section 8.1).

6.8.5   Influences on subsequent developments in politics in Uganda

History has created connections in Uganda between politics, varieties of Christianity and parts of the country, which has led to a number of socio-cultural schisms. Concentrating in this account on those that particularly involve the Baganda, an unresolved tension has continued to rumble on between the Baganda and other ethnic groups related to the behaviour of Baganda like Kakungulu, who imposed Kiganda culture forcefully on other parts of the country. The Banyoro formerly had another grudge against the Baganda, because two of its counties were allocated to Buganda under the terms of the Uganda Agreement of 1900 (see Section 5.4.3). However, this was rectified following a referendum in 1964, which returned them to Bunyoro.

The Baganda became increasingly disenchanted with the political behaviour of the British after the Second World War. They resented the way that the British had drawn them into their creation of the polity of Uganda and were especially alarmed when they learnt in 1953 that the British were intending to incorporate Uganda, along with Kenya and Tanzania, into a Federation of East Africa (see Sections 2.3.2 and 16.2) (Muteesa, 1967). What they particularly feared was that they would become politically dominated by the numerous White Settlers who had established themselves in the Kenyan Highlands. From the perspective of Kabaka Muteesa, the British had no right to impose these constitutional changes on Buganda without his express permission. The result was that tension built up between Muteesa and the British that led, in 1953, to him being bundled into a plane and deported to Britain. However, in 1955 he was allowed to return to his kingdom on the understanding that the British would drop the idea of an East African Federation, that he was to be a constitutional monarch, that the power of the Buganda parliament would be increased and that Britain would grant Uganda political independence in the future.

Britain began preparing Uganda to be a politically independent country during the 1950s and, as is normal in modern countries, political parties positioned along the left/right wing ideological spectrum began to emerge. However, thanks to Uganda's history, more significant for forging the characters of the main political parties were religion and tribal identity (Kasujja, 2023). Three political parties contested the first general election, which was held in 1962 just before the date of political independence. They were the Uganda People's Congress (UPC), which had a connection with Protestantism; the Democratic Party (DP), which had a connection with Catholicism; and Kabaka Yekka (KY), which advocated a high degree of political autonomy for Buganda (Kasujja, 2023). (*Kabaka Yekka* in Luganda means 'King Only'.)

After the pre-independence election in 1962 in Uganda, a political pact made between the UPC and KY resulted in an arrangement whereby Muteesa became the Head of State and Milton Obote (the leader of UPC) the Prime Minister. However, in practice, Muteesa and Obote found it impossible to collaborate, which resulted, in 1966, in Obote giving an order to the army to storm Muteesa's palace. About 3000 Baganda inside the palace were killed in the fighting (including A.H.'s father-in-law), but Muteesa managed to escape and, after a hazardous journey, reached Britain, where he was granted political asylum and later died.

With Muteesa gone, Obote declared himself executive president and abolished Uganda's four largest kingdoms, including that of Buganda. Power in the governance structure was centralised, the consequences of which for forestry are mentioned in Section 5.6. The general election of 1962 was the last held in Uganda when there was a peaceful and constitutional transfer of power from one ruler to another. It is ironic that the involvement of the British in Uganda has led to the destruction of Buganda's traditional system of governance, which was one of the attractions that drew them there in the first place.

6.9 Arrival of New Christian Influences in China in the 19th Century

China was a different proposition for 19th century Protestant missions than Uganda. As mentioned in Section 4.3, its culture had been profoundly influenced by the cultural revolutions of the Axial Age and it was no stranger to Christianity. The Nestorian (or Xi'an) stele, which is a limestone block engraved in Chinese and Syriac (the language of ancient Syria), records that Christianity was brought to China in 635 CE by the Church of the East (Russell, 2014). This was a branch of early Christianity that had its headquarters in Ctesiphon (near modern-day Baghdad). The stele was unearthed in about 1624 in Shaanxi Province, the ancient capital of which (Xi'an) stood at the eastern terminus of the terrestrial Silk Roads (see Section 3.5.2). The Venetian traveller, Marco Polo, found a Christian church already established in Yangzhou, when he arrived there in *c.*1282. (Yangzhou is situated on the northern bank of the Yangtze River where it connects to the Grand Canal – see Section 3.5.4.) The Italian Franciscan priest, John of Montecorvino (1247–1328), built a church opposite the imperial palace in Khanbaliq (now Beijing) and, during the 17th and 18th centuries, Jesuits were valued members of the imperial court (see Section 5.1).

The Chinese Government in the 19th century was wary about the intentions of the 19th century Protestant missionaries and forbade them from living and working outside the limited confines of Macau and the Thirteen Factories area of Guangzhou (see Section 5.3.4). However, in 1858 as a consequence of the terms of the treaties that concluded the Opium Wars, the missions obtained the freedom to travel and preach all over China. Responding with enthusiasm, Protestant missionaries (mainly British and American) set out in numbers for China, where they established mission stations, and Western-style clinics and schools.

One aspect of the missions that especially interested the Chinese was the opportunity to receive an education in the Western way and Western educational institutions forged ahead. A milestone was reached in the development of science in China in 1928 when Academia Sinica (the Chinese Academy of Sciences) was founded in Shanghai. However, along with enthusiasm for a Western education came an undercurrent of resentment at the humiliation that China had received in the Opium Wars and the one-sided terms (from the Chinese perspective) of the treaties of Nanking and Tientsin. These sentiments erupted in 1899 in the Boxer Rebellion, during which many missionaries and their Chinese converts were killed. Nevertheless, Western missionaries still came to China, their total reaching a peak in the 1920s, after which the disturbed state of China led to their numbers declining. The last Western missionaries were expelled in 1953, when the Communist Party of China came into power.

However, this was not the end of the story of Christianity in China. Since 1980, the Chinese Government has allowed a considerable degree of religious freedom and Christianity has experienced a dramatic revival. Nevertheless, the Chinese Government remains wary of religions whose leaders do not reside in China. For instance, it has reserved for itself authority over the appointment of Catholic bishops in China.

6.10   Environmentalism and Humanitarianism Go International

Since their origins in Europe at about the turn of the 19th century, environmentalism and humanitarianism have expanded in geographical influence and become international. The progress towards internationalisation of humanitarianism began in 1864, when 12 European countries agreed on the first Geneva Convention, which set down baseline standards of treatment expected to be provided to civilians in times of war and combatants no longer able to fight. It was a Swiss businessman, Henry Dunant (1828–1910), whom the world has to thank for the idea of the Geneva Convention. He was stimulated into campaigning for it when he observed at first hand the aftermath of the Battle of Solferino, which was fought in Italy in 1859 between the armies of the Austrian and French empires, the latter in alliance with the Kingdom of Sardinia. The battle had left 4000 soldiers of both sides dead or wounded on the battlefield without medical assistance of any kind. Today, the Geneva Conventions in their modern forms have been ratified by nearly every state.

The horrors of the two 20th century World Wars drove humanitarian concerns to the top of the international agenda. The League of Nations was founded in 1920 after the First World War and its replacement, the United Nations (UN), in 1945. The League of Nations had 44 signatory states, almost all countries in Europe or the Americas. Initially, the UN had 51 members, but many more joined as Europe's colonial territories gained their political independence.

A number of subsidiary bodies and connected organisations have been created to further the aims of the UN. Because, in actuality, the human and natural worlds are inextricably interconnected, many of these bodies are relevant to the pursuits of both humanitarianism and environmentalism. Those created between 1944 and 1948 included the World Bank and the International Monetary Fund (IMF), the World Trade Organization (WTO), the Food and Agricultural Organization of the United Nations (FAO), the United Nations Educational, Scientific and Cultural Organization (UNESCO) and the World Health Organization (WHO). The United Nations Development Programme (UNDP) was created in 1966 at the time when Europe was rapidly decolonising and the United Nations Environment Programme (UNEP) in 1972 as part of the upsurge in environmental awareness that was occurring at the time.

The seeds of the UN were sown in the recognition by Franklin D. Roosevelt (the president of the United States) and Winston Churchill (the British prime minister) that a moral justification was needed to unite the efforts of the Allies fighting the Axis powers in the Second World War. Additionally, in Britain, it had become politically necessary to offer a better future to the troops who were fighting for their country and the civilians on the home front who were backing them. In his State of the Union Address in January 1941, Roosevelt listed four fundamental freedoms that, he believed, every person on Earth should be able to enjoy. They were freedom of speech, freedom of worship, freedom from want and freedom from fear. In August of the same year, shortly before the United States started to supply significant quantities of military supplies to the British, Roosevelt and Churchill met in Newfoundland and jointly issued a statement (which later became known as the Atlantic Charter) that set out their joint goals and aims for the war and the post-war world. This charter formed the foundation stone upon which the UN was constructed and sounded the death knell of the British Empire. The United States had developed a distaste for British imperialism in the late 18th century, when it fought against Britain in its War of Independence. Because of the way in which it originated, the UN's headquarters was established in the USA, the land for its building being donated by the tycoon and philanthropist, John D. Rockefeller (see Section 6.6). Roosevelt's wife, Eleanor, chaired the committee that drafted the Universal Declaration of Human Rights, which was adopted by the UN in 1948 and forms the foundational document for the principles of international humanitarianism.

As mentioned in Section 6.6, environmentalism came on to the world stage later than humanitarianism. Landmarks in its institutional development have included: the founding of the International Union for Conservation of Nature and Natural Resources (IUCN, 1948) and of its Species Survival Commission (1949) and Commission on Protected Areas (1960); the founding of the World Wildlife Fund (WWF, 1961 – now called the World Wide Fund for Nature), initially a fundraiser and publicist for IUCN; the publication of a book *Silent Spring* by Rachel Carson, which drew attention to the extensive poisoning of birds of prey by persistent synthetic pesticides (Carson, 1962); the initiation of IUCN's Red List of endangered species (1964); the publication of two papers, *A Blueprint for Survival* and *The Limits to Growth*, pointing out that Earth's natural resources are not inexhaustible (Goldsmith and Allen, 1972); the launch of the UNEP (1972); the coming into force of the Convention on

International Trade in Endangered Species of Wild Fauna and Flora (CITES, 1975); the ratification of the Montreal Protocol (1987); and the ratification of the UN's Convention on Biological Diversity (CBD, 1992).

The purpose of the Montreal Treaty was to limit emissions of the man-made gases that were found to be destroying the ozone layer in the upper atmosphere and exposing terrestrial life to potentially lethal doses of radiation (see Section 2.1). As of today, the Montreal Treaty is the only environmental treaty that has been ratified by every single member of the UN. It is also notable for being a pioneer example of an international treaty that sets different conditions of compliance for different countries, depending on whether they are placed in the Global South or the Global North. (See Section 5.7, for an explanation of these terms.) The arrangement reached in the Montreal Treaty, which was seen as a reasonable compromise by all concerned, was to allow extra time for countries in the Global South to implement the treaty, recognising that they were less well positioned in terms of finance and expertise to do so (Sunstein, 2008).

6.11   The Convention on Biological Diversity

The Convention on Biological Diversity (CBD) has been ratified by all members of the UN except for Andorra, South Sudan, the United States and the Holy See. Those who have ratified CBD have agreed to do three things (Glowka et al., 1994) – to conserve the biodiversity that lies within their borders, to use their biological resources sustainably and to ensure that there is an equitable sharing of benefits when the allowing of access to the biodiversity of a country leads to the development of new commercial products. The third part of CBD was intended to be a mechanism whereby countries which have a high potential to develop new commercial products would share the benefits that they received from them with the countries from which the biodiversity originated, in particular to strengthen the conservation of biodiversity. Although the terms Global South and Global North are not mentioned in the convention, there is an implicit assumption that the countries that have proficiency at developing new commercial products are in the Global North and many of those rich in biodiversity are in the Global South.

Attempts to implement the third part of the CBD have run into many practical problems (Muzaffar et al., 2011). Such progress as has been made is contained in the text of the Nagoya Protocol on Access and Benefit-sharing, a supplementary agreement to the CBD reached in 2014 (www.cbd.int/abs, accessed 8 May 2024). In any case, the way that the third part of CBD was envisaged to work is largely out of date (Laird et al., 2020). It was originally thought that the research would be undertaken on physical samples taken from plants or other organisms. However, nowadays, much of the search for new useful products inspired by the biological world uses information on gene sequences that is already in the public domain or else is purchased by businesses from one another. Two contributions of Professor Pei Shengji and his colleagues towards making progress on access and benefit-sharing (ABS) are described in Chapters 13 and 14. They are characterised by taking a bottom-up, community-based, approach to deciding how best to distribute the costs and benefits involved, rather than the top-down approach that is more typical. The Chinese Ministry of Ecology and Environment, together with UNDP, has chosen these two examples as case studies to demonstrate at two Conferences of the Parties (COP) of the CBD what has been achieved with implementing ABS in China.

Around the year 2000, A.H. was asked by the Secretariat of the CBD to be a member of a small group of plant conservationists to develop the wording of a supplementary agreement to the CBD, entitled the Global Strategy for Plant Conservation (GSPC, which was ratified in 2002). The structure chosen for the GSPC included targets, an approach to management that was becoming popular at the time. In GSPC's case, there were 16 targets to be met by the year 2010. A.H. was assigned the job of clarifying the wording of Target 4, which, in its final form, reads 'At least 10 per cent of each of the world's ecological regions effectively conserved' (CBD, 2002).

A.H. set about this task by choosing a few sample countries to seek advice from experts within them on how they thought Target 4 should be worded. One of the countries he chose was Colombia, influenced by three considerations. One was that it contains more species of plants than any other country. Second, it had an excellent WWF country team eager to cooperate. Third, it is a difficult place to be effective in conservation, because it was beset with plant-related, drug-fuelled, violence (see Section 6.3.3). By then, A.H. knew from his own experience that social unrest and violence are common at places where biodiversity is concentrated (Hamilton *et al.*, 2000) – an association that, nowadays, is well-attested (Gaynor *et al.*, 2016). His reasoning was that whatever was decided at the UN should be workable in countries like Colombia, not just in countries in which the political climate is benign. He thought that, with the increasing pressure of people on the natural resources of the planet, the number of countries in which there are conflicts over access to natural resources is bound to rise.

The visit to Colombia gave A.H. the opportunity to meet some remarkable Colombians who were trying to make conservation work, sometimes at great personal risk. One of them was Maria Teresa Maya, director of Puracé National Natural Park. By taking a community-based approach, Maria was achieving some successes in protecting the park, despite it being situated in a drug-related hotspot (Ospina, 2006). It strengthened A.H.'s conviction that making progress in plant conservation is greatly dependent on the efforts of people committed to achieving practical results on the ground.

7 Human Knowledge of Plants

Abstract

Since humans are integrated systems of body, mind and soul (as most people historically have conceived it), so their knowledge is intimately associated with their memories, feelings, values and impulses to act. A description of the traditional plant-resource system of the Baganda is given to demonstrate how factual knowledge is associated with less tangible attributes, such as symbolic meanings, rituals and taboos. The contributions of human knowledge of plants to healthcare are discussed, including in relation to the origins of herbal medicine and of traditional medical systems, and the transition in Europe and Britain from traditional to Western medicine. The nature of scientific knowledge is considered, and mention made of some major scientific discoveries in botany. Also discussed are the influences of values and ideologies for determining the directions that have been taken in science. The significance of encounters between knowledge systems for conservation and development is discussed, with examples presented for botany, medicine and language.

7.1 Personal Knowledge and Its Botanical Component

Since the human is an integrated system of body, mind and soul (as most people historically have conceived it), so the knowledge about plants that a person possesses is not just a bundle of dry facts floating free (as it were) of other attachments. It is part of a living creature and intimately associated with memories, feelings, values and impulses to act. Each person's knowledge is unique, but, since some experiences are shared with people of similar cultural background, it can also be thought about as a component in knowledge systems. This is useful for understanding why people possess the particular knowledge that they do and when thinking about how to influence them for educational or other purposes.

When humans acquired the ability to use open-ended language, likely about 50,000 ya (years ago) (see Chapter 1), it enhanced their abilities to store knowledge about plants and draw upon it when thinking or communicating with other people. Those types of plants with which people were familiar would likely have been given names – handy mental pegs on which to attach information about them, such as how they could be recognised, where they occurred, their growth habits, their uses, how to harvest them, how to prepare products from them and to whom, if anyone, they belonged.

Research carried out on modern societies of hunter-gatherers, subsistence farmers and traditional pastoralists, all of whom have high dependencies on locally growing plants, has confirmed that they typically possess rich stores of linguistically related information about their local floras (Farnsworth and Soejarto, 1991; Schippmann et al., 2006). That the same was true of people living in similar societies in the past is confirmed by the occasional lucky archaeological find. Analyses of the stomach contents of prehistoric bog bodies, which are sometimes found entombed in peat in moist Northwest Europe, show that their diet could include an exceptionally wide diversity of plants (Nielsen et al., 2018). The remains and artefacts of an unfortunate Bronze Age man, nicknamed Ötzi, who perished high on the Alps 5000 ya and was frozen, revealed that his society knew how to use plants to make carrying equipment, clothing, shoes, weapons and fire, which ones were useful as food and medicines, and how to cultivate some on farms (Oeggl, 2009).

The safe transmission of knowledge about plants from generation to generation in small-scale societies is a survival necessity. The earliest years of people's lives are most critical in this respect, because this is when they acquire the norms of their societies about how to interact with the natural world – a process that has been called environmentalisation (Hamilton, 2001). This occurs at the same time as socialisation, the word used for the process whereby infants and growing children acquire the norms of their societies about how they should behave towards other people. One of the mechanisms that assists children in small-scale societies to learn about plants is the expectation that, from an early age, they will join in everyday family work involving plants, such as cultivating crops, collecting firewood and preparing food. Another is their inclusion in many of the rituals and ceremonies found in such societies, these, in effect, being mechanisms that confirm the place of each person in society, reinforcing social solidarity.

The environments that shape people's knowledge about plants in modern urban-industrial societies are quite different from those described above. In such societies, many families lack connection with long-standing traditions that have been passing on plant-related knowledge from generation to generation. Families can be composed haphazardly by historical standards, lacking the certainty of individual social position and continuity through time of traditional societies. People may have little direct contact with the plants on which their lives (in actuality) depend, and they are unlikely to be able to recognise many of them in the living state. Their knowledge of wild plants may come largely through the media, perhaps supplemented by occasional visits to beauty spots in the countryside. Their knowledge of the values of plants as food and healthcare may be largely media-derived, backed up by the information fed to them by the manufacturers and retailers of plant-derived products (see Section 6.3.2).

This said, there are certain subgroups of people within modern urban-industrial societies who are very knowledgeable about specialised aspects of plants. Among them are some involved in plant-related businesses (such as farming, food manufacturing and food retailing), have plant-related hobbies (such as natural history and gardening), work in education, or are employed in the plant-related pure and applied sciences (such as botany and forestry).

7.2   The Traditional Plant-Resource System of the Baganda

As mentioned in Chapter 1, three parts of the world were chosen as places of special concentration for the present book, one of them Uganda (especially Buganda). For this section, we have selected the Baganda and their homeland of Buganda to provide an example of a traditional plant-resource system. Some aspects of this system are material, involving such matters as the types of plants used as resources, where they are obtained, who uses them, the products made from them, their monetary values, and so on. The system also has a less tangible aspect, involving such matters as the feelings that the plants evoke, their symbolic meanings and the taboos associated with them. We have tried to open a glimpse into this less overt world in this account.

The Baganda's traditional knowledge of plants has evolved in their traditional homeland, which is an area about 20,000 km^2 in size situated in the northwestern hinterland of Lake Victoria (Map 3.2). The climate is warm and wet with two rainy seasons each year, a 'long rains' (*ttoggo mukazi*) from March to May and a 'short rains' (*ddumbi musajja*) from September to November. The relatively high altitude, which averages 1300 m, ensures that the climate is cooler than might be expected for a place situated on the equator. The natural vegetation is mostly lowland tropical forest (greatly reduced in extent nowadays – see Section 5.6). Papyrus chokes many of the wider valleys (see Section 15.4.2). The forests, swamps and other more natural habitats found in Buganda provide the Baganda, a farming people, with many resources harvested from wild plants. The climate is generally favourable for agriculture and some food is normally available on farms throughout the year. Even so, the Baganda have traditionally stored food for later use as seed in baskets or as dried slices of carbohydrate crops on racks above the cooking areas in their kitchens.

The British introduced the Baganda to commercial agriculture and the global market at the beginning of the 20th century (see Section 5.4.3). In spite of this, the Baganda's traditional knowledge of plants has survived to a considerable extent, especially in remoter places, and it is therefore possible to describe their traditional plant-resource system in the present tense. At its heart is a landholding (*ekibanja*) given by a father to a son on reaching maturity. Landholdings are not isolated socio-economic units, but are connected to one another through market transactions and the practice of farmers gifting plants and plant produce to their families and friends (see Section 8.4). The deep connection between the clans of the Baganda and the land is evident in the etymology of the word *ekibanja*. *Ekibanja* is concordant with *ebbanja*, which means 'debt'. The implication is that the gifting of land by a father to a son is a particular instance of a chief providing land to a clansman.

The components of a landholding include a house with outbuildings (including an outside kitchen and, these days, commonly a long-drop toilet), front and back courtyards (*oluggya lw'emiryango* and *oluggya lw'emmanju*), a home garden (*lusuku*), cultivated fields (*emisiri*) and fallow. There are no hedges or fences, but the boundaries of an *ekibanja* may be marked by shallow ditches (*ensalonsalo*) or planted at intervals with a certain type of small tree (called *oluwaanyi* or *omulamula – Dracaena fragrans*). The second name of this tree in Luganda is derived from the verb *-lamula* (meaning 'judge' or 'arbitrate'), which is a reference to it being extremely difficult to uproot – its presence potentially provides valuable evidence of where a boundary should be in the event of a dispute. The average size of a landholding (as at the beginning of the 21st century) was 1.4 ha, of which 22% consisted of the home garden (Edmeades and Karamura, 2007). It is probable that the average size of a landholding was larger in the past, when the human population was lower and there was less pressure on the land. Agricultural productivity has been declining in recent decades, one cause of which is believed to be an insufficient length of time given to fallow to restore the fertility of the soil (see Section 2.3.1).

The terminology used for some of the components of an *ekibanja* reflects the Baganda's traditional connection with the land. When a Muganda dies, the body of the deceased is traditionally buried in a clan graveyard on an *ekibanja*. The low pile of stones in the left foreground of Fig. 7.1 marks the site of a grave. The word used for such a graveyard (*ekiggya*) is concordant with that used for a courtyard of a house (*oluggya*), the significance of this being that it is in courtyards where the Baganda traditionally receive their visitors (see Section 4.5). The word used for one of the divisions of the clan (*essiga* – 'sub-clan') is also that used for one of the stones of a traditional hearth (see Section 6.8.1). The word used for a branch of a clan (*omutuba*) is also used for the name of the

Fig. 7.1. A *lusuku* near Mbarara, Uganda. Photo: A.H. (2010).

bark cloth tree (*Ficus natalensis*). The bark cloth tree has deep symbolic significance for the Baganda. For example, it is traditional for a father to plant a bark cloth tree to mark the boundary of the landholding that he presents to a son.

As is typical of home gardens in the wet tropics, that of the Baganda (*lusuku*) contains a variety of plant species of different growth habits and sizes. Its most distinctive feature, as a type of home garden, is its abundance of banana plants, of which there are always several different kinds. A photograph of a *lusuku* is shown in Fig. 7.1. (The photograph was taken in Ankole, which neighbours on Buganda.) The Baganda have a multitude of uses for the various parts of the banana plant. One of them is illustrated in Fig. 7.2, which shows baskets placed at the tomb of Kabaka Kalema at Mmende in Buganda. (Kabaka Kalema was the King of Buganda in 1888–1889.) This type of basket (*ekibbo*) is traditionally used in Buganda to present offerings, for example offerings of food to visitors to homes. *Ekibbo* baskets are made from strips of dried fibre (*obukeedo*) cut from the midribs of banana leaves and bound into spirals using bindings of *enjulu*, which is obtained from the stems of a large, ginger-like, plant (*Marantochloa purpurea*) found in moist places in rainforest.

Each of the four baskets at Kabaka Kalema's tomb is for receiving offerings connected with different parts of the spiritual realm, as the Baganda have traditionally perceived it. Nowadays, money is often placed in such baskets, but traditionally the offerings were of cowrie shells (*ensimbi*), roasted coffee beans (*emmwanyi*) and seeds of ensete (*ettembe*). Cowrie shells were a traditional currency in Buganda; coffee beans have various ritual uses, for instance being commonly offered to visitors to homes; and the seeds of ensete were once used as counters in the traditional board game of *omweso* (see Section 3.3.1). Note the pot, which is being used as a hearth (see Section 6.8.4). Kalema's tomb is in much better condition than some of the other tombs of past kings to be found in its neighbourhood. This is because Kabaka Kalema, who was a Muslim, died in exile during the religiously related chaos in Buganda during the late 19th century.

Fig. 7.2. Baskets placed in front of the tomb of Kabaka Kalema, Buganda. Photo: A.H. (2017).

This tomb was constructed in 1998 with the support of a Muslim country in Arabia for the repatriation of his remains (see Section 6.8.2).

Like the bark cloth tree, the banana plant (*ekitooke*) has deep symbolic significance for the Baganda. Its many varieties are classified into two major classes, 'male' banana plants (*ebitooke ebisajja*) and 'female' banana plants (*ebitooke ebikazi*) – a distinction that has nothing to do with biological gender (the plants in both categories bear both female and male flowers and, anyway, are sterile). Rather, the nomenclature is symbolic of the traditional structure of the family in Buganda (see Section 4.5). 'Male' banana plants are those types whose fruits are used for making beer (*omwenge*) and 'female' banana plants those used for making the Baganda's favourite food (*matooke*) (see Section 2.3.1). The same word in Luganda (*emmere*) is used both for food (in general) and for *matooke*. The home gardens of the Baganda always contain at least one 'male' banana plant, which is symbolically regarded as providing the 'female' banana plants with protection. When a woman gives birth to a child (*omwana*), the traditional practice is to present the placenta (*omwana ow'emabega* – literally meaning 'the child behind') to a 'male' banana plant, if the child is male, or to a 'female' banana plant, if female. When this is done, the placenta is provided with ritual protection by being covered with banana leaves. Uniquely, the same verb (*-zaala*) is used both for a woman giving birth (*-zaala omwana*) and for a banana plant giving rise to a new shoot (*-zaala ensukusa*). *-zaala* is not used in this way for any other type of plant.

Many other types of plants are found within the *ekibanja*, including within the *lusuku*. Three common trees, in addition to the bark cloth tree, are the incense tree (*omuwafu* – *Canarium schweinfurthii*), the mango tree (*omuyembe* – *Mangifera indica*) and a leguminous tree known as *omugavu* (*Albizia coriaria*). The edible fruit of the incense tree (*empafu*) was an important item of the diet of the ancestors of the Buganda when they lived in the Congolese rainforest (see Section 3.3.1), but nowadays in Buganda they are mainly eaten by children. The fragrant resin (*obubaani*) of the incense tree is burnt at the shrines of traditional deities and used to rosin the strings of musical instruments. Several varieties of mangoes are known, some of which have been in Buganda for a long time (the mango is native to the Indian subcontinent). One type (*kagoogwa*), which has long been present, has small fibrous fruits that are craved by pregnant women. *Omugavu* provides the wood favoured for smoking bark cloth, which is how it is cleaned.

Underneath the large trees in a *lusuku* is a layer of small trees to large shrubs. Three of its commonest species are robusta coffee (*emmwanyi* – *Coffea canephora*), *omululuuza* (*Gymnanthemum amygdalinum*) and *ekikookooma* (*Gymnanthemum auriculiferum*). Traditionally, the seeds of robusta coffee are dried and roasted, chewed as a stimulant (they contain caffeine) and used in various ritual ways, such as being offered to deities and visitors. The leaves of *omululuuza* form one of the commonest traditional treatments for malaria (see Section 7.3.2). The leaves of *ekikookooma* are soft and used as toilet paper.

It is common for home gardens to contain species in various stages of domestication. For instance, the *lusuku* frequently contains some species that are fully domesticated (e.g. the banana), some that are fully wild (e.g. plants that have grown from seed dropped by passing birds and deliberately not weeded) and others that lie somewhere along a spectrum in between. The ground layer of plants in a *lusuku* contains a number of herbaceous weedy species that are self-sown and deliberately left untouched during weeding. These species are eaten as edible greens (spinach), used in traditional medicine, or both. They include *doodo* (*Amaranthus retroflexus*), *doodo w'amaggwa* (*Amaranthus spinosus*), *ensugga* (*Solanum nigrum*) and *ejjobyo* (*Cleome gynandra*). They provide valuable sources of fibre and nutrients (Maundu *et al.*, 2002; Getachew *et al.*, 2013; Achigan-Dako *et al.*, 2014).

Each of the cultivated fields (*omusiri*) of an *ekibanja* is usually planted with a single type of annual crop that is required in quantity. Crops commonly grown nowadays include kidney bean (*ebijanjaalo* – *Phaseolus vulgaris*), peanut (*ebinyee-bwa* – *Arachis hypogaea*), maize (*kasooli* – *Zea mays*), sweet potato (*lumonde* – *Ipomoea batatas*) and cassava (*muwogo* – *Manihot esculenta*). All of these food plants were originally domesticated in the Americas and have, to a large extent, replaced crops of African origin, mostly since 1800 (see Section 3.3.1). Various types of vegetation can be present in the fallow part of a traditional landholding. They may include weedy abandoned fields (*ebisambu*), pastureland (*ettale*), bush (*ensiko*) and forest (*ekibira*). There can be extensive stands of single species of grass, for instance of spear grass (*essenke* – *Imperata cylindrica*) and a lemon-scented grass (*etteete* – *Cymbopogon nardus*). Spear grass is used by the Baganda in traditional beer-making and the lemon-scented grass is traditionally strewn on floors to suppress blood-sucking insects (see Chapter 13). (True lemon grass (*Cymbopogon citratus*), which has been introduced to Buganda, is a native of Southeast Asia.)

Home gardens, such as those of the Baganda, are common features of traditional agriculture worldwide (Huai Huyin and Hamilton, 2009; Galhena *et al.*, 2013; Yang Lixin *et al.*, 2014). They tend to be situated close to homesteads, to be closely tended (often mulched or manured) and contain a variety of plants of different statures, growth habits and uses. Much of the produce of home gardens is used in the home and, typically, it is women who are mainly responsible for their care (Dompreh *et al.*, 2020). The strong role of culture for determining the characteristics of home gardens was revealed by a survey in Jinping County, Yunnan, China (Huai Huyin *et al.*, 2011). This study found that each of the eight ethnolinguistic groups that have long lived in Jinping has its own form of home garden and that these forms have persisted for centuries, even when the villages of different ethnicities have been situated close to one another.

7.3 Plants and Healthcare in Traditional Societies

Traditionally, the family is the principal social institution responsible for holding plant-related knowledge of healthcare and transmitting it from one generation to the next. Often, it is women who are the chief knowledge-holders in this respect, because it is they who are typically most involved in the plant-related businesses of cooking, providing primary healthcare for the family, gathering wild plant foods and tending home gardens (Allen and Hatfield, 2004; Di Nicola, 2020).

It is normal for societies to have some people who are recognised as experts in healthcare and are consulted when assistance, beyond the knowledge of those in a home, is needed. In traditional societies, these traditional medical practitioners

(TMPs) are generally supplemented by traditional midwives (also called traditional birth attendants – TBAs) and expert bone-setters. Some TMPs may specialise, for instance in internal medicine, mental illness and ailments associated with local spiritual beliefs. Normally, TMPs do not request payment for their services, since they view their talents as having been bestowed on them by the divine. However, it is customary for those who consult them to offer them gifts – an arrangement that works well in subsistence economies, but less so when economies are monetised.

7.3.1 Health in hunter-gatherer societies

Homo sapiens has followed a hunter-gatherer way of life for at least 96% of its time on Earth (see Chapter 1) – a long evolutionary experience that has moulded its basic physiological and psychological characteristics. Researchers have discovered that modern hunter-gatherers are remarkably healthy in terms of non-communicable diseases, especially when compared with people in the West (Pontzer *et al.*, 2018). The lifestyle of the typical hunter-gatherer includes daily periods of physical exertion, combined with long periods of time spent resting and socialising (Bai XueJun and Li Hui, 2016; Andrews and Johnson, 2019). Overall, the amount of energy consumed is much the same as in the West, but the level of sustained physical exercise is, at times, much higher.

The diets of hunter-gatherers are diverse, but, generally, compared with Western people, more energy intake comes from plants (compared with animals), greater quantities of fibre, micronutrients and secondary metabolites are consumed, and there are decreased quantities of sugar and salt (Eaton, 2006; Mann, 2018). Fats and sugars are rare foods in nature and tend to be consumed avidly by hunter-gatherers when encountered. The sugar consumed by hunter-gatherers comes mostly in the form of honey (on average, 2–3% of food energy intake), rather than as refined sugar (on average, 15% of food energy intake in the West). Western people tend to be plagued by the 'diseases of affluence', such as obesity, type 2 diabetes and coronary heart disease – all associated with nutritionally poor diets and an insufficiency or imbalance in exercise.

Human health deteriorated when populations transitioned from being hunter-gatherers to farmers (Eshed *et al.*, 2004; Romer, 2012). One cause was having a less nutritious diet, since people were now eating large quantities of a few carbohydrate-rich foods and had insufficient intakes of secondary metabolites (see Section 6.5.2). Another was an increase in repetitive strain injuries, related to the back-breaking tasks of clearing land, hoeing fields, harvesting crops and pounding grain. There was also an increased susceptibility to communicable diseases, due to the greater ease of person-to-person transmission of pathogens with the higher populations of people that resulted from the adoption of agriculture (Dobson *et al.*, 2006).

7.3.2 The origin of herbal medicine

The chimpanzee and the human are closely related species that share a common ancestor that lived in Africa 5–6 million ya. Researchers have found that wild populations of chimpanzees use some of the same species of plants as medicines as do people living in nearby villages and that geographically separated populations of chimpanzees, like geographically separated populations of people, differ in their traditions of herbal medicine (Huffman and Wrangham, 1994). Evidently, chimpanzees, just like people, are passing on knowledge of herbal medicine from generation to generation. In Uganda, one species that both chimpanzees and people use as medicine is *G. amygdalinum*, which is one of the most common shrubs found on smallholdings in Buganda where it is used as an antimalarial (see Section 7.2). Another is *Aframomum angustifolium* (*ekitungulu* in Luganda), a tall, ginger-like, herb found in damp places in forests. It is the fruit of *A. angustifolium* (known as *ettungulu* in Luganda) that is the part taken medicinally. As mentioned in Section 5.1, at one time the seeds of *Aframomum* were much in demand as a spice in Europe. Laboratory analyses have demonstrated that the leaves of *G. amygdalinum* have antiparasitic properties and that the fruits of *A. angustifolium* are antibacterial (Koshimizu *et al.*, 1994; Kagoro-Rugunda, 2019).

Interestingly, the Luganda names for *A. angustifolium* (*ekitungulu*) and its fruit (*ettungulu*) are concordant with those used for onion (*akatungulu*) and garlic (*katunguluccumu*), the replacement of the prefix *eki-* by *aka-* indicating smallness in size. Neither onion nor garlic is native in Africa – their evolutionary origins lie in the steppe country of central Eurasia. Garlic is a well-known herbal medicine and laboratory research has shown that both it and onion have antibacterial properties (Garba *et al.*, 2013; Sharma *et al.*, 2018). What the linguistic evidence suggests is that, at some time in the past, the introduced crops of onion and garlic became recognised as substitutes for the fruits and seeds of the native *Aframomum*.

Comparison of plants used in traditional medicine in different parts of the world has shown that there are certain plant families, such as *Apocynaceae*, *Asteraceae* and *Lamiaceae*, that are consistently overrepresented, compared with their abundances in the local floras. These are also families that have been shown in the laboratory to be unusually rich in secondary metabolites, such as alkaloids (Saxena *et al.*, 2013). Given that human health deteriorated when people turned from a hunter-gatherer to an agricultural way of life (see Section 7.3.1), it may have been that it was at that time that a clearer distinction came to be made between plants conceptualised as foods and those conceptualised as medicines.

It is known that the invention of agriculture happened in several parts of the world independently of one another, beginning 12,000 ya (see Section 3.2). Since each place had its own distinctive native flora, it can be hypothesised that initially a plethora of local herbal traditions came into existence. This may help to explain why the number of plant species that have been used worldwide as medicines is so great (see Section 3.1). The total number of higher plant species that have been used medicinally has been estimated at 50,000–70,000, which represents 15–20% of the entire global flora (Schippmann *et al.*, 2006).

Some plants have psychoactive effects when taken into the human body, resulting in alterations of perception, mood, consciousness, cognition or behaviour. It is likely that these types of plants attracted early human attention. Table 7.1 gives examples of psychoactive plants, together with information on

Table 7.1. Examples of plants containing psychoactive substances.

Plant types and natural distributions	Traditional uses	Post-1500 CE developments
Tobacco (*Nicotiana tabacum*); not known in the wild; probably originated from a wild species on the lower eastern slopes of the Andes	A sacred plant of the highland Maya, used for both visionary and therapeutic ends (Groark, 2010); a trade item in pre-Columbian North America, used in pipe ceremonies for sealing treaties (Godlaski, 2013)	Introduced to Europe in late 16th century, recreational tobacco smoking then spreading rapidly; popular herbal medicine in 16th to 18th century Europe; was grown in the United States using slave labour; dangers to health of smoking tobacco realised from the 1960s; legislative control resisted by tobacco companies
Ayahuasca or caapi (*Banisteriopsis caapi*); native to Western Amazonia	A sacred plant, sometimes combined with other species, to prepare a hallucinogenic drink taken by shamans to diagnose and treat human ailments (Schultes and Rauffauf, 1990)	Shamanistic-type practices have been taken up by various non-indigenous groups in Brazil, a practice that can be resented by indigenous people; taking ayahuasca is part of the ritual of the syncretic religion of Santo Daime, founded in Brazil in the 1930s; use of ayahuasca decoction for religious purposes became legally permitted in Brazil in 1986
Opium poppy (*Papaver somniferum*); native to Southeastern Europe and Western Asia	Probably originally domesticated in the Western Mediterranean, being grown by 5500 BCE; probably grown for its edible seeds and oil (Salavert *et al.*, 2020)	Traded by the British East India Company from India to China in the 19th century (see Section 5.3.4); latex from unripe capsules yields codeine and morphine, the latter first purified in 1803; today, morphine is widely used as a painkiller; heroin first synthesised from morphine in 1874; at first considered a wonder drug, heroin was openly sold in North America until 1917, when its addictive properties became realised (Lewington, 1990)
Coca (*Erythroxylum coca* and *Erythroxylum novogranatense*); natives of Western and Northwestern Amazonia; frost-intolerant, not cultivated above 2700 m altitude	A sacred plant; leaves traditionally chewed with lime by Andean people and in Northwest Amazonia to supress tiredness and hunger (Schultes and Rauffauf, 1990; Dillehay *et al.*, 2010)	Cocaine first isolated from the leaves of the coca plant by Albert Niemann in 1859; therapeutic and recreational uses started soon thereafter; a highly addictive drug, its manufacture and use are illegal in most countries; cocaine was rarely taken in the United States before 1970, after which its use skyrocketed; today, the illegal cocaine trade is a major international business, associated with violent criminality

Continued

Table 7.1. Continued.

Plant types and natural distributions	Traditional uses	Post-1500 CE developments
Khat (*Catha edulis*); native to montane east and south Africa; possibly also to Yemen	Leaves traditionally chewed, mainly in company and mainly by men; mood change during khat-chewing sessions causes a transition from outgoing and sociable to introspective and mildly depressive (Al-Hebshi and Skaug, 2005)	Mainly used today by Muslim communities in Ethiopia, Somalia and (especially) Yemen. Khat use has expanded globally with air transport and immigration; an important economic plant, today a source of income for millions of Ethiopian farmers (Cochrane and O'Regan, 2016)

their traditional uses and their post-1500 stories. It is common for shamans – the seers of small-scale societies – to take psychoactive plants to help them float away in time and space to enquire about the causes of predicaments and what the future may bring. Shamans are viewed within their societies as those most in touch with the world of the spirits. Psychoactive plants have yielded a number of drugs used in Western medicine, some examples of which are included in Table 7.1. Some of the drugs, particularly heroin and cocaine, can be highly addictive, which has attracted criminality (see Section 6.3.3).

7.3.3 Traditional medical systems

Traditional medicine can be broadly classified into folk medicine and traditional medical systems (Pei Shengji, 2001, 2002). Folk medicine is characterised by being transmitted between the generations orally and through demonstration and by being associated with relatively small social units, such as families, local communities and local ethnolinguistic groups. In contrast, traditional medical systems are characterised as being associated with larger-scale societies, literacy, written pharmacopeia, formal ways of training medical practitioners, and formal theories about the meaning of human health and how good health can be acquired and maintained. In practice, folk medicine and traditional medical systems are not entirely distinct categories. It is common for folk healers to draw on knowledge from traditional medical systems and for practitioners of the latter to augment their knowledge with cures picked up from folk medicine.

The better-known traditional medical systems are listed in Table 7.2, together with information on their geography, history and associations with culture and religion. All, bar one, are Asian, which is the part of the world that has longest possessed literacy. Given the association of traditional medical systems with states and literacy on the Eurasian landmass, it seems likely that the literate pre-Hispanic civilisations in the Americas would have possessed them too (Marino and Gonzales-Portillo, 2000). The types of religion with which the traditional

Table 7.2. Traditional medical systems, showing geographical and cultural associations. An asterisk (*) indicates that the system is officially recognised by the Chinese Government. Note: homeopathy is not a traditional medical system and is not included; it was invented by Samuel Hahnemann (1755–1833).

Traditional medical systems	Geographical, historical and cultural associations, and numbers of medicinal plant species used
Traditional Chinese Medicine (TCM)*	China; associated with Daoism and Confucianism; the total number of species used is reported as 10,654, 1200 of them commonly (Pei Shengji, 2001). The figure of 10,654 includes *all* forms of traditional medicine found in China
Kampo	Japan; 217 kinds of drugs recorded (17th edition of the official Japanese pharmacopeia)
Bon	Tibet, associated with Bon, a pre-Buddhist religion that has close connections with shamanism in Central Asia; the foundational text is *'Bum bzhi* (Millard, 2005/2006)
Tibetan (also known as Sowa Rigpa or Amchi Medicine)*	Tibet; associated with Buddhism (Lama *et al.*, 2001); uses 3000 plant species (Yan Jinshen, 2017); the foundational text is *rGyud bzhi* (Millard, 2005/2006)
Mongolian*	Mongolia; associated with Buddhism; 879 plant species used in Mongolia (the modern political state); 1122 wild plant species used in Inner Mongolia (Chinese Autonomous Region)
Dai*	Dai people of Southern Yunnan, China; associated with Theravada Buddhism; 1260 medicinal plants used (Ma Xiao-jun *et al.*, 2017)
Ayurveda	Indian subcontinent, also Bali (Indonesia); associated with Hinduism (but with Buddhism in Sri Lanka); uses 1250–1400 plant species (Dev, 1999)
Siddha	Tamil-speaking areas of India; uses 328 plant species (Shiva, 1996)
Unani (known in China as Uyghur Medicine*)	Greco-Arabic system of medicine, today associated with Islam; common in the Indian subcontinent; uses 342 plant species (Shiva, 1996)
Western herbal medicine	Europe and the West; developed from Classical Greco-Romano medicine; traditionally associated with Christianity, especially monasteries; enriched later by herbal traditions from the Indian subcontinent, North America and elsewhere

medical systems are associated are world religions, rather than those associated with small-scale societies (see Section 4.3). This association is not surprising, given that both of them are associated with literacy and both are interested in the meaning of a healthy human and the origins of human ailments.

China has a rich body of ancient literature dealing with human health and herbal treatments. *Shennong Bencaojing*, the classic of Chinese herbal medicine, was written between 200 and 250 CE, although drawing on texts composed much earlier. Other major works are the 1500-year-old *Qi Min Yao Shu (Important Arts for the Peoples' Welfare)* by Jia Sixie, the works of Sun Simiao (581–682 – see Section 4.3.1) and the 26-volume *Bencao Gangmu (Classification of Materia Medica)* by Li Shizhen (1518–1593), the latter mentioning 1096 herbs used medicinally (Pei Shengji, 1984; Menzies,

2021). The Mawungdui Silk Texts, discovered in 1973 in a tomb sealed in 168 BCE in Hunan Province, map out the pathways on the human body that are still used in acupuncture (Harper, 1998).

Traditional medical systems have evolved down the centuries. Ideas have been exchanged between the systems and new remedies introduced from folk medicine. The routes of the Ancient Silk Roads, which were pioneered for exchanging goods, also proved fruitful for exchanging ideas (Frankopan, 2015). Past exchanges of knowledge help to explain why considerable similarity exists in the types of medicinal plants used in some of the traditional medical systems. This is the case with Unani and Western herbal medicine, and with the trio of Unani, Tibetan Medicine and Ayurveda. A deliberate attempt to garner the best medical information available from around the known world was made in the 8th century CE by Trisong Detsen, the King of Tibet. He called together medical experts from Tibet, China, Dolpo (now in Nepal), India and Persia to discuss how Tibetan traditional medicine could be improved (Lama *et al.*, 2001). Tibet is the home not only of Tibetan Medicine, but also of the older, pre-Buddhist, traditional medical system associated with the Bon religion and the ancient Himalayan kingdom of Zhangzhung (see Sections 3.4 and 3.5.2). Buddhism and Bon are today about equally represented at Dolpo, the site of the conservation case study described in Chapter 20.

All the traditional medical systems listed in Table 7.2 emphasise the importance of living a well-balanced life, which is conceptualised as maintaining an equilibrium between a few basic elements that are said to determine a person's fundamental constitution. These basic elements are variously described as humours, forces or dynamic states. It is the habitual practice of TMPs associated with these systems to examine their patients to determine how well these elements are in proper balance, both internally (in relation to the body, mind and soul) and externally (in relation to the social, natural and spiritual environments). However, where the systems differ is in the number of basic elements recognised. Traditional Chinese Medicine (TCM) makes reference to two, Ayurveda to three, Western herbal medicine and Unani to four, and Tibetan Medicine and aspects of Daoism to five. Figure 7.3 is a photograph of a Tibetan traditional doctor (amchi) taking the pulse of a patient in a village in Ladakh, India. Pulse reading is a finely tuned skill in Tibetan Medicine, much more being read into the pulse than is the case in Western medicine.

The duality seen in TCM has deep roots in Chinese philosophy, being found in the Yin Yang Theory of Daoism, which is that things exist as inseparable and contradictory opposites (Jiang Xinyan, 2013). It is found in Chinese food culture, as well as TCM, with foods as well as medicines being traditionally regarded as being either *ré* ('warming') or *liáng* ('cooling'), or a third category *ping* ('smooth' or 'even') in between. To take an example, the condition of constipation (*biànmi*) can be treated with a cooling food, such as *xiao mi* (millet), or a cooling medicine, such as *jin yin hua* (Japanese honeysuckle – *Lonicera confusa*). The fourfold way of analysing constitutional types in Western herbal medicine has origins in the philosophy and medicine of Ancient Greece. In modern times, it crops up in the psychology of Carl Yung (1875–1961) and the personality typing of experts in business efficiency (Hall and Nordby, 1973; Paul, 2005).

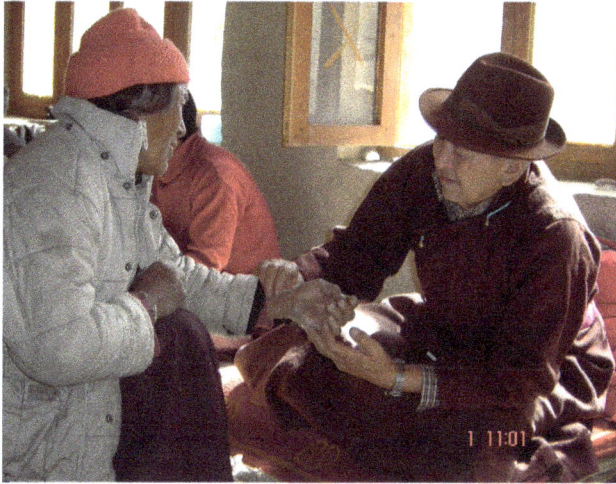

Fig. 7.3. Amchi reading the pulse of a patient in Ladakh, India. Photo: A.H. (2006).

7.3.4   The transition from traditional to Western medicine in Europe and Britain

The history of Western herbal medicine is closely linked to that of botany. Its foundational figures in the Greco-Romano Classical Age were Hippocrates of Kos (*c*.460–370 BCE), who is widely regarded as the father of Western medicine, Dioscorides (*c*.50–90 CE), who is widely regarded as the father of scientific botany, and the Roman army doctor Galen (129–*c*.216 CE). Much of the medical knowledge of the Greco-Romano Classical Age became lost to Europe during its Dark Ages (*c*.500–1300 CE), just as it lost much of the Classical knowledge of biology (see Section 6.5.1). However, the medical knowledge of the Classical Age was rediscovered and developed in the Islamic World during a period of intellectual flourishing (the Islamic Golden Age) dating to 800–1300 CE and, from there, some of the knowledge became reintroduced into Christian Europe during medieval times. It was particularly with the Christian monasteries and universities of medieval Europe that this learning became associated, monasteries being the hospitals of the day. Medical schools, such as those at the universities of Pisa and Padua in Italy and at Oxford in Britain, were the first institutions in the Western world to establish botanical gardens to display their knowledge of medicinal plants in the form of ordered beds (see Section 8.5). The botanical gardens of Pisa, Padua and Oxford were established in 1544, 1545 and 1621, respectively.

One use of the word 'herbal' is for a type of book containing the names and descriptions of plants, especially those that are used medicinally. Until recently, it was commonly supposed that early European herbals, such as *Herball, or Generall Historie of Plantes* by John Gerard (1597) and *The English Physician; and Complete Herbal* by Nicholas Culpeper (1652), were good guides to the medicinal uses of plants in late 16th century to 18th century England. However, this is not the case. The great majority of people at the time were illiterate and unable to read the herbals, and also too poor to afford to see the learned medical professionals who consulted them. They turned to local folk healers instead. Gerard's *Herball*, which was the most widely read book on botany between 1600 and 1800, is not even a good guide to the types of medicine being prescribed by the medical professionals. A careful assessment of the evidence relating to the types of plants used in medicine in Britain prior to the 19th century has revealed that about half of the medicinal plants mentioned in Gerard's *Herball* did not actually occur in Britain at the time of its writing and that much of its contents had been plagiarised from continental sources, notably *Cruydeboeck* (published 1554) by the Flemish physician and botanist, Rembert Dodoens (1517–1585) (Hatfield, 1999; Allen and Hatfield, 2004).

A better guide to the plants used medicinally by most people in 18th century Britain is probably provided by the information given in a small book, *Primitive Physic: Or, An Easy and Natural Method of Curing Most Diseases*, by John Wesley (1703–1791), the founder of the Protestant denomination of Methodism (Wesley, 1743). Wesley had deep concern for the common people and respect for their knowledge. He updated his manual several times during his lifetime, as he learnt more from the villagers about their medical practices as he toured the countryside on horseback on his proselytising rounds. An austere man, who did not believe in frivolity, Wesley noted in his journal, after he had visited Chelsea Physic Garden in 1749, how he did not see the purpose of 'heaping up' plants from around the world in the garden, just for the gratification of idle curiosity (see Section 5.3.4) (Minter, 2000). In his opinion, what the garden should be doing was to undertake research into how the plants could be used for the practical benefit of the people.

Scientific knowledge of the causes of disease and the values of treatments did not exist before the 19th century. The 'doctrine of signatures' was widely accepted in Europe, this being the belief that God has marked plants with signs to indicates their purposes. This belief could become incorporated into plants' names. For example, the lungwort (*Pulmonaria*), which has spotted oval leaves resembling diseased, ulcerated, lungs, was believed to be a cure for lung diseases, such as tuberculosis. Another common belief was that diseases could be caused by being possessed by evil spirits or the Devil. Exorcism ceremonies, intended to deliver people from evil spirits, are prominent today in Pentecostal churches and are still among the rites of the Anglican Church (Giordan and Possamai, 2020). In England, there was once a belief that being touched by a monarch had the power to cure people of scrofula, a form of tuberculosis. The last English monarch to engage in this laying-on-of-hands was Queen Anne (reigned, 1702–1714).

7.4 Scientific Knowledge

7.4.1 The nature of scientific knowledge

Science emerged in Europe as a recognisable cultural stream in the 16th century as part of its mid-2nd millennium cultural revolution (see Section 6.1). Its institutional progress has been marked by the founding of scientific societies, scientific journals and national academies (centres of national scientific excellence). The origin of national academies can be traced back in the West to a school established by Plato in Classical Greece in c.387 BCE. Aristotle, a philosopher whose ideas were influential in medieval Europe, attended the school for 20 years (see Section 4.3.2). Britain's national academy, the Royal Society, was founded in London in 1660 by a group of scholars and granted a royal charter by King Charles II in 1663. Joseph Banks, an influential figure in the development of economic botany and the British Empire, was the president of the society for 41 years (see Section 5.3.2). The California Academy of Sciences was founded in 1853 and the US National Academy of Sciences in 1863. China's national academy, the Chinese Academy of Sciences (CAS), was founded in 1928 (see Section 6.9). Its roots in Axial Age times consist of the same Classical Greek one as in the West and a national scholastic institute founded during the reign of Emperor Jing (157–141 BCE), one of the emperors of the Han dynasty (see Section 4.3.1).

Personal knowledge, as described in Section 7.1, consists of the things that a person knows to be true or, at least, takes to be true for the purpose of everyday living. It is an individual thing – something that each person acquires for her or himself during her or his lifetime. Scientific knowledge differs from personal knowledge in being associated with a formal system of knowledge creation. For most practical purposes – for instance, for determining whether papers submitted for publication to scientific journals are scientifically worthy – it can be taken to be those theories about the nature of reality that a consensus of recognised experts in a particular academic discipline agrees is the most accurate approximations to truth that have so far been placed in the public realm.

Scientific knowledge is an evolutionary form of knowledge that progresses through the presentation of hypotheses to explain facts (things taken to be true for the purpose of argument). Hypotheses gain strength from their powers of prediction. If predictions turn out to be wrong, then this too is useful, because it exposes the need to develop new hypotheses to explain the new observations or, alternatively, to question the truthfulness of the 'facts' that had previously been accepted as true. Scientific knowledge is an international form of knowledge to which anyone, in principle, can contribute.

Deductive logic, the process of reasoning outlined above, is relatively easy to grasp. It is an extension in linguistic and formally written form of the ability of organisms (plants as well as animals) to adjust what they do according to what they sense. More difficult to understand is the process of induction – how scientists come up with their original hypotheses, especially if they are wide-ranging. Wide-ranging scientific hypotheses, which are known as paradigms, gain credibility when scientists outside the relatively narrow scientific fields in which they have been incubated become convinced of their credibility (Mayr, 2001). When this happens, the results are paradigm shifts – momentous scientific revolutions that are characterised by the reorientation of scientific disciplines and the types of experimental and observational practices that they employ (Kuhn, 1962).

The processes that have led the great scientists in history to arrive at their transforming hypotheses remain largely a mystery. However, it is known that certain conditions help – that is, thorough immersion in scientific fields in a long-term disciplined way, a receptiveness to observations that fail to conform to current scientific thinking and a willingness to transgress the barriers that scientific disciplines tend to erect around their perimeters (Popper, 1976). Significant scientific advances have often been made by people patiently engaged in routine research and who, one day, receive a flash of inspiration (Biswas and Mani, 2008; Kourkouta *et al.*, 2018). Such flashes may come to people when their minds are wandering about in unfocused ways, for instance early in the morning while only half-awake. New hypotheses are vulnerable when first proposed, being products of the imagination, affronts to the existing scientific establishment and, as yet, unsupported by confirmatory research.

7.4.2 Scientific advances with special reference to plants

Early contributions to the development of botany included those of: Robert Hooke (1635–1701), on the cellular structure of plants; Stephen Hales (1677–1761), on transpiration; Jan Ingenhousz (1730–1799), on photosynthesis and respiration; and Jean Baptiste Boussingault (1801–1887), on nutrient cycling. The Swedish botanist, Carl Linnaeus (1707–1778), developed an improved system for categorising plants that proved invaluable for naming and cataloguing the flood of new types of plants discovered by Western plant hunters, as they fanned out across the world from the late 18th century (see Section 7.5.1). Linnaeus' system is based on the number of sexual organs in the flower. It is now known from genetic research that this gives only a crude approximation to the ways that plants are related to one another in terms of their evolutionary descent (Chase *et al.*, 2016).

In the 19th century, it was discovered that three elements – nitrogen, phosphorus and potassium – are required by plants in quantity and are taken into them through their roots. Nitrogen in its gaseous form (N_2) makes up 70% of the atmosphere, but it cannot be used directly by plants. It was therefore a scientific breakthrough when it was discovered by Martinus Beijerinck in 1901 that bacteria living in nodules on the roots of certain types of plants can convert atmospheric nitrogen into chemical forms, such as ammonia (NH_3), that they can use. Subsequently, in 1909–1910, two German chemists, Fritz Haber and Carl Bosch, who together combined academic excellence with industrial know-how, discovered how to manufacture ammonia from atmospheric nitrogen at scale. The large-scale application of synthetic nitrogen fertilisers in agriculture then began.

During the 19th century, chemists started to identify the chemicals (known as active principles) that give medicinal plants their therapeutic qualities. This led the way, first, to the standardisation of doses of herbal medicines and then, in some cases, to their replacements with synthetic substitutes. The first active principle to be extracted from a plant was morphine from the opium poppy (*Papaver somniferum*), which was achieved by the German pharmacist, Friedrich Sertürner, in 1803. Morphine is still used in medicine for the relief of pain (Table 7.1). The isolation of quinine from the tree *Cinchona* was first achieved by the French pharmacists and chemists, Joseph-Bienaimé Caventou and Pierre-Joseph Pelletier, in 1820. Quinine is the antimalarial compound that enables Europeans to live with less risk of death or serious disease in tropical Africa (see Section 5.2). Many medicines used in Western medicine today contain active ingredients (or less well-refined extracts) obtained from plants (Hollman, 1992; Roberson, 2009). An analysis of prescriptions dispensed from community pharmacies in the United States between 1959 and 1980 found that 25% of them contained ingredients of this type (Farnsworth et al., 1985). As mentioned in Section 8.1, if the ingredients used in medicines are obtained from wild plants, then the pressure of commercial collection can threaten their survival.

Chemists during the 19th century became increasingly adept at synthesising complex chemicals from simpler chemical ingredients and at creating chemicals unknown in nature. The discovery of how to synthesise heroin, which is a chemical with a structure similar to morphine but is not found in nature, was made independently by two people – one a Briton, Charles Wright, in 1874 and the other a German, Felix Hoffmann, in 1897. Hoffmann worked for the Bayer pharmaceutical company, which, at first, marketed heroin as an over-the-counter cough medicine to replace morphine. At the time, morphine was considered too addictive to be used in easily available medicines. Heroin's own highly addictive properties were not recognised until 1917, when its legal availability became restricted. Heroin-related criminality then began (see Section 6.3.3) (Hoszafi, 2001). Aspirin was the first pharmaceutical drug to be fully synthesised from inorganic compounds, which was in 1897, again by Hoffmann.

Petroleum is a naturally occurring organic and energy-rich liquid found trapped in geological strata. It is a product of the partial decomposition of plants and other photosynthetic organisms that, in the past, captured energy from the sun. Apart from the value of petroleum as a source of fuel, it has proved invaluable to industry during the last 150 years as a raw material for the manufacture of synthetic organic molecules. Some of these synthetic chemicals have replaced, to varying degrees, those natural substances produced by plants that have been influential in shaping world history during the last 500 years, as described in Chapter 5. They include food and drink sweeteners, medicines, textiles and rubber. Tea today is commonly sold in tea bags manufactured from petroleum, as are the synthetic colourings, preservatives and flavour enhancers popular with industrial manufacturers of food (see Section 6.3.3).

The plastics, rayon, polyester and nylon, which are produced from petroleum, were first used in the manufacture of textiles in 1905, 1928 and 1930, respectively. Currently, about 25% of the fibres used worldwide in textile manufacture consist of cellulose (the fibre in cotton), about 6% man-made cellulosic fibres (derived from wood pulp) and much of the rest is polyester (Savage, 2022). The first successful synthetic rubbers were invented independently in the United States and Germany during the 1930s, which was just in time for them to be used by both sides during the Second World War – supplemented, in the United States' case, by supplies of wild rubber from Amazonia thanks to the efforts of Richard Schultes (see Section 5.3.5). The tread on a modern tyre contains about equal quantities of natural rubber, synthetic rubber and carbon black, the last being a soot-like reinforcing agent that is produced through the partial burning of fossil fuel.

Petroleum has been a key ingredient catalysing the economic expansion of the world since the turn of the 20th century. It has been a catalyst for conflict, as more powerful states have fought with one another over supplies and drawn in others to fight on their behalf. Carbon dioxide, a by-product of its burning as a fuel, is the principal chemical responsible for anthropogenic climate change (see Chapter 1). Plastic microfibres and nanofibres released into the environment during the processes of producing, using and disposing of plastic-containing products are now abundant in the oceans and taken in by marine organisms with, as yet, not well-established effects (Beverley et al., 2019).

7.4.3   Influences on how science develops

Scientific research can be undertaken with different purposes in mind, as can be illustrated by the case of the invention of the breakfast cereal by two brothers surnamed Kellogg. One of the brothers was John Kellogg (1852–1943), the superintendent of a sanitorium in Battle Creek, Michigan (USA), and the other Will Kellogg (1860–1951), a bookkeeper in the same establishment (Rucker and Rucker, 2016). The first flattened, dried, flake of a cereal was made accidentally in the sanatorium's laboratory in 1894 and, thereafter, the brothers collaborated with one another in experimenting on how to produce such flakes at scale. However, in 1906, they parted company on disagreement about how to exploit their findings. John continued to pursue his primary interest, which was in the improvement of public health, and Will founded the business that is today the giant Kellogg's food company – manufacturers of breakfast cereals from a variety of grains (see Section 3.1).

Political ideology has played a hand in influencing the directions taken in scientific research. We mention in Section 6.2.2 how, for ideological reasons, the former Soviet Union favoured the Lamarckian theory of biological inheritance, setting back the work of Vavilov and other geneticists by 25 years. An example from the ideologically right-leaning side of modern politics is the corrupting effect that neoliberalism has had on academia through its setting of a premium on the acquisition of knowledge that enhances economic performance (Radice, 2013). Currently in the United Kingdom (2024), the government is using graduate income as a measure of the value of a degree. This favours medicine, engineering, technology and economics, but disadvantages the biological sciences and agriculture (de Vries, 2014). A knock-on effect of this has been that staff are now recruited and graded by some universities according to their abilities to attract external funds, produce graduates with marketable skills and achieve high, year-on-year, publication scores. The pressure placed on staff to publish has contributed to a weakening of academic standards, because, today, there are many scientific journals that require payment by authors to publish their articles and take a casual attitude to checking their scientific standards. The world today is awash with politicised or commercialised media peddling fake facts and fake science (Martin, 2021). There have been huge increases in sham scientific journals and in scientific fraud (Van Noorden, 2023).

It is ironic that, at the present time of crisis for the survival of the world's flora, there has been a worldwide decline in the emphasis being given in teaching and research to 'whole-plant botany' (Hershey, 1993; Lock, 1994, 1996; Uno and Bybee, 1994; BSA, 1995). This covers a range of subjects relevant to making progress in plant conservation, such as plant identification, taxonomy, phytogeography, vegetation science, ecology, ethnobotany and ethnoecology. The trend in universities has been for departments of botany and zoology to merge into departments of biology and for them to concentrate on microbiology and biotechnology, which are seen as growth areas in the economy and offering better prospects of employment (Hamilton et al., 2003; Carlson, 2016).

One of the consequences of the intrusion of political ideology into science is to raise the question as to whether there is more than one type of science. Lysenko was accused by the authorities in the former Soviet Union of pursuing 'bourgeois' (or 'Western') science. In the United States, 45% of the population oppose Darwin's theory of evolution for religious reasons, some of them adhering to the tenets of 'creation science', which is based on a literalist interpretation of the Bible (Jelen and Lockett, 2014). The definition of science is a political issue in India, where arguments about whether Ayurveda, a Hindu-associated traditional medical system, is a science have held back the development of an integrated system of healthcare to the particular disadvantage of the poor (see Section 7.5.3).

7.5 Encounters Between Knowledge Systems

In the concluding sections of this chapter, we give examples of what has happened when there have been encounters between knowledge systems that are relevant to the advancement of plant conservation. The first is the encounter between scientific and traditional botany in China, the second is that between Western and traditional medicine (also in China), the third that between Western and traditional medicine (in general) and the fourth that between languages, with special reference to that between English and Luganda. The medical examples are relevant to plant conservation because of the historical and continuing closeness of botany and medicine. The encounters between languages can be the most fundamental of them all, given the role of language in shaping the ways that people view the world and their interactions with nature and one another (see Addendum to Chapter 1).

There can be much to gain environmentally if the systems used for managing natural resources draw on the knowledge of different sections of society (Barber et al., 2014). We give an example of this is Section 9.2, where we discuss the benefits that can be derived for conservation of biodiversity and the sustainable use of natural resources when the management systems devised for tropical forest draw on the knowledge of vegetation scientists, hydrologists and local community members. Considered as distinct types of knowledge, each of scientific, local and indigenous knowledge has something special to offer. Scientific knowledge has the benefit of being a deliberately evolutionary form of knowledge that has the potential to produce an increasingly accurate picture of reality over time. Local knowledge has the benefit of being based on practical experience at a particular locality. Indigenous knowledge has the same advantage, plus the benefit of long familiarity.

7.5.1 The encounter between scientific and traditional botany in China

In the 19th century, botanists and gardeners from the West began coming to China to collect living or preserved specimens of plants to take back to their countries to study in laboratories and grow in gardens. Four of these plant hunters, as they became known, were born in Scotland. The first of them was William Kerr (1779–1814), who collected close to Macau and Guangzhou, the only places in China where Europeans were allowed to live at the time. The second, Augustine Henry (1857–1930), who was actually of Irish heritage, went to China after graduating as a doctor to work for the Chinese Maritime Customs Service (CMCS) (Nelson, 1986). The CMCS, which had been founded in 1854 (in between the two Opium Wars), had much wider responsibilities than its name might imply. Until 1929, it was staffed at senior levels entirely by foreigners, especially by Britons, but also by Germans, Americans, French and, later, Japanese. Its responsibilities included administrating both internal and external customs, running anti-smuggling operations, undertaking currency reform, representing China at world fairs and running some Western educational establishments. Working for the CMCS both as a customs officer and a doctor, Henry became fascinated by Chinese plants and Chinese medicine. A Kew correspondent (see Section 5.3.2), he developed the habit of using his spare time to wander around the countryside in the various parts of Central and Southwest China where he worked, collecting plants and asking the local people about their uses. Altogether, he shipped about 158,000 dried specimens of plants back to the Royal Botanic Gardens, Kew.

The remaining two Scottish-born early plant hunters were George Forrest (1873–1932), a freelance collector who was among the first Westerners to visit Yunnan, and Robert Fortune (1812–1880), a botanist who received a commission from the Horticultural Society of London in 1843 to collect plants in Southern China. Highly successful in his collecting work, Fortune was responsible for introducing many Chinese ornamental plants into British gardens, among them new types of azaleas, camellias, chrysanthemums, peonies, rhododendrons and roses. Fortune was appointed director of Chelsea Physic Garden in 1846, but in 1848 asked for leave of absence, so that he could accept a commission from the British East India Company to go to China and bring out seeds and seedlings of tea for the establishment of tea plantations in India (see Section 5.3.4).

Another plant hunter was the enigmatic Joseph Rock (1884–1962), an Austrian botanist and ethnographer. Rock lived in Yunnan during much of the period between 1922 and 1949, collecting plants and studying the culture of the Naxi, one of the ethnolinguistic minorities who inhabit Yunnan (see Chapter 21). Rock was particularly interested in the texts and ritual ceremonies of the Dongba, a term that refers to both the texts themselves and to the shaman priests who compose them. The Dongba use a unique type of pictographic writing reminiscent of the hieroglyphic script of Ancient Egypt.

Yunnan, situated at the southwestern extremity of China and the southeastern end of the Himalayas, is a place of remarkable diversity – topographically, biologically and culturally. Its altitudes range from 76 to 6740 m, its ecosystems range from lowland tropical rainforest to high alpine pasture, and it contains about half of all native plant species recorded from China, although accounting for only 4.1% of its total area (Zhu and Tan, 2022). Yunnan was a refuge area for moisture-loving plants during the last global ice age (Li Rong *et al.*, 2011) and, in common with similar places elsewhere in the world, has an exceptionally rich flora (see Section 15.5.2). The wide altitudinal range and rugged topography of Yunnan present ideal conditions for populations of both plants and people to diverge evolutionarily, including culturally in the case of people.

Two of the Chinese scholars involved in the early development of scientific botany in China were Zhong Guanguang (1868–1940) and Tsai Xitao (1911–1981) (Fig. 7.4). The stories of their lives, spaced about 50 years apart, illustrate how botany in Chinese academia transitioned from being a subject rooted in traditional Chinese scholarship with scientific overtones to one rooted in science with Chinese overtones (Menzies, 2021). Chinese scholars have been keeping records and producing books dealing with botanical subjects for a long time.

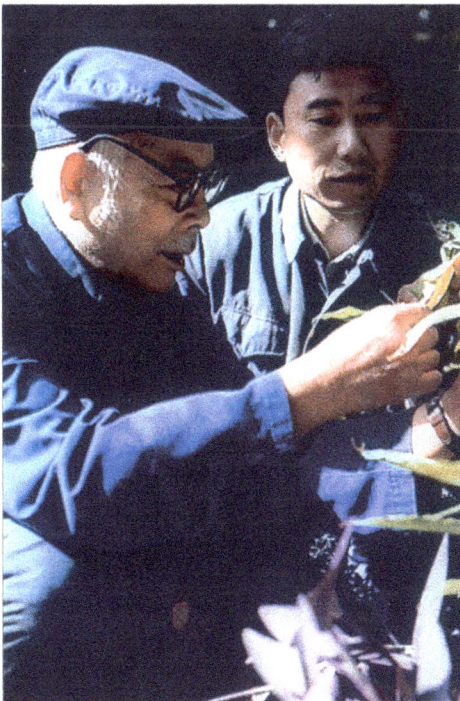

Fig. 7.4. Professor Tsai Xitao and Pei Shengji in 1974, China.

For example, 376 essays and books have been published on agriculture during the past 2000 years (Li Wenhua, 2016). In the 6th century CE, Jia Sixie systematically summarised Chinese knowledge of farming, forestry, animal husbandry and fisheries (Knechtges, 1997).

Zhong, the older of the two scholars, followed the traditional Chinese higher educational route and, in 1883, took the imperial civil service examination (see Section 4.3.1) (Menzies, 2021). However, thereafter, he followed his own path, taught himself scientific botany and become the first Chinese botanist to organise his own systematic plant-collecting expeditions in the Western plant-hunter style. Zhong was mindful of traditional Chinese learning and dedicated much of his career to investigating how best to match the Chinese and scientific names of plants, a subject which also intrigued Augustine Henry (Henry, 1893). Zhong was aware of the great wealth of botanical knowledge present within Chinese traditional culture and encouraged Chinese field botanists to talk with the herbalists and other villagers who were in close contact with plants in their working lives. In contrast to Zhong, Tsai Xitao had no contact with the traditional system of higher education in China, but loved the challenge of venturing into remote places, meeting the ethnic minorities who lived there and learning about their knowledge of plants. Like George Forrest and Joseph Rock, he was fascinated by Yunnan with its extraordinary flora and, like Zhong Guanguang, was a firm believer in the importance of botanists listening to the local people, respecting what they had to say and learning about their traditional knowledge. He believed in the universality of science and encouraged his students, one of whom was Pei Shengji (the author of the present book), to collaborate with foreign botanists and seek links with foreign botanical institutions (see Preface). Figure 7.4 is a photograph of Tsai Xitao demonstrating a specimen of ylang-ylang (*Cananga odorata*) to his student Pei Shengji. Ylang-ylang is a perfume-producing tree native of the Philippines. It is believed to have been introduced into China through its planting in the precincts of Buddhist temples in Xishuangbanna Prefecture.

Both Zhong Guanguang and Tsai Xitao contributed to the build-up of an institutional structure for scientific botany in China, in the process stimulating a growing appreciation within the Chinese leadership of the potential roles of botany in economic development. Milestones in its institutional development included the first scientifically organised botanical garden (in Hangzhou in 1927), the founding of the Botanical Society of China (1933) and the establishment in Kunming of the Yunnan Provincial Institute of Agricultural and Forestry Botany (1938). This last was subsequently brought under the CAS and renamed the Kunming Station of the Institute of Systematic Botany, with Tsai Xitao as its first director (from 1950 to 1958). Later, in the 1970s, the institution was once again renamed and became the Kunming Institute of Botany (KIB). Tsai was the driving force behind the founding of Xishuangbanna Tropical Botanical Garden (XTBG) (see Chapter 12). An ethnobotany department was established within KIB in 1987 with Pei Shengji as its first director. Thereafter, ethnobotany, both as a scholarly and applied subject, spread quickly to other institutions in China through the influence of Tsai's and Pei's students. The Chinese

Association of Ethnobotany (CAE) was launched in 2002 and the Asia-Pacific Forum on Ethnobotany (APFE) in 2004 (see Chapter 1).

7.5.2 The encounter between Western and traditional medicine in China

It was not until the second half of the 19th century that Western medicine, in the form that it had reached in the West, penetrated deep into inland China. The principal vehicles of its spread were Christian Protestant missionaries, mainly British and American, who took the opportunity to settle in China once they were allowed to do so under the terms of the treaties of Nanking and Tientsin (see Section 5.3.4). The first Western-style university was opened in 1895 (Peiyang University – now Tianjin University) and several medical colleges were founded by the missions at about the turn of the 20th century. The opening of Peking Union Medical College in 1906 was a key development. Built by the Chinese Government with the backing of a number of American and British missionary societies, in 1915 it managed to secure the support of the Rockefeller Foundation, which placed it on a firm financial footing (see Section 6.6) (Hsu, 2008).

When Mao Zedong launched the Cultural Revolution in 1966, he gave instruction to the Red Guards to 'Destroy the old and cultivate the new', referring to books, works of art and other records of China's cultural past (see Section 6.2.1). However, TCM was not among the 'old things' that Mao wished to destroy. Instead, it received a boost through Mao's decision to close all the country's universities and medical schools, and dispatch educated young people to 'go up to the mountains and down to the countryside to be educated by the poor farmers'. TCM was spared for two reasons. One was its association with ordinary rural people, who were among the keenest supporters of the new communist regime. Their support sprang from the gratitude that they felt for the introduction of social reforms that had the potential to lift them out of the poverty and indifference that they had experienced under the previous regime (Wood, 2020). The other was because the therapeutic value of TCM came to be appreciated by the Red Army during its Long March, a strategic retreat in 1934–1935 during the Chinese civil war (see Section 6.2.1). Largely out-of-touch of Western-style medical facilities, the soldiers and their retainers in the Red Army had little option but to resort to TCM.

Not only did Mao not suppress traditional medicine, he also actively supported the use of scientific research to further its development, as illustrated by another of his dictums: 'Chinese medicine and pharmacology are a great treasure house, efforts should be made to explore these and raise them to a high level'. One of the reasons for this advocacy was the success that Chinese scientists enjoyed in 1972 in isolating an active principle from a Chinese species that was administered to Zhou Enlai, the Chinese premier (see Chapter 12). When the chaos of the Cultural Revolution subsided in 1978 and orderly government became restored under the leadership of Deng Xiaoping, it was decided that, from then on, the Chinese national health system would be developed in a pragmatic way drawing on the best of both Western and tradi-tional medicine (Hsu, 2008). Dai, Mongolian, Tibetan and Uyghur medicine are all supported by the government in those parts of the country in which they are most relevant (Table 7.2).

It so happened that A.H.'s father was one of a group of British doctors who visited China in 1972 at the height of the Cultural Revolution. Being the first group of British physicians to tour Chinese medical facilities in 12 years, the visit provided a rare opportunity for an outside group of medical profes-sionals to see directly for themselves something of the current state of medicine in China. What they found astonished them and, on returning to Britain, some resolved to draw to the attention of a wider public what they had learnt. Four of the medical party, including A.H.'s father, succeeded in having a letter accepted for publication in the prestigious *British Medical Journal* (*BMJ*), a passage of which reads: 'We were greatly impressed with the standard of medical care, but above all, we were astounded by the use of acupuncture anaesthesia in major surgery. In all cases the patients were conscious, fully co-operative, and appear to suffer no pain' (Hamilton *et al.*, 1972). (Note: although acupuncture has long been used in therapeutic practice in China, its use as an analgesic in general surgery only began during the 1950s.)

The use of acupuncture as an analgesic is starkly different from the way that patients have been prepared for general surgery in Western medicine since the late 19th century, which has been by administering pain-killing drugs, such as morphine, and rendering them unconscious through the administra-tion of anaesthetics. Publication of the letter in the *BMJ* resulted in a torrent of responses. One of the most balanced and considered came from the head of the Cerebral Functions Unit at University College, London, who commented: 'Acupuncture anaesthesia presents a fascinating challenge to the "objectivity" of Western science and its application to man. It challenges cultural, scientific and political biases' (Wall, 1972) – the quotation marks around the word 'objec-tivity' are in the original. Other responses were more dismissive. Some medical scientists pointed out that the places where the needles are inserted into the human body in acupuncture (meridian points) have no neuroanatomical con-nections with anything known in Western medicine about how the body works. It was debated whether acupuncture could be considered to be scientific and, by implication, medically acceptable.

Another aspect of the Chinese health service that differed in 1972 from standard Western practice was the use of barefoot doctors to deliver primary healthcare and health education to isolated communities. Based on a concept announced by the Chinese Government in 1968, already, by the early 1970s, 1 million barefoot doctors had been engaged. The task of choosing who should become barefoot doctors was left to the communities, a policy that had the advantages that the people would likely choose people from their communities whom they knew to be trustworthy and knowledgeable and who spoke their languages. In fact, as Pei came to appreciate through his fieldwork in herbal medicine, many barefoot doctors had a good knowledge of herbal medicine and only needed training in some modern medical skills.

7.5.3   The relationship between Western and traditional medicine

Unlike academic botany, which is a 'pure' science, medicine is an applied science and, therefore, its success must be measured by progress made in the real world. The problem of how to deliver healthcare in places where there are few doctors, as arose for the Red Army during its Long March, is faced by the governments of many modern countries. The standard Western response has been to engage medical assistants, these being people who are provided with information and skills by those higher in medical hierarchies and assigned to work within the communities (Kachwano, 2022). This differs from the barefoot doctor approach in the positioning of authority, because it vests authority in external medical elites, rather than in the communities. A popular self-help manual within the Western tradition is *Donde No Hay Doctor* (*Where There is No Doctor: A Village Health Care Handbook*) (Werner, 1973). Subsequently translated into many languages and recommended by international organisations such as the World Health Organization (WHO), the manual provides advice to people on how they can look after their health in places where there are few doctors. It places special emphasis on the public health concerns of hygiene, a healthy diet and vaccination.

It makes sense for governments, faced with the responsibility of delivering universal healthcare, to be open to the possibility of drawing on all healthcare traditions represented within their countries (Cunningham, 1993). Debates on this topic customarily refer to Western medicine as 'conventional' and all other medical traditions as 'complementary and alternative medicine' (CAM). Even if politicians regard CAM as a second-best choice, bringing CAM into national healthcare systems can make pragmatic sense, given that an estimated 80% of the global population rely on CAM (rather than conventional medicine) to meet their primary healthcare needs (WHO *et al.*, 1993). Also, it should be noted, Western medicine, as actually delivered, can have major deficiencies. For example, investigations have revealed that 35% of samples of drugs from seven countries in Southeast Asia and 21 in sub-Saharan Africa failed chemical analysis – they were not what they claimed to be, in other words, fakes (Nayyar *et al.*, 2012). Some unscrupulous pharmaceutical companies have made huge profits out of marketing highly addictive opioids, such as fentanyl and oxycodone, resulting in large numbers of deaths from overdoses (Keefe, 2017; Ausness, 2022; White, 2022). The companies involved, some of which are household names, merge into outright criminal organisations, such as the Mexican and Colombian drug cartels mentioned in Section 6.3.3.

There are striking differences between countries in the relative availability of Western and CAM medicine. For example, the number of conventionally trained doctors per 1000 people and the annual health expenditure per capita are both about 30 times higher in Britain than Uganda, while there are about 30 times more traditional than conventional-trained doctors in Uganda (Hamilton *et al.*, 2003). In the 19th century, the flow of doctors and nurses between the two countries was entirely unidirectional – from Britain to Uganda. Today, the flow of doctors and nurses, and also care workers, is reversed. Ever since its creation in 1948, Britain's National Health Service (NHS) has never trained enough doctors to meet its staffing needs; it has always relied on importing doctors (Jalal *et al.*, 2019; Fagan and Bhutta, 2021). In effect, the costs of training doctors working in Britain has been transferred to other countries, including Uganda, a country which is about 30 times poorer. This is a bizarre turn of events, especially given that the British government announced in 1997 that its foreign policy was to be 'ethical' (see Section 5.7).

The extent of use and policies towards CAM vary considerably between countries. For example, CAM is used in regular medical practice much more in France and Germany than Britain (Fjær *et al.*, 2020). As explained above, state-supported medicine in China draws pragmatically on both conventional medicine and CAM. In Uganda, traditional medicine became marginalised during colonial times, when it was blanketly condemned by both the British administration and the Anglican Church and labelled as witchcraft, a policy that changed little at first with political independence (Tabuti *et al.*, 2003). However, nowadays, both the government and the Anglican Church are more open to traditional medicine, but it continues to be resisted by evangelical Christians.

In India, the debate as to whether and how CAM should be incorporated into its national health service has been politically charged and acrimonious (Ganesan, 2010; Islam, 2012). During colonial days, advocacy of Ayurveda, which is the most widely used type of CAM in India, became associated with Indian nationalism and, ever since, there has been a tension between conventional medicine and CAM. The lack of communication between conventional medicine and CAM in India has not served the country's citizens well, especially the poor. In rural areas, where most ordinary people continue to use CAM, doctors trained in conventional medicine can feel out of their comfort zones (Supe, 2016). In urban areas, many Ayurvedic practitioners have reportedly adopted practices from conventional medicine to gain social respectability (Patwardhan, 2014).

At a congress of the International Society for Ethnobiology (ISE) held at Montpellier, France, in 2012, Yildiz Aumeeruddy-Thomas of the Institut écologie et environnement (a French scientific institute) and A.H. organised a session entitled 'Maintaining resources for traditional medicine'. Because it was well-attended and there was reasonably good geographical coverage (Map 7.1), the session is thought to have given some indication of the current global state of research into medicinal plants, conservation and sustainable development (Hamilton and Aumeeruddy-Thomas, 2013). 'Maintaining resources', in the context of the session, referred to two things – one, the continuing availability of the plants to communities and, the other, the continuing existence of knowledge within the communities about how to use them. They are interconnected, since maintaining knowledge about medicinal plants is greatly dependent on their continuing use and being able to continue to use them is greatly dependent on their ready availability.

Everyone at the session agreed that knowledge of traditional medicine is declining worldwide. A principal reason for this, according to the experiences of those present, has been the spread of formal, Western-style, classroom education. This change in environment during their formative years has removed children for long periods of time from the environments in which their ancestors acquired

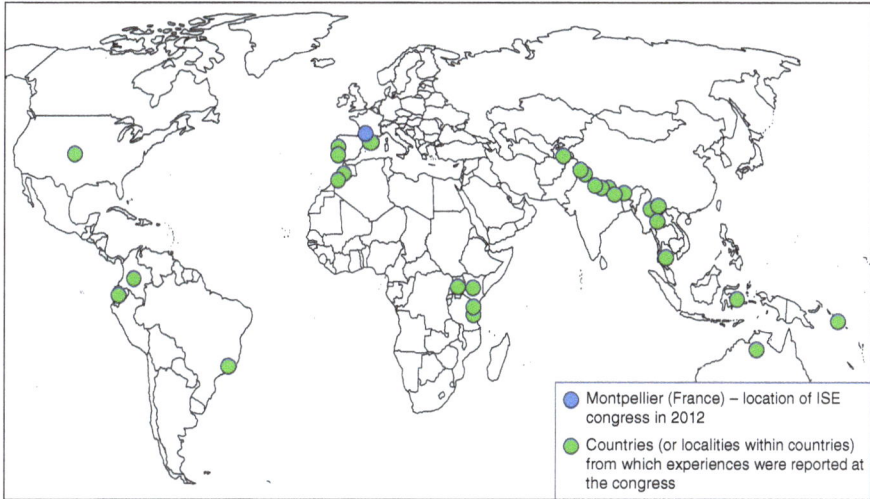

Map 7.1. Locations of experiences of traditional medicine shared at the 2012 Congress of the International Society of Ethnobiology. Note regarding the positions of country boundaries: see Disclaimer.

their traditional knowledge of healthcare and plants (see Section 7.1). There was general agreement that traditional medicine still has a useful role to play in primary healthcare, including for the management of chronic complaints and conditions related to local cultural beliefs, both of which are poorly served by Western medicine.

7.5.4 The encounter between languages, with special reference to English and Luganda

Pursuing plant conservation can require communication across language divides. Around the world, the languages spoken by people at community level are frequently different from the officially recognised languages of their countries and these can be different again from the languages that the countries use for international communication.

Of the world's estimated 7151 living languages (Ethnologue, 2022), only a handful are used as working languages in international affairs, such as in diplomacy, science, transnational business and development aid. The post-1500 CE expansionism of Europe and the West has ensured that all of these languages are of West European origin, the most commonly spoken being English, Spanish, French and Portuguese. About 1.5 billion people globally speak English. A few other languages, such as Arabic, Russian and Swahili, are used regionally. Where the European overrun of indigenous cultures has been particularly complete, as in virtually all of the Americas, one or another of the European languages has become accepted, more or less without question, as the national language of today's political states. Elsewhere, the independent states that emerged as the European empires were dismantled have generally retained the languages of their former colonial masters as their national languages.

Countries vary in their approaches to national language policy. This is not a trivial matter, but one that can relate to the fundamental developmental challenges that countries face. These challenges include the building of national unity among people of disparate ethnicities, the inclusion of all sectors of society in decision-making processes and the incorporation of indigenous knowledge into developmental initiatives (Heugh, 2018; Nankindu, 2020). The language policies of the three countries chosen for special attention in the present book (China, Britain and Uganda) reflect their histories and the present states of their political affairs.

China

In the case of China, which is a huge country with about 300 indigenous languages, a single language, Mandarin Chinese, was chosen in the 1930s to be the official national language, a decision that reflects China's historical concern about maintaining its national unity (see Section 4.3.1). A few other indigenous languages receive some official support. For medicine, they are those associated with the Dai, Mongolian, Tibetan and Uyghur traditional medical systems (Table 7.2). English is the main language of international communication and is taught at all levels in schools.

Britain

National language policy in Britain – or, rather, the lack of it – reflects the crisis of national identity that has arisen in the United Kingdom following the dissolution of the British Empire. Earlier, the empire and the English language had served as points of unity with respect to the way that British people regarded one another and looked out on the rest of the world (Leonard, 1997). The United Kingdom as a whole has no official language and neither do England and Scotland, which are two of its four regional parts. Wales has two official languages, and, in Northern Ireland, language policy has become a bitterly divisive issue, which has hampered the installation of a representative local government.

Uganda

There are about 41 indigenous ethnolinguistic groups in Uganda, living to a large extent in their traditional tribal areas (Simmons and Fenning, 2018). The most widely used of these indigenous languages is Luganda, with an estimated 7.4 million speakers. The great majority of people use the indigenous languages of their home areas in their everyday affairs, which, crucially (in relation to the subject of this book), include the plant-related activities of agriculture, forest utilisation, pastoralism, craft-making, the preparation and cooking of food, consultations with traditional healers and trading in local markets. In contrast, English is the language of government, the law, science, big business and development aid, as well as the required medium of instruction in schools at all but the most elementary levels. There is, in fact, one other official national language in Uganda – Swahili. However, there are few fluent speakers, mostly Muslims living in urban areas. For some in Uganda, speaking Swahili is associated with violence metered out on them by the army and police during times of political tension, which has given them a distaste for hearing and learning it (Leopold, 2021).

As in many countries, a gap has opened up in Uganda during recent decades between a rich upper stratum, which is largely self-perpetuating, and the bulk of the people (see Section 6.3.1). This divide is today becoming further cemented by a difference in home languages (Williams, 2011; Asante, 2020). It is reinforced at national political level by the requirement that all those who put themselves forward as parliamentary candidates must be competent speakers of the English language, which rules out most people. The language divide between the children of the elite and other people places them in an awkward social position, not knowing the everyday languages of the people among whom they are living.

A.H. has been involved in two collaborative efforts to help bridge linguistic divides in Uganda. One of them is a two-way Luganda–English dictionary (see Preface). The other, undertaken with James Kalema, a botanist at Makerere University, was the preparation of a *Field Guide to the Forest Trees of Uganda* (Kalema and Hamilton, 2020). The descriptions of the species in the guide are given in English (the main national language) and their scientific names in Latin (as required in science), but also included are 1104 names for the trees in 21 indigenous languages. It was hoped that inclusion of the vernacular names will make it easier for people to communicate across language divides when engaging in matters that relate to forests and trees. This list was initially compiled from the literature and from the authors' own experiences. Kalema and A.H. wanted to check the accuracy of the names, but for only about half the languages could they find people who had knowledge of both the vernacular and the scientific names, and then often for only a few species.

8

A History of Plant Conservation

Abstract

Long-standing societies often have beliefs and customs that protect certain features of their local natural worlds against their destruction for only passing benefits. Among them are taboos protecting certain types of plants and animals, the protection of groves of trees through association with local religious beliefs and the promotion of crop diversity through farmer-to-farmer exchanges of seed. Examples from Ethiopia and Cameroon illustrate how modern ideological and economic forces can be destructive of traditional conservation practices. A mounting awareness since the 1970s of the large numbers of traditional crop varieties and plant species in imminent danger of extinction catalysed the scientific community into devising a systematic, two-pronged approach for their conservation. One prong, *in situ* conservation, aims to conserve plants in their natural habitats, for instance through the establishment of legally protected areas. The other, *ex situ* conservation, aims to place plants in off-site facilities, such as botanical gardens and seed banks, so that they are removed from harm's way. The reintroduction of endangered species back into their natural habitats is encouraged, where feasible. The origins of legally protected areas in Britain, the United States and Uganda are described.

8.1 Traditional Conservation

Long-standing societies often possess beliefs and customs that protect certain features of their local natural worlds against their destruction for only passing benefits. Aspects of the natural world receiving protection in this way include certain species of plants and animals, individual specimens of trees, areas of more natural vegetation and prominent natural features, such as certain mountains and hills.

It is common in some parts of the world for small patches of indigenous forest to be preserved through their associations with religion. It is estimated that there are 150,000–200,000 such sacred groves in India alone (Wild and

McLeod, 2008). An example of a sacred grove in Uganda can be seen in Fig. 4.2. One place in which sacred groves have attracted the attention of conservationists is highland Ethiopia, an area that was extensively covered with montane forest until it started to be cleared extensively for agriculture in about 500 BCE (Darbyshire *et al.*, 2003). Nowadays, the landscape in these highlands is largely agricultural, with the result that the church forests have become very important places for the survival of the native flora and fauna (Aerts *et al.*, 2016). Some sacred groves in highland Ethiopia, particularly towards the south, are connected to African traditional religion, but an estimated 19,400 of them are associated with churches or monasteries of the Ethiopian Orthodox Tewahedo Church (EOTC). This is a type of Coptic Christianity with roots dating back to 1st century Egypt. The EOTC is hierarchical at its upper levels, but, at village level, the churches have close connections with the local communities, who supply their clergy.

Figure 8.1 is a sketch of the structure of a church and associated sacred grove of the EOTC, based on a visit made to Anchucho and a few other village churches by A.H. in 2002. At its heart is a small building which is only entered by the clergy and which contains a tabot (a replica of the Tablets of Law on to which the Ten Commandments of the Bible are said to have been originally inscribed). Around this building is a small garden containing ornamental and medicinal plants, such as *Artemisia afra* and *Ocimum*, and, surrounding this in turn, a large grass-covered courtyard, which is where the congregation assembles for services. The courtyard is contained within a palisade made of the wood of East African cedar. The sacred grove on the outside of the palisade

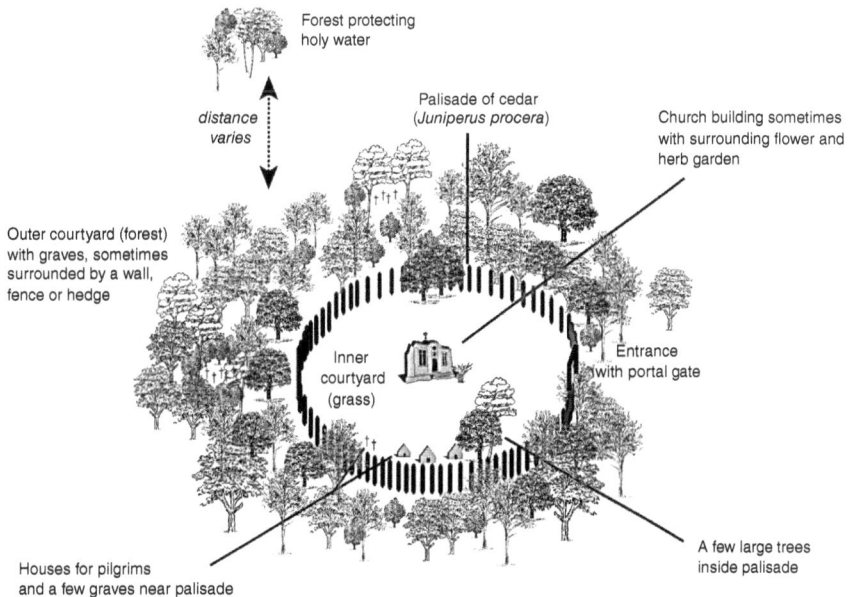

Forest protecting holy water

distance varies

Palisade of cedar (*Juniperus procera*)

Church building sometimes with surrounding flower and herb garden

Outer courtyard (forest) with graves, sometimes surrounded by a wall, fence or hedge

Inner courtyard (grass)

Entrance with portal gate

Houses for pilgrims and a few graves near palisade

A few large trees inside palisade

Fig. 8.1. Architecture of a Coptic church with sacred grove, Ethiopia. Reproduced from: Hamilton and Hamilton (2006).

contains species of trees that are typical of drier types of African montane forest. Medicinal plants are common in the undergrowth. There are scattered graves in the forest, many planted with a small, red-flowered species of shrubby euphorbia.

A.H. was accompanied on his brief visit to the EOTC village churches by Desalegn Desissa Daye, an Ethiopian ethnobotanist, and Pierre Bingelli, a Swiss ecologist. They told him that some of the plant species associated with the church have religious or healing significance. For example, the East African cedar (which is actually a juniper – *Juniperus procera*) is considered locally to represent the biblical Cedar of Lebanon (*Cedrus libani*). Christian crosses for use in the church are made of the wood of the olive tree (*Olea europaea*), the wild form of which is common in the forests. The olive tree and its oil are often mentioned in the Bible. One aspect of the ecology of these groves, seemingly little studied, is the extent to which people are moving plant species between the groves. If this is an active tradition, it would be of significance for devising strategies to enhance the conservation of the biodiversity of the groves.

It can be seen on Fig. 8.1 that, in addition to the forest near the church, there is a small grove of trees a short distance away labelled 'Forest protecting holy water'. Figure 8.2 is a photograph of this grove at Anchucho. On his visit to Anchucho, A.H. confirmed that the grove contains a spring, which, he was told, was consecrated. Its water was being used in religious rites, such as baptism, in the church. In wider geographical perspective, associations between religion, sources of water and forests are common. Examples from China are mentioned in Chapters 12, 14 and 21. In Britain, there are thousands of

Fig. 8.2. Sacred grove protecting a holy spring, Anchucho Church, Ethiopia. Photo: A.H. (2002).

springs and wells historically associated with Christian saints; their waters are traditionally believed to possess special healing properties (Davies and Robb, 2002). It is still a common practice in Britain for people to toss coins into holy wells to bring them good luck.

These associations between religion, water and forests give a clue as to how traditional conservation practices may have originated. Since water is a resource essential for human existence, it is not difficult to imagine how, through observation and experience, people in different parts of the world have discovered that the presence of a cover of forest in catchments tends to ensure that water supplies are more reliable. The association with religion is valuable for helping to protect the forests, because then, according to people's thinking, a failure to safeguard the forest will incur not just the anger of people, but the wrath of the gods.

In Buganda, there is a form of traditional conservation associated with the clans of the Baganda (see Section 6.8.1) (Roscoe, 1911). Each clan (ekika), of which there are about 50, has a main totem (omuziro) and a secondary totem (akabbiro), most of them animals. Examples are the civet (effumbe), the edible grasshopper (enseenene) and the haplochromis fish (enkejje). Members of clans are forbidden to eat their totem animals, thereby affording them some protection. However, since there is nothing to prevent one clan's totemic animals from being eaten by members of other clans, the protection afforded is relatively weak.

Exceptionally, eight of the totems and secondary totems of the Baganda are plants, most of them species useful to people in one way or another. One of these plants is ekkobe (the bud yam – Dioscorea bulbifera), which, according to oral tradition, has been in Buganda for a long time. Scientifically, it is known to have originated in Asia, which suggests that it may have been among those crops carried to Africa by Austronesian-speaking people during the early centuries CE (see Section 3.3.3). Another clan has envuma as its totem. This is the fruit of the water chestnut (Trapa natans), a species of aquatic plant with floating leaves found in warmer parts of Africa and Eurasia. The large fruits, which are considered a delicacy, are distributed in nature through being released by their parents to float about on the surface of the water before eventually sinking down and germinating in the mud on the lake floor. Possibly, the discovery of such a surprising form of food would have fascinated the ancestors of the Baganda when they settled on the margins of Lake Victoria during the 1st millennium BCE.

Although traditional conservation beliefs and practices can be resilient, they can prove inadequate for protecting nature in the face of hostile ideological and religious forces. For example, a communist government, that held power in Ethiopia in 1987–1991, was ideologically opposed to religion of any sort, which weakened the support traditionally provided to forests by the Coptic Church. In China, the Cultural Revolution witnessed direct assaults on both religion and traditional conservation (see Chapter 14). Today, under the influence of the forces of globalisation, younger villagers in Ethiopia tend to be less concerned about the conservation of the church forests than their parents (Reynolds et al., 2017). Evangelical Christianity, which is a growing force in

Ethiopia especially towards the south (where the Coptic Church is relatively recently established), can be dismissive of the traditional beliefs that have historically safeguarded the forests (Mergo, 2012; Freeman, 2013; Fantini, 2015). It can be similarly damaging to traditional conservation practices in some other parts of the world (see Section 6.3.1).

The introduction of a monetary market for commodities harvested from wild plants can break the hold of traditional beliefs and practices. An example is the introduction during the last 50 years of an international market for extracts from the bark of an African montane forest tree, *Prunus africana*. Extracts from the bark are used in modern Western herbal medicine in continental Europe to treat benign prostate hyperplasia (Komakech *et al.*, 2017). The commercial harvesting of the bark started first in Cameroon, when a French-owned company, Plantecam Medicam, received a licence from the government to collect the bark for export to Europe (Cunningham *et al.*, 1997). Commercial harvesting subsequently spread to other African countries, including DR Congo, Ethiopia, Kenya, Madagascar, São Tomé and Uganda. In Cameroon, Plantecam Medicam intended to introduce a system for harvesting the bark which, it thought, would be sustainable. This involved stripping the bark from opposite quadrants of the lower trunks of the trees, then leaving the trees alone to allow them to recover, then stripping the bark from the other two quadrants, then leaving those to recover, then back to the first two quadrants, and so on. A.H. has seen no evidence that even the second stripping phase of this supposedly sustainable system has ever been reached.

In 1985, the Government of Cameroon issued licences to additional companies to collect the bark, but with insufficient controls in place to ensure that they did so responsibly. The result was a competitive scramble to collect as much bark as possible. Figure 8.3 is a photograph of Cameroonian conservationist, Fonki Mbenkum, and A.H. standing by the side of a *Prunus* tree at Kovvifem in the Bamenda Highlands. The tree has been stripped of its bark right down to its smallest branches. The matter came to A.H.'s attention in 1992 when he was visiting Cameroon on behalf of WWF. As a result, the People and Plants Initiative (PPI) launched a campaign, spear-headed by Tony Cunningham, to place the harvesting of the bark on a sustainable basis (see Chapter 10 for a description of PPI). The campaign succeeded, at least with regard to having *P. africana* included in 1995 in Appendix II of the Convention on International Trade in Endangered Species of Wild Fauna and Flora (CITES) and in causing the species to be among those that are regularly selected for agroforestry planting in montane Africa (Cunningham *et al.*, 1997, 2002).

Figure 8.4 is a photograph taken in 1992 of agricultural land on Mt Oku in the Bamenda Highlands that had recently been created through the clearance of montane forest. Notice the scattered remaining forest trees and the logs lying about, here and there, on the newly tilled ground. In this part of the Bamenda Highlands, the harvesting of *Prunus* bark by Plantecam Medicam started in 1972 and the new wave of licensed companies arrived in 1985. Reportedly, there was a link between the indiscriminate stripping of the *Prunus* trees and the forest clearance (Stewart, 2003a,b). Up to the time of the indiscriminate stripping, the forest had received protection from a local taboo, but, when the

Fig. 8.3. *Prunus africana* tree stripped of its bark, Bamenda Highlands, Cameroon. Photo: A.H. (1992).

local people observed that the bark-harvesters were not receiving retribution from the guardian spirits of the mountain, they scrambled to clear the forest to add land to their farms.

8.2   Legally Protected Areas

Nowadays, written laws, created by governments, determine the activities which are allowed or not allowed within states. Here, we outline how networks of protected areas have originated in Britain, the United States and Uganda. For this discussion, we take a protected area to be a legally designated part of the Earth's surface in which conservation of biodiversity is one of the priorities of management.

Fig. 8.4. Agricultural land recently created from forest on Mt Oku, Cameroon. Photo: A.H. (1992).

8.2.1   Origin of protected areas in Britain

Part of the foundation of Britain's modern protected area system consists of areas of land that were declared during medieval times as royal forests (see Section 5.4.1). As an example of what happened to these forests, Epping Forest, on the outskirts of London and which had been declared a royal forest in the 12th century, was handed over by Queen Victoria, the reigning monarch, to the City of London Corporation in 1878. It was stipulated that, henceforth, it was to be a 'people's forest', dedicated to their use and enjoyment for all time. Another example is the New Forest in Southern England. This was proclaimed a royal forest in about 1079 and still belongs to the Crown, but with its management devolved to Forestry England, which is the government agency responsible for managing English forests. Today, the New Forest is managed for multiple purposes, one of which is conservation of biodiversity.

Two cultural movements, the romantic movement and humanitarianism, played big roles during the 19th century in the creation of Britain's protected area system (see Sections 6.1 and 6.6). Star figures in the romantic movement, such as the poets William Wordsworth (1770–1850) and Samuel Taylor Coleridge (1772–1834), extolled the virtues of contact with nature as being a way to discover one's true self. Humanitarians, such as Octavia Hill (1838–1912), who was one of the founders of social housing in Britain, lobbied for the creation of accessible places in the countryside, where those who laboured in the grime and gloom of the cities could refresh their souls. A concrete achievement of these movements was the creation in 1895 of the National Trust, an organisation with a remit to safeguard the country's natural and cultural heritage. National Trust properties receive an exceptionally high degree of legal protection, because they are held inalienably for the nation and cannot be sold or mortgaged. Today, the National Trust is one of Britain's largest landowners, responsible for managing many areas of important natural habitat.

Direct action has sometimes sharpened political lobbying in Britain to gain improved access for the public to the countryside. The 'mass trespass of Kinder Scout' was an event in 1932, organised by members of the Young Communist League, when hundreds of people illegally invaded Kinder Scout, a plateau area of moorland in the Pennine Hills of Northern England. At the time, Kinder Scout was being used by members of the privileged classes for grouse-shooting. (The grouse is a type of 'game' bird.) Kinder Scott is only about 25 km from Manchester, a city that, from the early 19th century, had been a major manufacturing centre. The Kinder Scout mass trespass is thought to have been instrumental in causing the government to pass legislation in 1949 allowing the establishment of National Parks. The first National Park in Britain, the Peak District National Park, was established in 1951 on the land where the Kinder Scout trespass had taken place.

If Britain had had a free hand to develop its protected area network according to the principles of conservation biology, without the baggage of history, the end result would have looked quite different. This said, a network of rationally selected national Nature Reserves does exist in Britain, the first of which was declared in 1951. Their sites were selected primarily with the aim of capturing a representative cross-section of the natural and semi-natural habitats present in the country (Tansley, 1946). Meanwhile, all the organisations responsible for managing land mentioned above now include conservation of biodiversity among their management objectives, regardless of the original reasons lying behind their acquisition of land. The network of legally defined protected areas in Britain has been further extended by individuals and organisations, such as local governments, landowners and the military, founding their own protected areas.

8.2.2 Origin of protected areas in the United States

The United States, like Britain, had its 19th century nature romantics, two of the most influential being Henry Thoreau (1817–1862) – a writer who advocated the benefits of living close to nature and learning from the ways of the indigenous people – and John Muir (1838–1914), a Scottish-born naturalist and botanist, who emigrated with his family to the United States at the age of 11 and became a leading proponent for the preservation of wilderness. Thoreau was an introverted person, who shared Humboldt's view that a true understanding of nature can only be achieved by spending time immersed in its reality. His best-known book, *Walden*, published in 1854, describes the years that he spent living quietly in a small cabin by the side of Walden Pond in Massachusetts (Thoreau, 1854). Muir, an outgoing and adventurous character, became entranced with Yosemite in the Sierra Nevada of California when he first visited in 1868, setting off his long-term campaign for it to be granted federal protection.

The first national parks in the United States (and the world) were Yellowstone (1872) and Yosemite (1890) – both of which are in the mountainous Western United States and were conceived by those Americans of European extraction who first discovered them as areas of pristine wilderness (Schama, 1995). Theodore Roosevelt (US president, 1901–1909) was impressed by Muir's vision of wilderness and was instrumental in creating many national parks, bird reserves and game preserves in the United States. In 1905, he established the US Forest Service, which today manages huge areas of public land, especially in the mountainous west. In 1970, President Richard Nixon established the Environmental Protection Agency with a mission to protect the human and environmental health of the nation (Jelen and Lockett, 2014).

Roosevelt was a naturalist and big-game hunter, as well as a politician, which accounts for his interest in establishing game preserves – places set aside for game-hunters to hunt animals. After he had completed two terms of presidency, Roosevelt packed his bags and set off on game-hunting expeditions to Amazonia and East Africa. His library is known to have contained a book by John Henry Patterson, a surveyor on the Uganda Railway, inviting people like him to visit East Africa and shoot its 'game' (see Section 5.3.1) (Thompson, 2010). Roosevelt's attitude to nature and the environmental policies he pursued were influenced by the environmental disaster that had befallen the Appalachian Mountains in the Eastern United States in the 19th century. There, poorly controlled expropriation of land by European settlers, combined with massive unsustainable exploitation of timber, had resulted in large-scale loss of catchment quality, flooding and erosion (Yarnell, 1998). It was hoped that the newly formed US Forest Service would be able to prevent mountainous Western United States suffering the same fate as the Appalachians.

Much of what is now the Western United States was settled in the 19th century by immigrants from Europe, who headed west from the east coast along trails pioneered by the Lewis and Clark Expedition of 1804–1806 (see Section 6.5.1). They took advantage of the Louisiana Purchase of 1803 (when the United States bought a huge area of land in North America from France) and the United States' victory in the Mexican–American War of 1849. The discovery of gold in California in 1848 soon resulted in fights breaking out at Yosemite in the Sierra Nevada between American miners and the indigenous Ahwahnechee. The Ahwahnechee, who were hunter-gatherers, stood little chance against the Americans and, after their defeat, those who survived were shipped out of Yosemite and left to fend for themselves.

Contrary to the American imagination, Yosemite was not a wilderness, in the sense of being in a natural state uninfluenced by people. It has been inhabited for over 10,000 years and its ecosystems had long been manipulated by its inhabitants to serve their purposes, for instance through the periodic burning of areas of vegetation to favour those plants and animals in which the people were interested (Gassaway, 2009). One species of interest to the Ahwahnechee was the California black oak (*Quercus kelloggii*) (see Section 3.1) (Fig. 8.5). This oak was not named after the Kellogg brothers of breakfast cereal fame, as might be supposed (see Section 7.4.3), but after Albert Kellogg, the first resident botanist to live in California and a founder member of the California

Fig. 8.5. Leaves and acorns of black oak, Yosemite, California, USA. Photo: A.H. (2013).

Academy of Sciences. Albert Kellogg (1813–1887) was born in Connecticut and became a doctor. Like Augustine Henry, who hunted plants in China (see Section 7.5.1), Alfred was fascinated by botany and spent much of his time exploring the vegetation and collecting specimens of plants in the various places where he worked. In 1849, he joined the California gold rush, travelling to California by ship via the Straits of Magellan, and found employment with the Connecticut Mining and Trading Company in Sacramento. The first scientific collections of the black oak were made by John Newberry (1822–1892), who worked for the Pacific Railroad Survey and was another doctor-cum-botanist. The person responsible for scientifically naming and describing the species was Karl Hartweg (1812–1871), a German botanist who collected plants in the Americas for the Horticultural Society of London (see Section 5.3.2).

Central Valley, the place where Sacramento is situated, is one of California's principal topographic features. To its west lie the Coastal Ranges fringing the

Pacific Ocean and, to its east, the Sierra Nevada. The coastal redwood (*Sequoia sempervirens*), the tallest living thing on Earth, is found in the Coastal Ranges and the giant redwood (*Sequoiadendron giganteum*), which is the Earth's most massive tree, grows in the Sierra Nevada. Both of these huge trees are sustained by the cool, wet, winter weather and the frequent, blanketing and moisturising, summer fogs.

The climate of Central Valley is dry, ranging from hot-Mediterranean in the north to dry-desertic in the south. The first Europeans to come to California, which was in the 16th century, considered Central Valley to be unsuitable for European habitation and the only agriculture attempted by American settlers in it prior to the 1930s used rainfed techniques to precariously raise crops of annual cereals, such as barley and wheat (Mitchell *et al.*, 2016). However, during the 1930s, a vast system of reservoirs, canals, aqueducts and pumping stations started to be constructed to bring water into the valley from the Sierra Nevada and the Cascades (a range to the north of the Sierra Nevada). Central Valley was transformed into one of the most agriculturally productive places on Earth.

8.2.3   Origin of protected areas in Uganda

The two motivational forces that contributed towards the founding of Uganda's protected area network were the interest of the British upper classes in hunting 'game' (see Section 2.3.2) and that of the colonial government in setting aside areas of land for environmental protection and timber production (see Section 5.6). At the time, hunting game and appreciating nature were not seen as mutually incompatible activities.

The population of Uganda rose rapidly during the 20th century, from about 2 million in 1900 to 5 million in 1950 and to 24 million in 2000. The result was a decrease in the areas of more pristine habitats and this, combined with the possibility of earning revenue from tourism, resulted in the colonial administration deciding in the early 1950s to create Uganda's first National Parks. The largest of the parks, Queen Elizabeth and Murchison Falls, both gazetted in 1952, have proved to be excellent places for tourists to view wildlife safely, while being driven around in comfort in vehicles. The vegetation of these parks consists largely of open savannah, without too much clutter of trees and bushes to obscure panoramic views of herds of antelope, prides of lion and the other types of wildlife that interest tourists.

Images of spectacular amounts of wildlife in national parks propounded by tourist agencies have reinforced a common perception in the West that Africa is a wild place full of wild animals (Adams and McShane, 1992). The perception is misleading. Just as with Yosemite in the United States, the places where Africa's savannah national parks are now situated are not places of long-existing, unsullied, raw nature. In fact, ecosystems in Africa have been subject to major modifications imposed by people for even longer than those in the Americas (see Chapter 1). For example, the sharp boundary between forest and savannah often seen in Africa today is believed to be an artifact of the burning of savannah vegetation by people that has been happening for a very long period of time (Hamilton, 1989). In the 19th century, prior to the arrival of people from the West, the places where Queen Elizabeth and Murchison Falls national parks are now situated lay within the domains of well-organised feudal kingdoms, the most powerful of which was the Kingdom of Bunyoro-Kitara, whose economic base rested on the large-scale herding of cattle (see Section 3.3.1).

Two deadly diseases, arriving in aggressive form at around the turn of the 20th century, were responsible for weakening the hold of the feudal kingdoms of Western Uganda. The first of these diseases to come was rinderpest, a disease of cattle caused by a virus with an evolutionary origin in the steppes of Central Asia. The initial introduction of rinderpest into sub-Saharan Africa was in 1895–1896 by means of a cattle train that was accompanying an Italian army that was set on conquering the Empire of Ethiopia (Pearce, 2000). Italy had decided to conquer Ethiopia and make it a colonial territory, because of its resentment at not receiving enough of Africa, in its opinion, when the continent was divided up between the European powers in 1884–1885, and because it was conveniently close to Eritrea, which it had been given, to launch an invasion (see Section 5.3.5).

From its initial base in Eritrea and Ethiopia, rinderpest spread rapidly through Eastern and Southern Africa, altogether causing the loss of an estimated 80–90% of the cattle. In Uganda, the resulting reduction in browsing pressure on savannah vegetation allowed woody plants, formerly suppressed by the heavy browsing of cattle, to grow freely and spread. The result was the creation of an ideal bushy habitat for tsetse fly, a blood-sucking insect that is responsible for transmitting the other deadly disease that arrived in force in Uganda at about the turn of the 20th century. This was trypanosomiasis, which is a disease that affects both cattle and humans and is known in the latter as sleeping sickness. In 1900, the newly established British colonial administration ordered the evacuation of people from the most heavily tsetse-infested areas, which resulted in the creation of extensive areas of countryside almost devoid of people (Imperato, 2005; Picozzi et al., 2005). Some of these evacuated areas were later declared as Forest or Game Reserves and, in Murchison Falls' case, eventually as a National Park.

Among the areas ordered to be evacuated were all of the Ssese Islands and the land on the mainland that was close to the shore of Lake Victoria. An order was also given to destroy all large canoes on the Ssese Islands, one of the places where they were traditionally manufactured (Roscoe, 1921). As mentioned in Section 3.3.1, these canoes were used by the Baganda for transporting people and goods on the lake. The intention of this order was to make it difficult for the people to sneak back to their homes when the authorities were not looking. A side-effect was to cause the loss of some of the detailed knowledge about how to construct these unique vessels.

8.3   Biosphere Reserves and Related Models

In 1972, the United Nations adopted the World Heritage Convention with the aim of conserving the world's natural and cultural heritage. The United

Nations Educational, Scientific and Cultural Organization (UNESCO) laid some of the groundwork for the convention in 1971 by launching its Man and Biosphere (MAB) programme with the aim of establishing a scientific basis for ways to improve human relationships with the natural world. To support MAB, UNESCO began compiling a list of Biosphere Reserves, which are places considered to provide good examples of what can practically be achieved. A protocol for the recognition of Biosphere Reserves was established, involving, first, the nomination of sites for inclusion on the list by the countries that contain them and, second, checks on the nominated sites to ensure that they meet some quality criteria. Several of those places in which we have worked mentioned in Chapters 10–21 have been nominated or accepted as Biosphere Reserves, for instance the East Usambara Mountains in Tanzania (see Chapter 16).

Within the context of MAB, UNESCO has drawn attention to the importance of Sacred Natural Sites (SNSs) for their strength in linking together natural and cultural heritage. SNSs are places, such as the church forests of Ethiopia, where the natural world has been traditionally protected through its association with religion (Lee and Schaaf, 2003; Verschuuren *et al.*, 2022). Buddhism is a particularly nature-friendly religion, as demonstrated by the examples given in Chapters 14 and 21. One aspect of this nature-friendliness is the exceptionally large numbers of specific types of plants traditionally present in the precincts of Buddhist temples (Wang Xinyang *et al.*, 2020). Two of the most famous of these species are the sacred fig (*Ficus religiosa*) in lowland parts of India and Nepal and the ginkgo tree (*Ginkgo biloba*) in China and Japan. The Buddha is believed to have attained enlightenment under a sacred fig.

Based on experience in conserving Biosphere Reserves, a general spatial model has been proposed to guide their development (Fig. 8.6) (Hadley, 2002). This is based on the division of an area into zones prescribed for different types of human activities. At the centre is a core area in which human activities, apart from monitoring and research, are kept to a minimum. Surrounding this core is a buffer zone, in which more types of human activities are allowed,

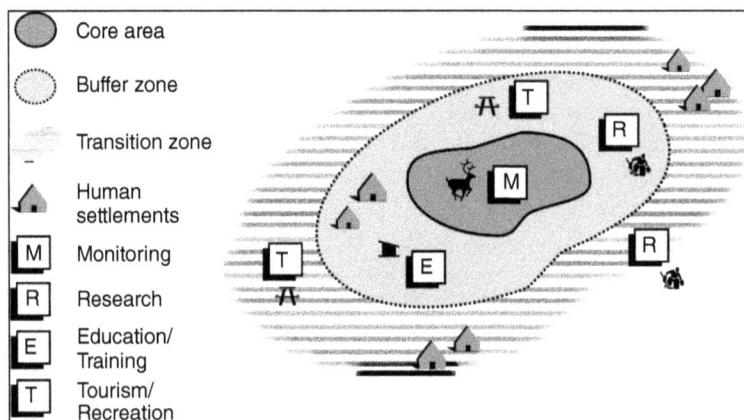

Fig. 8.6. Biosphere Reserve model.

especially ones that are supportive of biodiversity conservation in the core. The outer margin of the buffer zone grades outwards into the general landscape, as typical for the region. This model is not intended to be proscriptive, but rather to be implemented creatively to accommodate local geographical and other restraints. One of its virtues is that, however sketchily conceived, it provides a common platform for people having different primary interests in conservation and development to cooperate with one another to find ways of managing the landscape that are considered to be satisfactory compromises by all concerned.

During the 1980s, the International Union for Conservation of Nature and Natural Resources (IUCN) and the World Wildlife Fund (WWF – now World Wide Fund for Nature) promoted a version of the Biosphere Reserve model known as the Integrated Conservation and Development Project (ICDP). IUCN mounted a project of this type on the East Usambara Mountains (Tanzania) in 1987, as a follow-on to the conservation planning work of A.H. described in Chapter 16 (Stocking and Perkin, 1991). Likewise, in 1989, WWF started to support an ICDP project at Bwindi Impenetrable Forest in Uganda in connection with its declaration as a National Park (see Chapter 18). However, when an evaluation of the success of the ICDP approach was carried out during the 1990s, it was found that many of the projects were struggling to meet both their conservation and their development goals, a major reason being the lack of much connection between the two (McShane, 1999; Wells and McShane, 2004).

One of the common deficiencies of ICDP projects was a failure to pay sufficient attention to the complexities of tenure. Justine Vaz, who ran a project in the early 2000s in Sabah in Malaysia to find how best to protect biodiversity-rich rainforest, has pointed out that local communities are not homogeneous as regards their interests in the forests or in their aspirations for development (Vaz, 2006). There may be some people who are interested in traditional ways and traditional knowledge, but there may be others who are not. In Chapter 9, we recommend the use of applied ethnobotanical research as a way of exploring the complexities of how communities interact with their local natural worlds.

8.4   Plant Conservation and Traditional Farming

Traditional farming contributes to plant conservation in several ways (Tuxill and Nabhan, 2001). One of them is the tendency of the farmers to maintain a wide variety of types and varieties of crops, each valued for a reason. A diversity of crops helps farmers to stave off catastrophe, in the event of some of them failing (Wilson, 2023). Another common traditional conservation practice is for farmers to exchange seed or planting stock of crops with one another, as described for varieties of banana plants in Uganda in Section 2.3.1. The consequence of many generations of farmers following such practices can be seen in the patchwork of small fields that covers much of the Ethiopian Highlands (Fig. 8.7). This is a landscape renowned among agricultural scientists for its high diversity of cereals, both in terms of the number of species (wheat, finger millet, teff, etc.) and the number of landraces (Benin et al., 2007).

Fig. 8.7. Agriculture in the central Ethiopian Highlands. Photo: A.H. (2002).

Research into traditional farming practices in Ethiopia has revealed that the farmers can be keen observers of nature. For instance, some of those who grow sorghum are interested when new types of sorghum, previously unknown to them, appear spontaneously in their fields (Teshome *et al.*, 1999). These farmers are aware that the presence of wild sorghum growing near their domesticated sorghum makes it more likely that new varieties will be discovered and, for this reason, are content to tolerate a certain amount of wild sorghum growing on the margins of their fields.

8.5   From Home Gardens to Botanical Gardens

As mentioned in Section 7.2 in connection with the *lusuku*, home gardens are common features of traditional agricultural systems in many parts of the world. Their characteristics vary, according to the natural environment, local cultures and the interests of individual owners. Traditional doctors everywhere tend to grow an exceptional diversity of species in their home gardens, a reflection of their intrinsic interest in plants and in having supplies of medicines situated conveniently close to their homes. Figure 8.8 is a photograph of a traditional doctor in her home herbal garden in India. The use of pots for the garden, as here, can be common in urban environments. In China, it is traditional for people of all nationalities to include ornamental plants in their home gardens – a practice that, over time, has led to the evolution of a large number of varieties of garden plants (Pei Shengji, 1984). The collection of seed or vegetative stock from these plants was one of the attractions that drew Western plant hunters to China in the 19th and early 20th centuries (see Section 7.5.1).

Fig. 8.8. Traditional doctor in India with her herbal garden. Photo: A.H. (2007).

Modern gardens describing themselves as botanical gardens should have a firm foundation in science, including involvement in activities supportive of biodiversity conservation and the sustainable use of plant resources (Pei Shengji, 1984). Depending on the garden, suitable activities may include raising public awareness, education, holding systematically organised *ex situ* collections of plants, involvement in practical conservation projects with communities, and horticultural experimentation aimed at the domestication of over-harvested wild plants. Since 1987, Botanic Gardens Conservation International (BGCI), a UK-registered charity, has been working with botanical gardens around the world to support their efforts to become more involved in conservation (www. bgci.org, accessed 10 May 2024).

A number of historical roots can be traced for the modern botanical garden. One of them consists of those gardens of medicinal plants attached to medical schools that were created in Europe during the 16th and 17th centuries (see Section 7.3.4). Another consists of those national botanical gardens created around the world from the mid-19th century onwards to further the interests of industrial economies, examples being Kew Botanical Garden in Britain (see Section 5.3.2) and Xishuangbanna Botanical Garden in China (see Chapter 12). Prior to its transfer to government ownership, Kew had its roots in gardens belonging to the British monarchy. Elsewhere in the world, other monarchs over the course of history have also created magnificent gardens for their own pleasure, to impress their subjects and to awe foreign envoys. For example, such a garden called Shan-Ling-Yuan was constructed by the family of the Han emperor during the Qing-Han dynasty in 128 BCE. In China, there is a long tradition of gardens designed to align with the principles of Ancient Chinese philosophy, notably to emphasise the harmony that should exist between people and nature (see Section 4.3.1). Such gardens, which are designed to mimic actual landscapes on a micro scale, incorporate rocks, water features, architectural features and special viewing points, in addition to being places for the display of plants.

8.6   The 'Standard Approach' to Plant Conservation

Systematic efforts to conserve global plant diversity began after the Second World War. An early stimulus came from the discovery by agricultural crop breeders involved in the green revolution that the traditional landraces of crops upon which they depended as sources of genes were becoming scarce (see Section 5.4). As a result, a new organisation, the International Board for Plant Genetic Resources (IBPGR), was founded in 1974 with a remit from the United Nations to coordinate an international response to the conservation of crop genetic diversity. Some field collections of crop landraces already existed, such as those for bananas mentioned in Section 2.3.1. For crops that have seed capable of long-term dormancy, such as cereals, the most efficient *ex situ* way to conserve their genetic diversity is in seed banks, which are collections of dry seeds, normally stored under reduced temperatures to help maintain their long-term viability. The advantages of seed bank storage, compared with conserving crop diversity in field collections (as with bananas), is that little space is needed, the cost is lower and the samples can be handily available to agricultural scientists for their crop-breeding experiments. Initially, IBPGRI (now, incorporated into Bioversity International) concentrated its efforts on conserving the landraces of the major world crops. Its work has subsequently expanded to cover minor crops, culinary herbs, wild crop relatives and the *in situ* conservation of crop genetic diversity on farms.

Systematic efforts to conserve the global diversity of *wild* plant species began after the founding of the Species Survival Commission of the IUCN in 1949. The need to sharpen the focus on those species which urgently need conservation attention led to the use of the colour red – emblematic of urgency or danger in Western culture – being used for lists of endangered species (Red Lists) and books with associated information about them (Red Data books). The first global Red Data Book for Plants was published in 1978 (Lucas and Synge, 1978) and the last in 1997 (Walter and Gillett, 1997), after which IUCN's Species Survival Commission moved its Red List of threatened plant species online (www.iucnredlist.org/, accessed 10 May 2024). Species included in this database are classified into categories of threat according to their degrees of global endangerment (IUCN, 2012). The assessments of threat are made by members of volunteer groups of experts known as Specialist Groups (SGs), some of which are taxonomically based (e.g. Orchid SG), some geographically based (e.g. Mediterranean Plant SG), some use-based (e.g. Medicinal Plants SG) and some cross-cutting (e.g. Invasive Species SG).

An approach, referred to in this book as the 'standard approach', was developed in the 1980s to guide IUCN and WWF's work in plant conservation (Synge, 1988; Given, 1994). It recognises two linked forms of conservation for endangered species – *in situ* conservation and *ex situ* conservation (Fig. 8.9). A third type of conservation, *circa situm* conservation, is sometimes recognised as

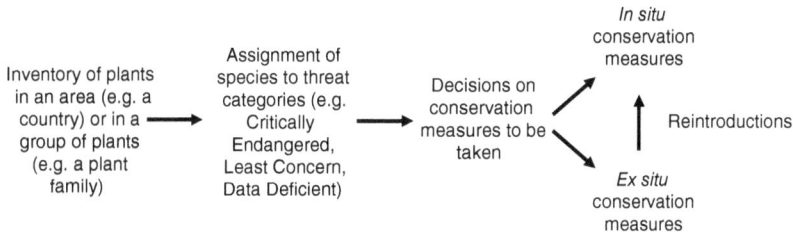

Fig. 8.9. Logic of the 'standard approach' to plant conservation.

a category in between (Hawkes *et al.*, 2001). *In situ* conservation approaches attempt to maintain or restore species in their natural habitats, in principle as populations that are genetically viable. Tools commonly used to support *in situ* conservation are the creation of protected areas and the passing of regulations designed to prevent the unsustainable harvesting of resources of wild plants. *Ex situ* conservation approaches involve maintaining plants, seeds or tissues of species in living collections outside their natural habitats. *Ex situ* collections are typically situated in botanical gardens, arboreta and experimental forestry and agricultural stations. *Circa situm* conservation includes such practices as the retention of 'wild plants' when land is cleared for agriculture and the growing of 'wild plants' in gardens, which is a common practice of traditional doctors. The reintroduction of species from *ex situ* collections back into their natural habitats is encouraged with the standard approach.

If there is a choice, *in situ* conservation is preferable to *ex situ* conservation, because it allows populations of species to continue to evolve in response to environmental change, it might prove possible to save the whole of an endangered population (rather than just a genetic sample) and because the expense is likely to be less (Hamilton and Hamilton, 2006). Another advantage of *in situ* conservation is that it links into efforts to achieve other aspects of sustainable development, such as conserving animals, securing water supplies and maintaining traditional botanical knowledge. Anyway, for many species, *ex situ* conservation is not a practical proposition. For example, it is technically difficult to store the seeds of many species of tropical rainforest trees in seed banks, because they cannot survive the desiccation that is normally required for seeds to remain viable long-term. Many countries with tropical forest lack the resources of money and professional manpower needed for systematic *ex situ* efforts.

The scientists responsible for delimiting plant species and assigning them to Red List categories often have close links with *ex situ* conservation facilities, such as botanical gardens and seed banks. This makes it relatively easy for them to encourage the managers of these facilities to include the species in which they are interested among their collections. In contrast, pursuing *in situ* plant conservation can be a complicated business. It may mean having to work with different sections of society separately (e.g. men and women), a range of community groups and outside business interests (e.g. those interested in different types of plant resources), various cultural organisations (e.g. those connected with education and religion) and a number of government departments (e.g. those connected with agriculture, forestry and local government).

In situ conservation is not a one-off job but requires continuing monitoring and research to ensure that the management prescriptions for plants remain optimal (see Section 9.8).

Creating systemically organised *ex situ* collections of wild plants in seed banks is relatively new (Hawkes *et al.*, 2001). The Royal Botanic Garden, Kew, in Britain, led the way in 2000 when it opened a Millennium Seed Bank on its estate at Wakehurst near London. Kew's seed bank contains seed from a number of other countries, in addition to Britain. Sensitive to its historical role in the British Empire (see Section 5.3.2), it has been careful to be ethical in the agreements it makes with those countries that provide it with seed (see Section 6.11) (Williams, 2022). In China, The Germplasm Bank of Wild Species in Southwest China was opened at the Kunming Institute of Botany in 2005. Figure 8.10 is a photograph of one of its staff preparing seed for storage. By 2020, this germplasm bank contained over 10,000 accessions.

In recognition of the threats of extinction that are facing many species, animals as well as plants, researchers have attempted to identify key localities

Fig. 8.10. Processing seed in The Germplasm Bank of Wild Species in Southwest China. Photo: A.H. (2007).

in the world that, according to their analyses, should be priorities for conservation efforts. Two that have been influential are the Global 200 priority ecoregions of WWF-US and the Biodiversity Hotspots of Conservation International (Olson and Dinerstein, 1998, 2002; Myers *et al.*, 2000). Two proposed for plants only are the Centres of Plant Diversity of WWF-International and IUCN and the Important Plant Areas of Plantlife International (WWF and IUCN, 1994–1997; Anderson, 2002; Plantlife International, 2004). They received international recognition in 2002, when they became the foundations of Targets 4 and 5 of the Global Strategy for Plant Conservation, which are 'At least 10 per cent of each of the world's ecological regions effectively conserved' (T4) and 'At least 75 per cent of the most important areas for plant diversity of each ecological region protected, with effective management in place for conserving plants and their genetic diversity' (T5) (see Section 6.11).

There has been a tendency in the standard approach to plant conservation to concentrate on certain types of vegetation and landscapes, and neglect others. Forest tends to be disproportionately favoured, as has been noted for Mt Kinabalu (see Chapter 17) and Uganda (Osmaston, 1968). In Britain, 'weeds' have been traditionally neglected by conservationists, but, more recently, it has been discovered that some of them are among the most critically endangered species at country level (Storkey and Westbury, 2007).

The case study described in Chapter 17 for Mt Kinabalu in Sabah, Malaysia, revealed inadequacies in the Red List data that were currently available for the species of plants that live there. Analyses revealed major differences between the quantity and quality of data that had become available from the efforts of the more than 200 scientific botanists who have researched its flora during a 147-year period and those of a few villagers using ethnobotanical methods working part-time for 6 years. For example, the scientific botanists collectively discovered 48 species and infraspecific taxa and ten genera of palms on the mountain, while the community ethnobotanists discovered 74 species and infraspecific taxa and 19 genera of palms during a much shorter time. Furthermore, much more information on the distributions, conservation statuses, local names, local uses and local methods of management had become available from the efforts of the community ethnobotanists than from the combined efforts of all the scientific botanists.

Also revealing of the inadequacies of current endangered species data for a locality was a recent analysis of that available for the tropical forest tree species of Uganda (Kalema and Hamilton, 2020). Three separately complied lists of endangered species were available for the analysis. One was IUCN's Red List of threatened plant species, which included 172 of the 452 tropical forest trees known to occur. Another was a recently published book, *Conservation Checklist of the Trees of Uganda* (*TOU*), which included all of the species (Kalema and Beentje, 2012). The third was an assessment of *national* conservation status for 42 species (WCS, 2016). Despite differences in coverage and methodology, it is striking that only three of the species classified as endangered are on all three of the lists. One reason for the discrepancies was that the IUCN and WCS lists contained many species judged to be endangered by the international timber trade, while *TOU* relied heavily on geographical criteria, such as 'small total range size'. In any case, both *TOU* and WCS were in agreement that many of the species are in grave danger of extinction at the national level, because of the severe degrees of forest reduction and degradation that are taking place (see Section 5.6). How far the species are similarly endangered in nearby African countries is difficult to know, given the difficulty of carrying out fieldwork in war-torn eastern DR Congo and Southern Sudan.

9 Principles and Practice of Ecosystem-Based Plant Conservation

Abstract

Ecosystem-based plant conservation (EBPC) is an approach to conservation of plant species and genetic diversity that seeks to take account of all contributions that plants make towards maintaining an ecologically sustainable world. Anyone combining a special interest in plants with wishing to contribute towards sustainable development, and who is prepared to make an effort to see practical results achieved, can contribute. It is place-centred and makes use of the concept of ecosystem services, which are the benefits that people receive from the existence of ecosystems. It seeks to strengthen the links between the promotion of biocultural diversity at the local level and efforts at the global level to achieve an ecologically sustainable Earth. An evidence-based approach is recommended for the development of policies at higher levels in political, socio-economic and cultural systems.

9.1 What is Ecosystem-Based Plant Conservation and Who Are the Plant Conservationists?

Ecosystem-based plant conservation (EBPC) is an approach to conservation of plant species and genetic diversity that seeks to take account of all contributions that plants can make towards maintaining an ecologically sustainable world. The actions that people take in its support may affect plants directly or indirectly. For example, farmers, as people who are in direct contact with plants through their work, can contribute to EBPC through adopting environmentally friendly farming practices, while consumers – people whose influences on plants are by definition indirect – can contribute through purchasing products that contain botanical ingredients that have been produced in environmentally friendly ways. People can contribute towards EBPC in more than one way. For example, farmers are consumers, as well as producers. Making progress with EBPC will be a learning journey for all those who try to contribute. Some idea of our own journeys in this respect (the authors of this book) should be apparent

© Alan Hamilton and Pei Shengji 2024. *History and Future of Plants, Planet and People*
(Alan Hamilton and Pei Shengji)
DOI: 10.1079/9781789248944.0009

from the accounts of our lives in the Preface and the programmes and case studies presented in Chapters 10–21. We hope that others, armed with their own experiences, will be able to make improvements on what we offer here.

We propose that a practical definition of a plant conservationist is someone who combines a special interest in plants with a desire to contribute to sustainable development and who is prepared to make an effort to see practical results achieved. All people have something to offer, but what they are able to do practically will depend on their circumstances. Those with limited public influence may be able to achieve something locally, while those who occupy powerful positions within society have the opportunities to directly influence the contributions of many other people. Scientists have special responsibilities, related to the strength of the scientific method for investigating reality and their privileged access to the stores of information that have been accumulated through scientific research.

An ecosystem-based approach to conservation is not a new idea. It was stipulated under the Convention on Biological Diversity (CBD) as the primary framework to be used for its implementation (see Section 6.11) (CBD, 1992). The precise meaning of an ecosystem approach *sensu* CBD is debateable, but, broadly, it encourages conservationists to think in terms of systems and to bear in mind that people are major elements of ecosystems in nearly all cases. Compared with other taxonomic groups of organisms, an ecosystem-based approach is particularly appropriate for plants, because of their keystone roles in the functioning of both ecosystems and economies, and the many ways in which they are culturally connected to people (Hamilton, 2013). Plants form a substantial part of the scenery that forms the backdrop to people's lives. Plants, as a taxonomic group, are relatively easy for people to influence in favour of conservation and sustainable development, compared to some other taxonomic groups, such as insects and fungi. Their physical sizes are on a scale relatively similar to that of the human and their fixedness in place avoids the complications that arise with trying to manage mobile organisms.

A set of 12 guiding principles has been drawn up under the CBD to guide conservationists in their adoption of an ecosystem approach (Table 9.1). However, progress in developing the approach has been slow. For example, none of the 16 targets of the Global Strategy for Plant Conservation (GSPC) agreed in 2002 (see Section 6.11) mentions the concept of ecosystem services, a fundamental aspect of the ecosystem approach.

9.2 The Concept of Ecosystem Services

The concept of ecosystem services is helpful for analysing how people relate to plants and for implementing EBPC. Defined as the benefits that people receive from the existence of ecosystems, the four categories generally recognised are shown in the left-hand column of Table 9.2 (Constanza *et al.*, 1997; Alcamo and Bennett, 2003).

Provisioning ecosystem services
These refer to the material products that people receive directly or indirectly from plants. Examples of products received directly from plants are vegetables

Table 9.1. Principles of the ecosystem approach to conservation, as drawn up for the Convention on Biological Diversity. Source: CBD (2004).

1.	The objectives of management of land, water and living resources are a matter of societal choice
2.	Management should be decentralised to the lowest appropriate level
3.	Ecosystem managers should consider the effects (actual or potential) of their activities on adjacent and other ecosystems
4.	There is usually a need to understand and manage the ecosystem in an economic context. Management should reduce market distortions that adversely affect biological diversity, align incentives to promote biodiversity conservation and sustainable use, and internalise as feasible costs and benefits in the given ecosystem
5.	Conservation of ecosystem structure and functioning, in order to maintain ecosystem services, should be a priority target of the ecosystem approach
6.	Ecosystems must be managed within the limits of their functioning
7.	The ecosystem approach should be undertaken at the appropriate spatial and temporal scales
8.	Objectives for ecosystem management should be set for the longer term
9.	Management must recognise that change is inevitable
10.	The ecosystem approach should seek the appropriate balance between, and integration of, conservation and use of biological diversity
11.	The ecosystem approach should consider all forms of relevant information, including scientific and indigenous and local knowledge, innovations and practices
12.	The ecosystems approach should involve all relevant sectors of society and scientific disciplines

Table 9.2. The four categories of ecosystem services and examples of subcategories. The subcategories have been chosen because of their connectedness to plants.

Categories	Subcategories with connections to plants
Provisioning	Agricultural produce
	Produce from wild plants
Regulating	Delivery of fresh water supplies
	Regulation of water flows
	Climate moderation
	Soil formation and stabilisation
	Pollination
	Pest and disease control
Cultural	Worldviews
	Child-raising and education
	Sectoral interests (e.g. an interest in plant conservation, involvement in farming, engaging in artistic endeavours that alert people to conservation, etc.)
Supporting	These are the basic properties of ecosystems that enable the delivery of the more tangible benefits received from the other three categories. Those having connections with plants include: (i) their abilities, as primary producers, to manufacture organic from inorganic substances; (ii) their roles as formers of habitats – properties related to their size and fixedness in place; (iii) their contributions to the recycling of nutrients

and firewood; examples of products received indirectly from plants are milk and hides from cows, and honey from bees. If the full range of products originating from plants is taken into account, then much of the cover of plants on the planet – from carefully managed land to wild places – is today providing products that people use.

Because people obtain so many products directly or indirectly from plants, integrating efforts to conserve the diversity of plants into the productive landscape is one of the major challenges that plant conservationists face (Hannah and Hansen, 2005; Ashley et al., 2006; Manning et al., 2009). Facets of this challenge cover gardening, agriculture, forestry, pasture management and the harvesting of produce from wild plants. As of now, the challenge of making the production of products from plants ecologically sustainable is huge. Industrial agriculture is one of the greatest threats to the conservation of biodiversity (see Section 5.4.2). Extensive areas of savannah and other rangelands are being ecologically degraded (Perrings, 2000; He Fangliang, 2009). The harvesting of wild plant produce is often undertaken in ecologically unsustainable ways. The harvesting of wood for fuel (fuelwood and charcoal) is an example of the latter. An estimated one-quarter to one-third of all the wood that is being harvested worldwide for wood fuel is being harvested unsustainably (FAO, 2016, 2017). In Uganda, where 90% of the people depend on plants for fuel, 90% of the trees that are felled for products are felled to supply fuel (Kabogozza, 2011).

Regulating ecosystem services
The nature of the cover of plants on the land, whether in 'wild' places, farmland or urban areas, has a strong regulatory influence on the functioning of ecosystems. Ecosystem services that are influenced by the nature of the plant cover include the delivery of fresh water supplies, the regulation of water flows, climate moderation, soil formation, soil stabilisation, pollination, and pest and disease control. Plants moderate the climate on several geographical scales (IPCC, 2007; Nicholson et al., 2009; Haslett et al., 2010). Locally, trees provide people with shade when the sun is blazing hot and shelter when it rains. At intermediate scales, the type of plant cover on the land has a strong influence on flows of energy and water through ecosystems. The recycling of water, falling as precipitation back into the atmosphere through evapotranspiration, can carry water, originally evaporated from the oceans, far into the hinterlands of continents (Eltahir and Bras, 1994).

Figure 9.1 is a sketch showing how energy and water flow through (on left) an intact tropical forest and (on right) an overgrazed grassland that has replaced the forest following its clearance by people. The widths of the arrows are approximately equivalent to the relative quantities of energy and water moving through the pathways, based on typical values for Uganda. Notice the greater photosynthesis and evapotranspiration, and reduced reflection and overland runoff, with the rainforest. Conversion of rainforest to overgrazed

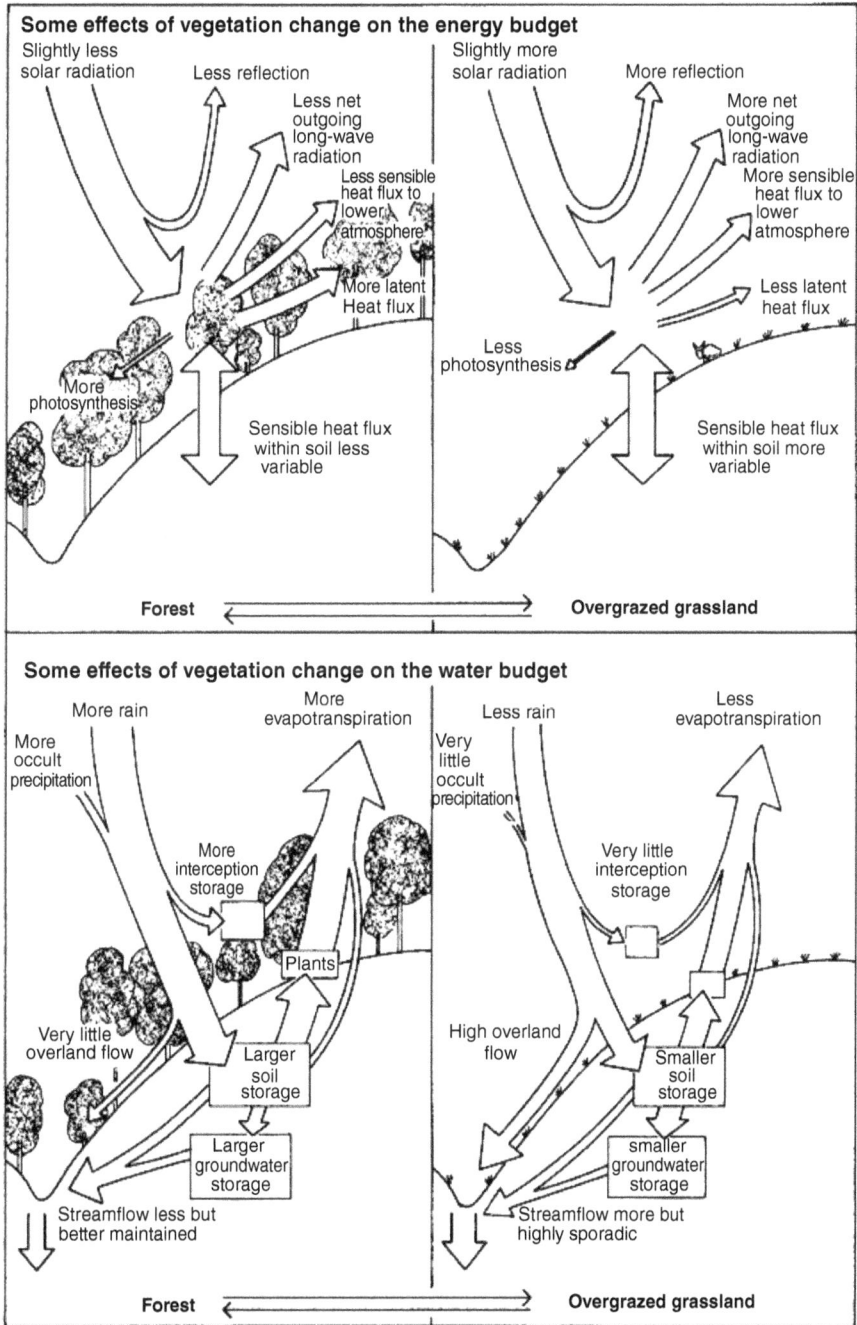

Fig. 9.1. Energy and water flows in a tropical rainforest (left) and replacement grassland (right). Reproduced from: Hamilton (1984).

grassland is a common ecosystem transition in Uganda today, as in other parts of the tropics. The environmental consequences of such transitions extend beyond the immediate sites where they are experienced. At the regional level, they result in less favourable agricultural climates, decreased security of water supplies and increased risks of downstream flooding. Destruction of tropical forest increases the carbon dioxide load in the atmosphere and contributes to global warming.

We wish to stress the value of the 'water argument' for plant conserva-tionists, an argument that derives its power from the essentiality for human beings of having regular access to supplies of fresh water. Its importance for plant conservation is demonstrated by our case studies for China, Mexico, Pakistan and Tanzania (see Chapters 12, 16, 19 and 21). Scientific hydrolo-gists have sometimes struggled to come up with clear advice on which types of vegetation and which types of land use are best for delivering water-related ecosystem services. They have been weighed down by the complexities of hydrological systems and by the difficulties of measuring how water is stored and how it flows in real-world circumstances (McCulloch and Robinson, 1993; Blackie and Robinson, 2007). This is a field in which traditional knowl-edge can be extremely helpful (see Section 8.1) (Londono et al., 2016). We suggest that a useful rule-of-thumb for land managers who wish to combine the conservation of biodiversity with the delivery of water-related ecosystem services is to retain or restore the vegetation cover to as close to its natural state as possible.

Cultural ecosystem services
The three subcategories of cultural ecosystem services included in Table 9.2 were selected because of their importance for determining the quality of the plant-related services that are delivered to people by ecosystems. The impor-tance of worldviews in influencing how people think and behave is discussed in Chapter 4 and the importance of child-raising and education in Section 7.1. There are many sectoral interests that have a bearing on the availability and quality of provisioning and regulating ecosystem services. The three chosen for inclusion on the table are: (i) an interest in plant conservation itself; (ii) an occupation that is directly concerned with managing plants (farming); and (iii) a set of cultural activities that are only indirectly connected to plants (artistic endeavours).

9.3   A Place-Centred Approach

Places are at the centre of attention with EBPC, a place being defined for this purpose as any area of the Earth's surface relevant to the practical manage-ment of plants. Places may be of different sizes, nested within one another or overlapping. A farm provides an example of how management areas can be nested. A farm is a management area in its own right, while, within it, it may contain several spatial units, each with its own management regime. For example, on a typical British farm, the spatial units are liable to be the individual fields, the home garden, hedgerows and patches of woodland. Individual farms are connected to wider systems which also have an influence on how they are managed. In Britain, these include government regulations, market forces, local cultural expectations and the interests of those people who actually own the farms. With respect to the latter, in Britain, it is common for farmers not to be the owners of the farms that they farm. About a third of the land is thought to be owned by the traditional aristocracy and an increasing proportion by agrifood businesses (see Section 5.4.2).

One advantage of place being at the centre of attention is that it acts as a point of convergence for the aspirations of all sections of society that see themselves as having a stake in its future. Initially, such sections of society may be unaware of one another's existence or, if they are aware, dismissive or uninterested. Several of our case studies demonstrate the value of EBPC-thinking and follow-up actions for establishing arrangements that bring together the interests of different sections of society and raise the prospects of them being able to live amicably together in the future. For instance, at Bwindi Forest in Uganda, initial hostility between local communities and a newly declared National Park became transformed when ethnobotanical research revealed the types of commodities in the forest in which the communities were interested, followed by a process of rural development facilitated by the humanitarian organisation CARE (see Chapter 18). The result was the development of a management system that better accommodated the interests of both the local communities and the National Park and reduced the levels of suspicion and hostility.

EBPC is applicable to everywhere where plants naturally grow or, if denuded of plants through human activities in the past, where they have the potential to naturally do so. With EBPC, there is less reliance on protected areas than with the standard plant conservation approach outlined in Section 8.6. This is useful for the conservation of plant diversity, because there are many endangered plant species that are not found in protected areas, the populations of those that are present within them can be small and of doubtful long-term viability, and anthropogenic climate change is changing the potential ranges of species. The implication of the latter is that networks of protected area designed to have an optimal configuration at any one time are liable to become suboptimal in the future.

The entire landscape of the Earth can be considered conceptually to consist of individual local places, at any one of which EBPC can be pursued. An ideal pattern of plants in the landscape can be envisaged, balancing the delivery of all types of plant-related contributions to conservation and sustainable development, set one against the other, while making provision for the use of some areas of land for purposes that contribute little or nothing to these purposes, as with completely built-over areas and much industrial agriculture. This thinking is resonant of that of Patrick Geddes (1854–1932), a pioneer British town planner and one-time professor of botany at the University of Dundee in Scotland, who devised the slogan 'Think Globally, Act Locally' (Geddes, 1915).

9.4   An Approach that Recognises Linkages between Biological and Cultural Diversity

Terrestrial plant diversity is geographically related, the product of the ways in which plants have evolved and become distributed over the years and the histories of their environments (see Section 2.1). To a large extent, success in conserving global plant diversity will depend on the extent to which elements of the natural geographical patterns of plants can be retained, these being the patterns that plant species would have had, had the human species never existed.

Recognition of the threats that people pose to the survival of biodiversity underscores the protective thinking apparent in the standard approach to plant conservation described in Section 8.6. This approach places emphases on protected areas and *ex situ* conservation, the first to keep threats at bay and the second to provide sanctuaries where species can be kept out of harm's way. With EBPC, conservationists are encouraged to counterbalance such protective measures by identifying aspects of human societies and cultures that are *positively* aligned with the conservation of plant diversity (Maffi, 2007). For example, every place is seen as requiring people who are interested in its particular plant diversity and prepared to make an effort to see it retained. Our experience (the authors of this book) is that such people can be found almost everywhere and at all levels of society.

9.5   The Value of an Ecosystem Services Framework for Planning, Implementing, Monitoring and Evaluation

An ecosystem services framework is a table listing ecosystem services on the vertical axis and 'place' on the horizontal, the latter divided into 'conditions at locality' and 'wider systems influencing locality' (Table 9.3) (Hamilton et al., 2017). Such frameworks can be useful at different stages of EBPC projects – planning, implementation, monitoring and evaluation. At the planning stage, they can help ensure that significant ecosystem services have not been overlooked and that the individuals and institutions associated with the various services have been identified. The precise details of how ecosystem services frameworks are constructed will depend on the circumstances. The example shown in Table 9.3 is a general one, designed to draw attention to conditions that are favourable both to the conservation of plant diversity and to the delivery of other plant-related contributions to sustainable development. Supporting ecosystem services have been omitted. They are the same as those in Table 9.2. An example of an ecosystem services framework filled up with data for an actual project is given in Chapter 21. It shows the initial conditions at the project site, the actions taken by the project and the results of these actions.

Local people – defined as those who do (or could) influence the plants directly – are limited in what they can do on their own accounts, due to constraints

Table 9.3. Ecosystem services framework. Adapted from: Hamilton *et al.* (2017).

		Conditions favourable to the conservation of plant diversity and to the delivery of other plant-related contributions to sustainable development	
Ecosystem services (categories and examples of subcategories)		Conditions at locality	Wider systems influencing locality (the enabling environment)
Provisioning	Agricultural produce Produce from wild plants	Land used for the supply of biological products managed in ways that are ecologically sustainable	Policies of government and industry and, also, the decisions of consumers, favour environmentally friendly production systems
Regulating	Delivery of fresh water supplies Regulation of water flows Climate moderation Soil formation and stabilisation Pollination Pest and disease control	Landscape managed to maintain vegetation types that deliver water supplies, protect against floods, moderate the climate (on local to global scales) and contribute to the formation and stabilisation of soils Diversity of plant species maintained in production areas and across the landscape, especially species known to deliver these services	Government and industry policies promote delivery of these services in environmentally friendly ways
Cultural	Worldviews	Local worldviews favourable to conservation of plant diversity and sustainable development; supportive local institutions	Prevalent religious, philosophical and ideological belief-systems supportive of environmentally friendly behaviour
	Child-raising and education	Local child-raising practices and educational opportunities favour acquisition of attitudes, knowledge and skills favourable to plant conservation and sustainable development; relevant information available locally	National policies supportive of acquisition of attitudes, knowledge and skills favourable to the acquisition of environmentally friendly behaviour
	Sectoral interests	Sectoral interests favourable to plant conservation and sustainable use exist; may include individuals and organisations concerned with natural history, gardening, local heritage, craft-making, the arts, etc.	Many types of sectoral interests can potentially be supportive of plant conservation – among them, various scientific specialities

imposed on them by wider political, socio-economic and cultural systems. For example, the laws of the land determine what is possible legally, the socio-economic system of a country the types of economic opportunities available locally, and a country's ideological or religious norms what is acceptable socially. It is because wider political, socio-economic and cultural systems can be so influential at the local level that those plant conservationists who are potentially in positions to influence them should do so when opportunities arise. Depending on their fields of influence, this might involve (as examples) making recommendations for changes in the law, revisions to the ethical codes of businesses or suggestions for changes in educational curricula in schools.

For the purposes of analysis and practical action, people can be considered to be divided into two groups for the purpose of pursuing EBPC. One group is composed of those who do (or could) influence the plants of the place directly; the other is composed of those who do (or could) influence the plants, but only indirectly. This breakdown is included in the filled-in ecosystem framework for an actual conservation project included in Chapter 21 (Table 21.1). In general, farmers, gardeners, forestry workers, pastoralists and harvesters of wild plant produce are among those in the first group, while lawmakers, big businesspeople, consumers, the headquarters staff of government agencies and lab-based scientists are among those in the second. Virtually everyone is a member of the first group, since virtually everyone lives somewhere. However, the psychological strengths of the ties that bind people to their local natural environments can vary enormously. For indigenous people, a rootedness in place is basic to their identity (see Section 4.2), while, for the more mobile members of urban-industrial societies, the details of the natural environments of the places where they happen to live at any one time may be of only passing interest (see Section 7.1).

9.6   The Usefulness of Applied Ethnobotany

We suggest that field teams involved in plant conservation will often benefit if they include at least one ethnobotanist among their number (Hamilton *et al.*, 2006; Pei Shengji *et al.*, 2009b). Ethnobotany is the interdisciplinary subject concerned with the relationships between people and plants and, as such, straddles the boundary between the sciences and humanities. On its margins, ethnobotany grades into a number of other interdisciplinary subjects whose names begin with the prefix 'ethno-', such as ethnoecology, ethnomedicine, ethnopedology and ethnozoology. In the People and Plants Initiative (described in Chapter 10), we called the application of ethnobotany to the conservation of biodiversity and the sustainable use of plant resources 'applied ethnobotany'. This can be seen as a new historical phase in the development of ethnobotany, a subject that had evolved from an early purely descriptive phase, involving making lists of plants used by communities, to an analytical stage, involving trying to establish the determinants of these uses, such as how they fit into the geographical settings of the people and their cultural beliefs (Ford, 1978;

Pei Shengji, 2003). As a discipline applied to conservation and sustainable development, the connections between ethnobotany, as a research subject, and the management of plants can involve making uses of the bundles of rural development techniques known as participatory rural appraisal (PRA). The key elements of PRA include having interdisciplinary project teams and community involvement in setting and implementing work plans. The work involves drawing on both traditional and scientific knowledge and skills.

Many aspects of the relationships between people and plants can be investigated ethnobotanically. Among them are: the names given to the plants and how they are classified; the distribution of botanical knowledge within communities and how this knowledge is transmitted between people; the values and uses of plants; the places where plants are found and local knowledge about their abundance, ecology, ownership and management; whether products from plants are exchanged or sold and, if so, in what ways; the influences on plants of parties external to the communities, such as government departments and plant-based businesses; and the beliefs, traditions and stories attached to plants and their connections with traditional conservation (see Section 8.1).

The work of the ethnobotanist is challenging. It involves gaining some familiarity with the knowledge and ways of thinking associated with a number of academic disciplines, two of the most central of which are whole-plant botany and social anthropology (see Section 7.4.3). Even so, a high level of formal education is not necessarily required, provided that there is an openness to learning and a preparedness to reach out across cultural boundaries (Shinwari and Khan, 2001). Serving as an apprentice to an experienced ethnobotanist can be very helpful. Ethnobotanists require interpersonal skills. They must take people coming from cultural backgrounds different from their own seriously, listening to what they have to say (Tuxill and Nabhan, 2001).

Ethnobotanists have privileged access to local, traditional and indigenous knowledge, which places them in positions of trust as regards how they make use of it (see Section 6.11). Fortunately for them, the International Society of Ethnobiology (ISE) has provided advice (ISE, 2006) and further guidance can be found in one of the publications of the People and Plants Initiative (Laird, 2002). ISE stresses that ethnobotanists should not become complacent in their ethical thinking and urges them to continually re-evaluate their ethical understandings as they continue with their work. Sarah Laird has pointed out that, to be successful, relationships of any kind, whether personal or business, are ultimately dependent on trust. No amount of legal detail will be able to cover all eventualities or substitute for the flexibility required when trying to achieve results in the real world. Ethical questions most commonly arise in ethnobotany in relation to medicinal and aromatic plants and the related interests of some pharmaceutical and botanical companies in acquiring indigenous knowledge to develop new products. The responses on this subject in projects in which we have had some engagement can be seen in Chapters 14 (rubber in China), 13 (fleagrass in China), 17 (Mt Kinabalu in Malaysia), 20 (Dolpo in Nepal) and 21 (Ludian and the Meili Snow Mountain Range in China).

9.7   Stages in an Applied Ethnobotany Project

There is much still to learn how to best apply ethnobotany to conservation of biodiversity and the sustainable use of plant resources. We (the authors of this book) have been involved in only limited arenas of conservation, P.S.'s mainly with the ethnic minorities of Yunnan and A.H.'s mainly in the contexts of WWF's interest in protected areas and Plantlife's in medicinal plants. The suggestions given in the rest of this chapter should be seen in the light of this limited experience. We have also drawn on recommendations for approaches and methodologies in ethnobotany in *Ethnobotany, A Methods Manual* (Martin, 1994) and *People, Plants and Protected Areas* (Tuxill and Nabhan, 2001) (see Chapter 10). Examination of the ways that ethnobotany has been applied in plant conservation and sustainable development projects in China (Chapters 12–14, 21) and in other parts of the world (Chapters 17–20) should give some idea of the present state of development of the subject.

Before embarking on fieldwork, it is recommended that ethnobotanists make contact with institutions and organisations with advice to offer and with whom ongoing collaboration might prove useful as the work continues. Among them could be government departments, academic institutions and herbaria, businesses, trade unions and non-governmental organisations involved with the communities. In parallel with this, studying the scientific literature to find out what is already known about the people and plants of a locality and the relationships between the two will provide a valuable foundation for the work. A knowledge of the histories of people and plants and the relationships between them, as they relate to a locality, should be helpful for understanding the ways in which political, economic, social and cultural systems are shaping the relationships between the two today.

During this introductory phase, including when initial introductions are made to communities, it is important not to raise unrealistic expectations of what a project might be able to achieve and to be aware that initially held views about the conservation and development issues present at a site may be misleading. Local people may not share the same perspectives on conservation and development as outsiders. For example, it can be that the concepts associated with the standard approach to plant conservation, as described in Section 8.6, have little meaning to the local people. This can apply, for instance, to the distinctions made in the standard approach between wild and non-wild plants, protected and non-protected areas, *ex situ* and *in situ* conservation, and botanical gardens and gardens of other types. It is common for the species of plants that have caught the attention of compilers of Red Lists of endangered species not to be the ones that field projects will need to concentrate on to make progress on the ground. These key species are more likely to be those that are of special interest to the communities.

It is a good idea, if circumstances permit, to slip gently into the fieldwork through steadily building up relationships with local people, discovering which types of plants are of interest to them and learning about how they think and talk about their botanical interests. One approach that has been found helpful in trying to gain an initial understanding of local people/plant relationships is to 'wander about in the woods' with knowledgeable local people (Tuxill and Nabhan, 2001). Janice Alcorn has suggested that the simple question, 'What good is this plant?', can be a key to opening up dialogue and revealing which issues that are of local interest and concern (Alcorn, 1995). A graphic depiction of this is given on Fig. 9.2, which shows connections between the questions that might be asked of knowledgeable local people and related subject areas in academia and the professions. More formal scoping surveys during the initial stages of projects are useful for narrowing down the focus of the work on to subjects that require more concentrated attention, as demonstrated by several of the projects described in Chapters 12–21. They should cover the full range of vegetation and land-use types present in the project area, from closely cultivated agricultural land to places that seem to be totally wild. Scoping surveys should help in understanding the influences of outside systems on local people/plant relationships, for example those involving government departments, businesses and cultural organisations.

9.8 Project Teams and Two Interconnected Cycles of Action and Reflection

The establishment of a core group of people who are prepared to dedicate time to the project is useful. Preferably, they should cover those fields of special

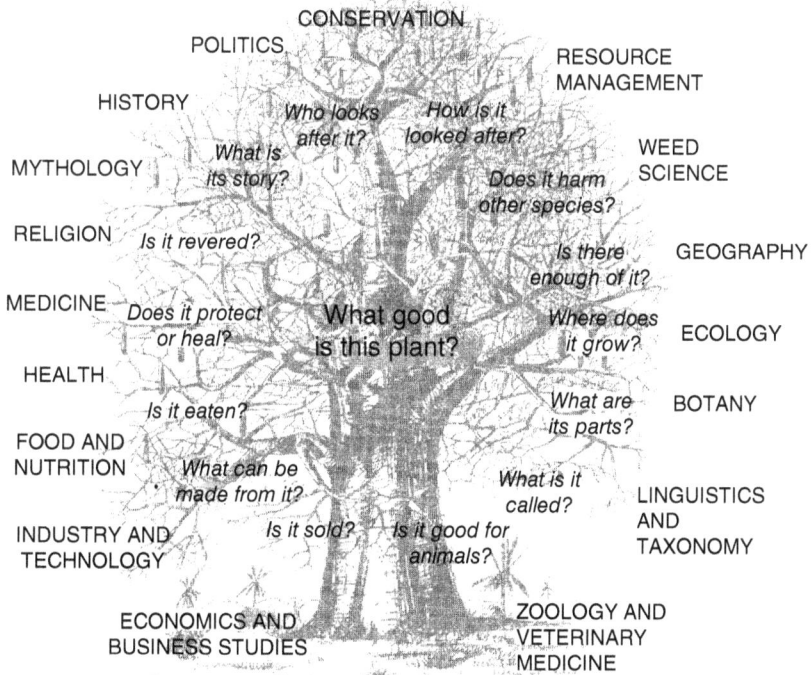

Fig. 9.2. What good is this plant? From: Hamilton and Hamilton (2006).

interest that are revealed to be of importance as the work proceeds. Experience has shown that including local 'key knowledge-holders' in core project groups can be critical to success (Tuxill and Nabhan, 2001). These are community members who are recognised locally as being particularly knowledgeable about plants. Very often, they have received little in the way of formal education, but instead have acquired their expertise through their own interest in plants, their curiosity and long practical experience.

Research into the relationships between people and plants undertaken for the purposes of conservation and sustainable development is not a one-off matter. It is to be expected that new questions for research will arise as gaps or errors in existing knowledge become apparent and as the external environment evolves. Similarly, it is to be expected that management systems for plants will likely require adjustments from time to time in the light of new knowledge and environmental developments. The relevant types of research and management are known as 'action research' and 'adaptive management', respectively. They can be depicted graphically as cycles of activities and reflection, as shown in Figs 9.3 and 9.4. The two systems are linked through the recommendations

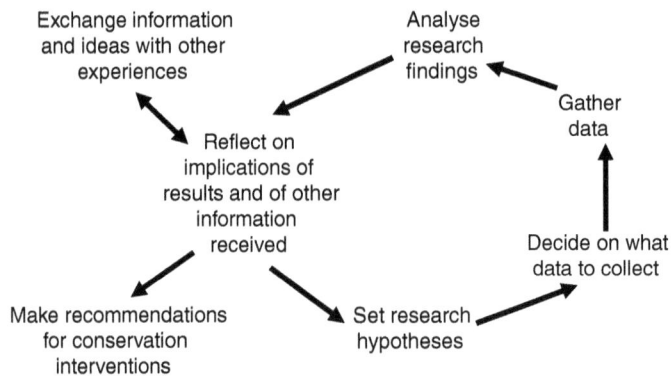

Fig. 9.3. Activities in the action research cycle.

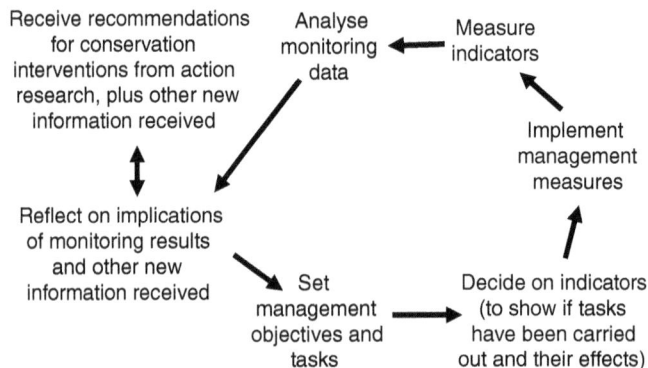

Fig. 9.4. Activities in the adaptive management cycle.

for management generated from the action research and the recommendations for research generated by questions arising during management.

Both action research and adaptive management require the close engage-ment of communities at all stages – setting the research or monitoring questions, collecting and analysing the data, reflecting on the implications of the results, and drawing up recommendations for further research or for conservation and development actions (Sheil et al., 2006). In the past, it has been common in research projects concerned with the management of natural resources for the research to be undertaken by scientific specialists working to a large extent in isolation, with little involvement of communities. The recommen-dations of these specialists regarding management may be presented to the communities in a near-final state to allow some slight adjustments. This can prove confrontational, because it can give the impression to communities that external parties have already decided what is going to happen, even though it is they whose lives will be most disrupted by the decisions that have been reached.

In Section 7.5.1, we describe what happened in China during the encoun-ter between Chinese scholarly botany and the scientific botany that became introduced from Europe from the beginning of the 19th century. Each of these types of botany was associated with particular ways of learning for its practi-tioners and each was associated with a body of accumulated knowledge that had been built up over time. A vast amount of information about plants had become recorded in writing by Chinese scholars for 2000 years. Scientific botany had a vibrancy associated with the scientific way of advancing human knowledge, including its openness in principle to considering new hypotheses (see Section 7.4.1). In an analogous way, making progress in a plant conser-vation project at a locality can benefit from the knowledge and skills of both communities and scientists. Communities can be expected to have much more detailed knowledge about certain aspects of the plants in their neighbourhoods than scientists will ever be able to discover working on their own. On the other hand, scientists have access to the vast store of scientific knowledge that has been built up over the years and to the research techniques that have been developed by scientists.

In the case studies of botany and conservation in Chapters 13–21, there are many examples of how the various social groups associated with particular localities differ in their interests and knowledge of plants and of how making progress in developing management systems for conservation and sustainable use depends on taking account of this cultural diversity. As examples: at Mt Kinabalu in Malaysia, ethnobotanical research revealed that the communities collectively possessed much more knowledge about the types of palms pres-ent on the mountain and about their distributions and conservation statuses than was possessed by botanical scientists (see Chapter 17); at Bwindi Forest in Uganda, resolution of the tensions that had been generated by the creation of a new National Park only started to become possible after ethnobotanical research had revealed the types of plants in the park in which the people were interested (see Chapter 18); at Ayubia National Park in Pakistan, the unwillingness of the authorities to acknowledge the need to accommodate in the management system the interests and knowledge of women about plants resulted in their failure to seize on an opportunity that could have contributed towards reducing the rate of forest loss on the foothills of the Himalayas (see Chapter 19).

9.9 Value of an Evidence-Based Approach for Policy Development

Ecosystems are complex and the outcomes of interventions made for conservation purposes can be difficult to predict (Lester *et al.*, 2010). For these reasons, we suggest the adoption of an evidence-based approach in EBPC for the identification and promotion of best practice (Pei Shengji *et al.*, 2010; Hamilton, 2011, 2012; Hamilton *et al.*, 2017). This suggestion is made in the light of the success that such an approach has enjoyed in the practice of medicine in recent years – medicine being a subject analogous to conservation, in that both are scientifically influenced arts dealing with complex systems and in both of which practical results are required. An evidence-based approach involves periodic reviews of the evidence relating to the success or failure of attempts to deal with particular issues in practice, reviews which can then be followed up by the dissemination of recommendations on best practice for wider adoption or which, alternatively, can be treated as scientific hypotheses for further testing. Both quantitative and qualitative methods can be used to judge success, as appropriate to the case. If applied properly, an evidence-based approach should not lead to 'cookbook' solutions, but rather to the integration of the experience of the individual practitioner with the best external evidence, just as in medicine (Sackett *et al.*, 1996).

An early use of an evidence-based approach in conservation was the testing by WWF of a conservation model known as the Integrated Conservation and Development Project (ICDP) that was popular in the 1980s (see Section 8.3) (McShane, 1999; Wells and McShane, 2004). We ourselves have used an evidence-based approach to evaluate one of the international conservation programmes in which we have both been involved, the Medicinal Plants Conservation Initiative, the question asked being 'How can the conservation of biodiversity and sustainable development best be achieved, based upon the interests of communities in medicinal plants?' (see Chapter 11) (Hamilton, 2008; Pei Shengji *et al.*, 2010).

10 The People and Plants Initiative (PPI, 1992–2005)

WRITTEN IN THE FIRST PERSON BY ALAN HAMILTON

Abstract

The People and Plants Initiative (PPI) was a programme of global capacity-building in ethnobotany, as applied to conservation of biodiversity and sustainability in the use of plant resources. The three partners in the programme, which ran from 1992 to 2005, were WWF, UNESCO and the Royal Botanic Gardens, Kew. The areas of capacity-building covered included the mentored training of national ethnobotanists in selected countries in Africa, Asia and the Pacific, training courses attended altogether by hundreds of people, the founding of national ethnobotanical societies, the development of curricula in ethnobotany in universities and colleges, and the production of manuals and other materials, which were made freely available to a targeted list of several hundred people in the Global South. A new non-governmental organisation, People and Plants International, was founded by the initiative and this is continuing to take the subject forward at the international level.

The Plants and People Initiative (PPI) was a programme launched by World Wildlife Fund (WWF – now World Wide Fund for Nature) in 1992 with the aim of increasing global capacity in applied ethnobotany, the scientific discipline concerned with the applications of ethnobotany to the conservation of biodiversity and the sustainable use of plant resources. Three people introduced me to the subject of ethnobotany, which is a word that I had not consciously heard before joining WWF in 1989. One of them was Richard Schultes, a member of an international advisory panel that had been established by the International Union for Conservation of Nature and Natural Resources (IUCN) and WWF to guide their work in plant conservation (Synge, 1988). I first met Schultes at one of the advisory panel's meetings and, realising that he had much useful to tell me, decided to visit him at his place of work at Harvard University in the United States to benefit more from his advice. Schultes (1915–2001), who has been called the father of modern ethnobotany, had gained his scientific reputation from his research into the botanical knowledge of the indigenous people of Latin America (Schultes and von Reis, 1995; Bussman, 2012; Soejarto, 2012). His work had included studying the hallucinatory drugs and arrow poisons prepared by the indigenous people of Colombia (Schultes and Rauffauf, 1990). In 1941, he was ordered by the American Government to revive the long-dormant wild rubber-collecting industry of Amazonia to boost the Allied war effort in the Second World War (see Section 5.3.5).

The other two people who introduced me to ethnobotany were Tony Cunningham and Gary Martin. Tony is a South African botanist who was preparing a report for WWF entitled *African Medicinal Plants: Setting Priorities at the Interface between Conservation and Primary Health Care* (Cunningham, 1993) and Gary Martin is an American ethnobotanist, who had worked in Mexico and Bolivia and was preparing a manual for WWF on basic techniques in ethnobotany (Martin, 1994). Later, once PPI had got underway, Tony Cunningham and Gary Martin became the key people responsible for guiding its activities in Africa and Southeast Asia-Pacific, respectively. Their skill in knowing how to work constructively with rural communities on matters of conservation and related development was an eyeopener for me.

For my part, by the time that I joined WWF in 1989, I was aware from my experiences in botany and conservation in Uganda, Mexico and Tanzania that insufficient attention was being paid by the governments of those countries to involving local communities constructively in forest management (see Chapters 15 and 16). After I had joined WWF and was editing manuscripts received for inclusion in the Centres of Plant Diversity project (WWF and IUCN, 1994–1997), it became obvious that there was an urgency about finding ways to better involve communities in conservation everywhere (Hamilton, 1997). It seemed to me that launching a programme to increase global capacity in applied ethnobotany would be a useful thing for WWF to do.

Tony Cunningham, Gary Martin and I came up with the outline of the structure of PPI during an afternoon spent together in my garden in Godalming in Britain in 1991. From the start, we targeted the tropics and subtropics for special attention, on the basis that this is where plant diversity is most concentrated, where communities are most dependent on locally growing plant resources and where resources for conservation are most limited. Ethnobotany can be used with different purposes in mind. In PPI's case, the applications of interest were those that promoted conservation of biodiversity and sustainability in the use of plant resources (see Section 9.6). We identified the need for capacity-building in applied ethnobotany at three levels of society – the local one, at which people come into direct contact with plants; the national one, because of the centrality of the state in modern systems of governance; and the global one, in recognition of the interconnectedness of the modern world and the value of sharing knowledge across national boundaries. The cover of one of the brochures produced for PPI is shown in Fig. 10.1. It shows Amchi Tangual Lama of Pungmo Village, Dolpo, Nepal, carrying *bonmak* (*Aconitum lethale*), the root of which is used in Amchi Medicine (after detoxification) to treat fever (see Chapter 20).

We decided to divide the work of PPI into three Regional Projects and a Global Project. The regions were Africa, the Hindu Kush-Himalayas and

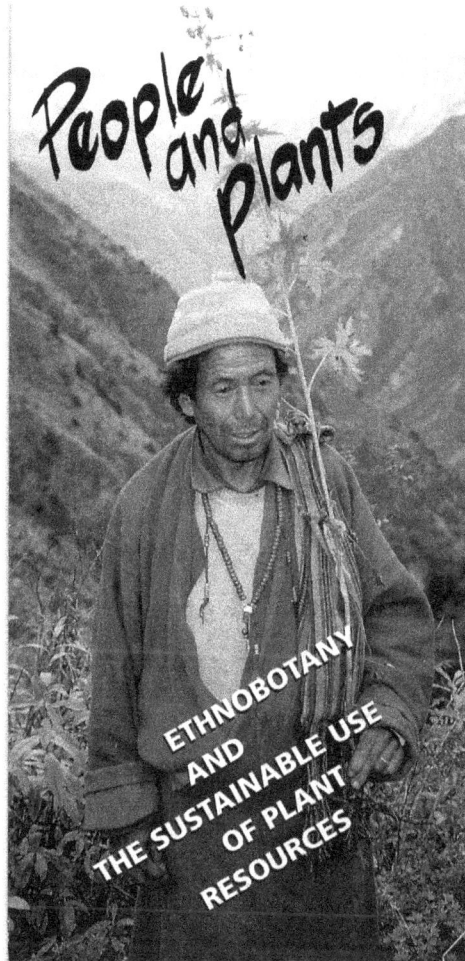

Fig. 10.1. Brochure of the People and Plants Initiative.

Southeast Asia-Pacific. Within each of these, one to a few countries were selected for integrated national capacity-building programmes in ethnobotany. These programmes included the mentoring of aspiring young ethnobotanists, especially through their engagement in applied research on local conservation projects, organising national training workshops, supporting the establishment of local ethnobotanical associations, and encouraging the teaching of ethnobotany in universities and colleges. These countries of special concentration were Kenya and Uganda in Africa, Nepal and Pakistan in the Hindu Kush-Himalayas, and Malaysia in Southeast Asia-Pacific (Map 10.1). They were all chosen because the WWF staff working within them expressed an interest in collaborating with PPI and could identify specific field sites where, they thought, the involvement of PPI would be useful for advancing the conservation work. The field sites initially selected were Mt Kinabalu on the island of Borneo in Malaysia (see Chapter 17), Bwindi Impenetrable Forest in Uganda

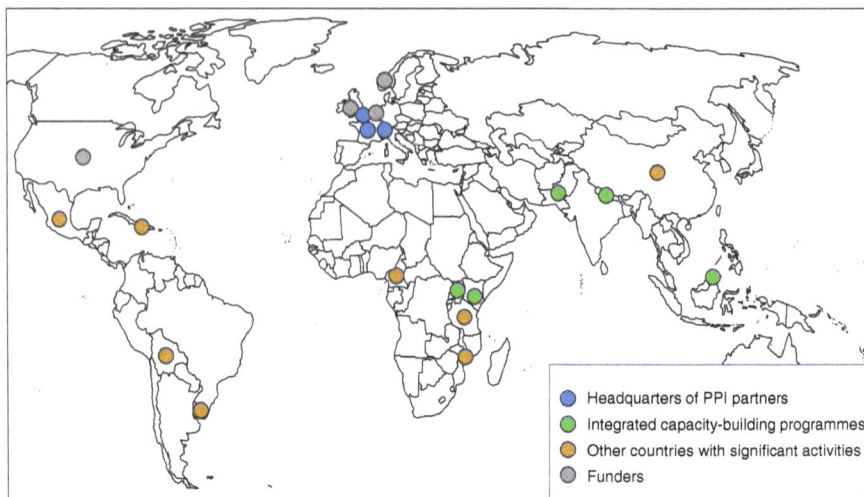

Map 10.1. Geography of the People and Plants Initiative. Note regarding the positions of country boundaries: see Disclaimer.

(see Chapter 18), Ayubia National Park in Pakistan (see Chapter 19) and Shey Phoksundo National Park in Nepal (see Chapter 20).

Later, after the United Nations Educational, Scientific and Cultural Organization (UNESCO) had become a partner of WWF in PPI, Mt Kenya National Park in Kenya was added as an additional site for a field project. All these places have protected areas, which reflects WWF's interest in this particular conservation tool and the difficulties that the countries were facing in establishing effective working relationships between National Parks and local communities (see Section 8.2). It can be seen from Map 10.1 that all the funders of PPI were based in countries in the Global North and that quite a number of other countries in the Global South, in addition to those mentioned above, were involved in significant ways in PPI. The work undertaken in Cameroon was about tackling the over-harvesting of the bark of the montane forest tree *Prunus africana* (see Section 8.1).

The work of PPI at the field sites fell into a number of general themes. A common one was to use ethnobotanical approaches and methods to identify which types of plant resources were of interest to the communities, followed by working with the communities to try and ensure that their uses of plant resources was sustainable. Since there were protected areas present at all of the sites, a variant of this was to help the communities and the protected area authorities work out how the interests of the communities in certain plant resources could best be accommodated within the protected areas' management plans. Where wild plant resources were being harvested by commercial collectors, then studies were undertaken to try and find out what was actually happening and sometimes to assist the parties concerned to find ways of making the harvesting more sustainable. In some cases, PPI worked with the communities and others to identify alternatives to over-harvested plant resources. The conservation of traditional botanical knowledge was a development issue at nearly all of the sites. PPI became involved, in particular, with the conservation of traditional healthcare knowledge and, in some cases, with its integration with Western healthcare to provide improved primary healthcare services at community level.

Soon after WWF had launched PPI, Gary Martin approached Malcolm Hadley of the Man and the Biosphere Programme (MAB) of UNESCO to discuss the possible involvement of UNESCO in PPI. Malcolm, who worked in the secretariat of MAB, was receptive, doubtless because MAB's approach to conservation was well-aligned with the thinking of PPI (see Section 8.3) (Hadley, 2002). Later, the Royal Botanic Gardens, Kew was asked if it, too, would like to become a partner in the programme. Personally, I had some misgivings about including Kew, because, unlike WWF and UNESCO, it is a national (rather than international) organisation and I knew that it was regarded with suspicion in some parts of the world, related to its activities during the British colonial era (see Section 5.3.2). On the other hand, Kew was probably the best equipped botanical institute in the world in terms of some types of relevant technical expertise and it was conveniently situated close to the office of WWF-UK in Britain, which is where I was stationed.

The key drivers of the Regional Projects were the Regional Coordinators and the National Coordinators, the latter normally being members of the staff of local WWF offices. Tony Cunningham remained as WWF Regional Coordinator for Africa throughout the programme. The Regional Coordinator for the Hindu Kush-Himalayas was Yildiz Aumeeruddy-Thomas, Pei Shengji later taking over the role for Pakistan. The Regional Coordinator for Southeast Asia-Pacific was initially Gary Martin and later me. Robert Höft was UNESCO's Regional Coordinator for East Africa and Nepal. Malcolm Hadley of UNESCO was responsible for introducing Yildiz, Pei and Robert to the programme.

One of the main tasks of the Regional Coordinators was to mentor trainee local ethnobotanists. It was considered a high priority for PPI to train up competent applied ethnobotanists at the national level so that work of the PPI-type could continue after PPI's demise. The 86 young ethnobotanists who received mentored training under PPI were required to undertake (typically) two years of individual field research within their own countries, the subjects of their work being selected to be of relevance to solving issues of current conservation concern. Most were registered for Masters' Degrees at local universities and had local supervisors. Of the 86 young professionals trained under PPI, 17 undertook research at Mt Kinabalu, 18 at Bwindi Forest, 19 at Ayubia and 20 at Shey Phoksundo. The staff of WWF-UK who worked with me on backroom activities included Ros Coles (Administrator), Susanne Schmitt (Assistant Plants Conservation Officer), Robert Wilkinson (Fundraiser), Clive Wickes (WWF Conservation Officer) and Chris Wilde (Accountant).

The Global Project of PPI was charged with coordinating the whole programme, fundraising, and commissioning and distributing capacity-building materials. Some of PPI's publications are still accessible on the websites of People and Plants International (https://www.peopleandplants.org/what-we-do/knowledge-exchange-tools, accessed 18 May 2024) and UNESCO (search

'People and Plants', https://unesdoc.unesco.org/search/, accessed 18 May 2024). Among the published materials were eight book-length manuals on various aspects of ethnobotany and related subjects (Table 10.1). In addition to English versions, several of them were produced in other language editions, the principal ones being Spanish and Chinese, which were chosen because of their large number of their speakers. The editor of the English-language versions of the publications was Martin Walters and those responsible for translating the English versions into other languages were Ana Elena Guyer, Laura del Puerto and Hugo Inda (for Spanish); Pei Shengji and Huai Huyin (for Chinese); and Maryati Mohamed (for Bahasa Malaysia). Provision was made for several hundred copies of the manuals and other publications to be made available for distribution free-of-charge to a carefully targeted list of recipients in the Global South. I was personally keen for this element to be included, because I knew from personal experience in Uganda how difficult it can be for scientists in the Global South to access the literature relevant to their professions.

The outcomes of PPI are listed in Table 10.2 (Hamilton, 2004b). On reflection, I think that a principal ingredient in the successes of PPI was the existence of a core team of people united by shared values and having complementary skills. Professor Amyan Macfadyen, a former colleague of mine at the University of Ulster and president of the British Ecological Society, once told me, based on his experience in the International Biological Programme (IBP), that a period of about 10 years is about the right length of time for an international programme in applied ecology to run. IBP was an ambitious 'big science' programme, that was mounted in 1964–1974 to explore the biological basis

Table 10.1. Manuals produced by the People and Plants Initiative. Key to languages: B, Bahasa Malaysia; C, Chinese; E, English; S, Spanish.

Author(s)	Title (in English)	First published (languages)
Gary Martin	*Ethnobotany: A Methods Manual*	1995 (E, S, C, B)
Quentin Cronk and Janice Fuller	*Plant Invaders: The Threat to Natural Ecosystems*	1995 (E, S)
Michael Berjak and Jeremy Grimsdell	*Botanical Databases for Conservation and Development*	1999 (E)
John Tuxill and Gary Nabhan	*People, Plants and Protected Areas: A Guide to In Situ Management*	1999 (E, S, C)
Anthony Cunningham	*Applied Ethnobotany*	2001 (E, S, C)
Bruce Campbell and Martin Luckert	*Uncovering the Hidden Harvest: Valuation Methods for Woodland and Forest Resources*	2002 (E, S)
Sarah Laird (ed.)	*Biodiversity and Traditional Knowledge: Equitable Partnerships in Practice*	2002 (E)
Patricia Shanley, Alan Pierce, Sarah Laird and Abraham Guillén (eds.)	*Tapping the Green Market: Management and Certification of Non-Timber Forest Products*	2002 (E)
Alan Hamilton and Patrick Hamilton	*Plant Conservation: an Ecosystem Approach*	2006 (E, C)

Table 10.2. Capacity-building achievements of the People and Plants Initiative.

Aspects of capacity-building	No.	Details
Young professionals	86	All from countries in the Global South; typically, receiving 2 years of training under local supervisors plus mentoring by PPI Regional Coordinators; most of them were registered for MScs, a few for PhDs
Training workshops	~40	3- to-5-day training workshops, generally field-based, on selected ethnobotanical themes, each with ~25 participants
Courses developed	25	Ethnobotanical courses in 23 universities and colleges, in eight countries in the Global South
Ethnobotanical groups	10	One international non-governmental organisation (People and Plants International), two regional groups, seven national groups
Conservation manuals (books)	17	Each distributed free-of-charge to ~600 targeted individuals in developing countries; the languages were English, Spanish, Chinese, French and Bahasa Malaysia (see Table 10.1)
Working papers and other publications	22	Case studies, newsletters, local ethno-floras, etc.; in English, Spanish, French, Bahasa Malaysia, Urdu, Nepali and Tibetan
Training videos	7	Mostly about field techniques in applied ethnobotany, most filmed by Tony Cunningham
Websites		Some PPI materials continue to be accessible on: People and Plants International (https://www.peopleandplants.org/what-we-do/knowledge-exchange-tools, accessed 18 May 2024) and UNESCO (search 'People and Plants', https://unesdoc.unesco.org/search/, accessed 18 May 2024)
Integrated capacity-building programmes	6	Each including fieldwork, training, awareness-raising and policy promotion. Countries/regions: Kenya, Nepal, Pakistan, Southeast Asia (mainly Sabah, Malaysia), South Pacific, Uganda
Other countries or regions with significant activities	7	Activities included: campaign for sustainable trade in medicinal plants; development of ethnobotanical curricula; ethnobotanical inventories. Countries: Bolivia, Cameroon, China, Dominican Republic, Mexico, Mozambique, Tanzania, Uruguay
World Heritage Site	1	Bwindi Impenetrable National Park declared a World Heritage Site in 1994

of productivity and human welfare. Macfadyen's conclusion was based on his observation that such programmes are founded and initially driven forward by the vision and energy of a few people, but, after a time, dynamism falls away, routine takes hold, and interpersonal frictions emerge. And so it proved with PPI.

11 The Medicinal Plants Conservation Initiative (MPCI, 2004–2008)

WRITTEN IN THE FIRST PERSON BY ALAN HAMILTON

Abstract

The Medicinal Plants Conservation Initiative (MCPI) was a programme of Plantlife International in 2004–2008 which explored the question of how the conservation of biodiversity and sustainable development can best be achieved, based on the interests of communities in medicinal plants. The programme involved National Partners in two countries in Africa and four in Asia working with local communities to try and achieve practical results in the field. Four events were held to share experiences between countries. Analysis of the results revealed that the involvement of three social elements – community groups having an interest in medicinal plants, project teams from conservation organisations and policy makers – is helpful for advancing the work. The motivations for community involvement include people's interests in healthcare, the opportunities for sustainable sources of income from the sale of medicinal plants and affirmation of local cultural identity.

11.1 The Design of the Programme

In 2004, I was appointed manager of a new Plant Conservation and Livelihoods Programme by Plantlife International, a small, British-based, non-governmental organisation (NGO) dedicated to plant conservation, especially through *in situ* means (see Section 8.6). Jonathan Rudge, a New Zealander with experience of working in conservation in Nigeria and Russia, had recently been appointed as Plantlife's first International Director and my job was to assist him by establishing the new programme in practice, especially in those countries in the Global South in which I had gained some experience of conservation during my time at World Wildlife Fund (WWF – now World Wide Fund for Nature) (see Chapter 10). Initial funding to support the livelihoods programme had been obtained from the Allachy Trust and further funds were provided later by the Rufford Foundation, a UK-registered charity.

© Alan Hamilton and Pei Shengji 2024. *History and Future of Plants, Planet and People*
(Alan Hamilton and Pei Shengji)
DOI: 10.1079/9781789248944.0011

'Livelihoods' in a conservation context is a vast subject and Jonathan and I decided to narrow down the programme's scope by concentrating on a single category of plant resources. We chose 'medicinal plants' on the basis of the strong ties that can link people to them, potentially providing them with motivations to save medicinal species and their habitats. These motivations stem from their connections to health, a major human interest, the opportunities that they can provide for making money by selling them on the market and the close connections that they often have with cultural identity (see Table 7.2). Many more species of plants are used medicinally than for any other single purpose (see Section 3.1), which ties in with Plantlife's prime reason for existence – the conservation of the global diversity of plant species.

We also decided to restrict the geographical scope of the programme to two regions with which I was relatively familiar, namely East Africa and the Hindu Kush-Himalayas. The majority of people in these regions rely on local herbal medicine to support their primary healthcare, often using plants that are obtained locally. We next contacted relevant national NGOs and botanical institutions in the chosen countries to ask if they were interested in applying for grants on a competitive basis and serving as Plantlife's National Partners. The grants were for sums of up to £5000 (US$6000) – which may not sound much to the Western ear but was a considerable amount of money for some of Plantlife's potential partners. We stipulated that the money must be used on practical projects with particular rural communities, the aims of the projects being to include strengthening the conservation status of local medicinal plants. Since most species of medicinal plants are wild-collected in East Africa and the Hindu Kush-Himalayas, one obvious way to achieve this would be by increasing the sustainability of harvesting wild plants.

We requested that, if possible, National Partners should feed the results of their work into the national policy level, so as to contribute to the development of national policies on conservation, healthcare, business development and local governance. This was a big ask, given that community-based projects can require time to produce significant developmental results and Plantlife was offering a maximum of only four years of financial support. However, for the most part, Plantlife's National Partners already had pre-existing relationships with both specific communities and government policy makers, which made it relatively easy for them to start work at community level at short notice and to be heard in policy circles.

The NGOs that agreed to be National Partners and the titles of their projects are shown in Table 11.1. The localities of the field projects are shown on Map 11.1. In terms of cultural outlook, some of the communities possessed worldviews typical of small-scale societies. Also represented were the world religions of Buddhism, Christianity, Confucianism, Hinduism and Islam. Politically, the governments of the involved countries varied from left- to right-leaning. Geographically, most of the communities involved in the programme were situated in remote parts of their countries, far removed from the centres of political and economic influence.

Table 11.1. National Partners of Plantlife in the Medicinal Plants Conservation Initiative, and their projects.

National Partners	Projects and communities involved
Applied Environmental Research Foundation – AERF (India)	Capacity-building for linking medicinal plant conservation and sustainable livelihoods in Western Himalayas, Uttarakhand
Ashoka Trust for Research in Ecology and the Environment – ATREE (India)	Development of a strategy for participatory conservation of medicinal plants in the Darjeeling and Sikkim Himalayas
Ethnobotanical Research and Advocacy – JERA (Uganda)	Community-based cultivation of commercially used medicinal plants and their integration in home healthcare in Bunza Village, Mpigi District
Ethnobotanical Society of Nepal – ESON (Nepal)	Community-based conservation and sustainable utilisation of potential medicinal plants in Rasuwa
India (Ladakh Society for Traditional Medicines)	Development of a methodology on medicinal plant conservation to strengthen Amchi Medicine in Ladakh
Kunming Institute of Botany – KIB (China)	The development of methodologies for conservation of medicinal plants based on field-level application at Ludian, Yunnan
National Museums of Kenya – NMK (Kenya)	Building capacity for community-based conservation of medicinal plants. Three case study communities: (i) Bondo and Kisumu East Districts (Nyanza Province); (ii) Nyeri District (Central Province); (iii) Mbeere District (Eastern Province)
Tooro Botanical Garden (Uganda)	Farmer-based First Aid Herbal Toolkit for the Rwenzori Region
Uganda Group of the African Network of Ethnobiology – UGANEB (Uganda)	Conservation and sustainable use of key malaria medicinal plants in Rakai District
WWF-Pakistan (Pakistan)	Promotion of sustainable harvest of medicinal plants at Miandam, Swat

11.2   Events Held to Share Experiences

In addition to the community-based projects of the Medicinal Plants Conservation Initiative (MPCI), four events were mounted by Plantlife and its National Partners to share experiences and draw out general conclusions (Table 11.2). One of the events included a meeting in Kathmandu (Nepal) to identify Important Plant Areas (IPAs) for medicinal plants in the Hindu Kush-Himalayas. For Plantlife, an IPA was (and is) an important conservation concept. This was partly because it had been selected in 2002 to be the lead organisation responsible for developing and promoting Target 5 of the Global Strategy for Plant Conservation (see Section 6.11). Target 5 reads: 'At least 75 per cent of the most important areas for plant diversity of each ecological region protected, with effective management in place for conserving plants

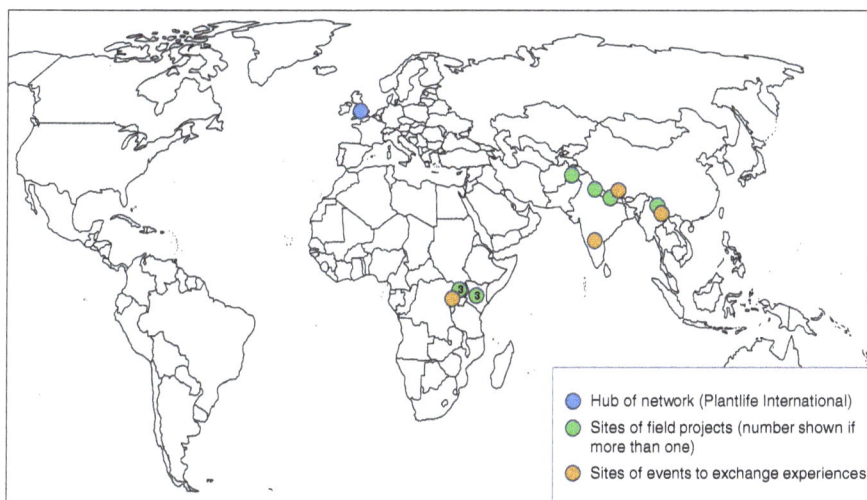

Map 11.1. Geography of the Medicinal Plants Conservation Initiative. Note regarding the positions of country boundaries: see Disclaimer.

Table 11.2. Events to exchange experiences in the Medicinal Plants Conservation Initiative.

Events	Organisers (locality; date)
Regional workshop on identification and conservation of Important Plant Areas (IPAs) for medicinal plants in the Hindu Kush-Himalaya	Jointly organised by ESON and Plantlife (Kathmandu, Nepal; September 2006)
International training and capacity-building on medicinal plants conservation and sustainable utilisation – based on Indian experience	Foundation for Revitalisation of Local Health Traditions – FRLHT (Bangalore, India; October 2006)
Field consultation of the International Standard for Sustainable Wild Collection of Medicinal and Aromatic Plants (ISSC-MAP)	FRLHT (Bangalore, India; 2006-2007)
China/India/UK dialogue on conservation of Himalayan medicinal plants	KIB and Plantlife (Kunming, China; April 2007)

and their genetic diversity [by 2010]'. Prior to the meeting in Kathmandu, Plantlife had requested its National Partners in the Hindu Kush/Himalayan Region to select knowledgeable local botanists to identify IPAs for medicinal plants within their countries. When doing so, Plantlife drew their attention to the three criteria that Plantlife advocated should be used for identifying an IPA: (i) the presence of threatened species of global or regional concern; (ii) exceptional botanical richness for its biogeographic zone; and (iii) the presence of threatened habitats (Anderson, 2002).

It became clear at the meeting in Kathmandu that the use of Plantlife's three suggested criteria to identify IPAs for medicinal plants in the Hindu

Kush-Himalayas had serious limitations. One of them was the unevenness with which the distribution of plant species was known. What has happened historically is that those botanists who are interested in the floras of these ranges have tended to visit the same places as one another, resulting in vast tracts elsewhere remaining unexplored botanically. Another limitation is that the species that appeared on the Red List of the International Union for Conservation of Nature and Natural Resources (IUCN) as endangered were heavily biased towards species in international trade, especially if trade extended to the West. These species tend to be widely distributed across the Hindu Kush-Himalayas, in contrast to the many more species of plants that are used medicinally only locally or regionally, some of which are local endemics. Many of the latter are restricted to higher altitudes, where they are vulnerable to extinction through the exceptionally rapid climate change that the Hindu Kush-Himalayas are now experiencing (see Section 21.3).

However, the most serious criticism of the use of Plantlife's quantitative criteria for recognition of IPAs for medicinal plants is the fact that we are here dealing with a category of plants which is defined by cultural interest and there are no cultural criteria included among Plantlife's criteria for identifying IPAs. This omission was, in fact, pointed out at the meeting in Kathmandu by Professor Pei Shengji, who attended as one of the representatives for China. Later, he followed up the IPA-identification exercise in Kathmandu by developing field projects at Ludian and the Meili Snow Mountain Range in Yunnan, in the process incorporating cultural components into the conservation methodologies (see Chapter 21).

The most ambitious of the exchanges of experience held under the auspices of MPCI was a course organised in 2006 by the Foundation for Revitalisation of Local Health Traditions (FRLHT), a not-for-profit organisation founded by Darshan Shankar and based near Bangalore in India. By 2006, FRLHT had developed an integrated programme of conservation, health security and livelihood support centred around medicinal plants, which had spread across the Southern Cone of India (Mrudula, 2002). The word 'local' in the title of FRLHT was a deliberate choice by Darshan to emphasise the diversity and cultural importance of folk medical traditions in India, over and above the traditional medical systems of Ayurveda, Siddha, Tibetan and Unani (see Section 7.3.3). According to my knowledge, FRLHT in 2006 was unmatched anywhere in the world as an organisation trying to achieve conservation and sustainable development with a focus on medicinal plants. The course was organised by Dr G.A. Kinhal, a forester who had been seconded from the Indian Forest Service. The attendees on the course included nine from East Africa, five from India, ten from other countries in Asia and me. The international transportation costs for the participation of non-Indians attending the course were met directly by Plantlife.

The course was well-organised and included demonstrations of the full range of methodologies that had been developed by FRLHT. One of these was the Medicinal Plants Conservation Area (MPCA), which was conceived as an area of natural or semi-natural vegetation set aside specifically for the *in situ* conservation of medicinal plants. By 2006, 34 MPCAs had been established in the Southern Cone of India, mostly within Forest Reserves. According to

FRLHT, all had received support for their establishment from local people. FRLHT's intention in establishing the MPCAs was to create a network that would eventually cover most of the Indian medicinal flora. The ideal size of an MPCA had been worked out by FRLHT's scientists as being 80–400 ha, based on the principles of population genetics and ground realities.

Associated with many of the MPCAs were Medicinal Plants Development Areas (MPDAs), which were designed to support local communities in developmental matters relating to medicinal plants. Figure 11.1 is a photograph of one of the MPDAs, showing nursery workers, seedlings of medicinal plants produced for local distribution and, in the background, some of those attending FRLHT's course. Another methodology was the promotion of home herbal gardens, which were conceived as being small gardens near homesteads equipped with about 15 species of medicinal plants chosen for their medicinal, cultural and ecological appropriateness. A similar idea was being promoted in Uganda by Tooro Botanical Garden, another of Plantlife's National Partners in MPCI.

Following on from FRLHT's course, a meeting was held at the Kunming Institute of Botany (KIB) in China to allow a more detailed discussion of conservation issues relating to medicinal plants, based on experiences in China, India and Britain. The two participants from India were Drs G.S. Goraya and G.A. Kinhal from FRLHT and the two from Britain were Jonathan Rudge and me. Discussions at the workshop resulted in four conclusions being reached, based on practical experiences in India and China. First, it was agreed that, in both countries, it was folk healers who constitute the social group that is the most knowledgeable about medicinal plants and who were most concerned that these resources are conserved. From this, it was concluded that it is essential for

Fig. 11.1. Medicinal Plants Development Area, Southern India. Photo: A.H. (2006)

folk healers to be included in conservation initiatives based on medicinal plants for them to have much chance of success. Secondly, it was agreed that both in China and India, local health traditions were being rapidly eroded and that it was urgent to document their associated ethnobotanical knowledge before it disappears. Thirdly, in both China and India, industries involved in the medicinal and aromatic plant (MAP) sector were generally insensitive to conservation. It was thought that conservation initiatives based on the certification of herbal products were only likely to have much chance of success where there are substantial sales to ethical markets in the West. Only then, it was thought, would sufficient money be generated to cover the necessary extra costs. Finally, it was agreed that the development of traditional medicine to meet the challenges of modern times would benefit greatly if traditional medical systems received official recognition. In 2007, such recognition was patchy. In China, five traditional medical systems had been recognised (see Table 7.2), while in India Ayurveda was officially recognised, but not Tibetan Medicine (see Section 7.5.3).

11.3   A Model for Biodiversity Conservation Based on Community Interests in Medicinal Plants

MPCI concluded with an analysis of the work undertaken to try and draw out general conclusions. The approach used was to use the information obtained during the programme to answer the question, 'How can the conservation of biodiversity and sustainable development best be achieved, based on the interests of communities in medicinal plants?' The result of the analysis was a conservation model composed of three social elements and recommendations for the relationships between them (Hamilton, 2008; Pei Shengji et al., 2010). The three social elements were community groups, project teams and policy makers.

It was assessed that the primary interests of the community groups in medicinal plants do not necessarily have to relate to healthcare, although they have to have some interest. The primary concerns of the 12 community groups included in the programme's sample included agriculture, tree-planting, forest management and women's affairs, as well as healthcare directly. Their motivations to act in favour of the conservation of the species were judged to be about equally divided between a concern to maintain resources for local healthcare and a concern to have a source of sustainable cash income. The third potentially motivating factor mentioned at the beginning of this chapter, 'the close connections that medicinal plants often have with cultural identity', was too closely intertwined with local healthcare, local religious beliefs and local worldviews to be easily distinguished as a separate analysable factor.

Regarding project teams, it was judged that the values and motivations of members of project teams are critical. In particular, there must be a genuine interest in the plants and the concerns of the communities, and respect for the local cultures. A degree of competence in ethnobotany is required, as well as the knowledge of how to draw upon the additional expertise and skills as

Fig. 11.2. Village ethnobotany workshop, Rasuwa, Nepal. Photo: A.H. (2007).

are required for particular projects. Figure 11.2 shows two members of the project team in Nepal in discussion about medicinal plants with villagers in Rasuwa District, which is in the foothills of the Himalayas north of Kathmandu. In this case, the team consisted of members of the Ethnobotanical Society of Nepal (ESON) and representatives of the Manekor Society Nepal, a local NGO dedicated to the alleviation of poverty in Rasuwa, and the Federation of Community Forestry Users Nepal (FECOFUN), which is an umbrella organisation that covers the whole of Nepal.

Plantlife's National Partners in MPCI generally had good ties to national policy makers, which made it easy for them to introduce the results of their efforts into policy circles. Depending on the country, fields of policy relevant to meeting the objectives of MPCI could include those relating to indigenous people, local autonomy, national healthcare, natural resource management and the orientations of research institutes. Three types of organisations were judged to be well-placed to carry forward work of the MPCI-type at the landscape scale, namely faith-based organisations, women's associations and indigenous people's groups. All can potentially be in favourable positions to transfer lessons learnt from experiences in one community to others in the neighbourhood.

12 The Saga of Rubber, Part 2 – The Story in China and the Founding of Xishuangbanna Tropical Botanical Garden

Abstract

An indigenous rubber industry was created in China as a response to the United States imposing an embargo on the sale of rubber to China in 1949. The research needed to achieve this development eventually led to the founding in 1959 of Xishuangbanna Tropical Botanical Garden as a permanent centre for research into tropical crops and for studying traditional ethnobotanical and ethnoecological knowledge. The expansion in the area of plantations of rubber and other tropical crops in Yunnan was so rapid that it was feared at one time that all the remaining tropical forest would be lost. More effort started to be put into conservation of tropical forest and research into traditional knowledge after 1979, when their values for developing the national economy, healthcare and agriculture became better appreciated by the Chinese leadership. Also realised was the value for China of maintaining scientific exchanges with the international scientific community.

In Section 5.3.5, we told the stories of the booms in the exploitation of the pará rubber tree (*Hevea brasiliensis*) in Amazonia during the second half of the 19th century and of the African rubber tree (*Funtumia africana*) and the African rubber liana (*Landolphia owariensis*) in the Congo in 1885–1908; also, how Britain managed to establish its first *Hevea* plantations in the Malay Peninsula in 1895. In Section 7.5.1, we described how, in the 19th century, scientific botany came to China and mentioned the contributions of Tsai Xitao and the Kunming Institute of Botany (KIB) to how it subsequently developed. Here, we tell how these two strands of history came together in China, how a domestic Chinese rubber industry was born and the contributions of Pei Shengji to these stories.

In 1955, after his schooling and taking a course in agriculture at Chengdu Provincial Agricultural School, Pei Shengji was instructed to join a team investigating the vegetation of China for on-the-job training in botany. Being included in this team gave Pei an exceptional opportunity to become acquainted with the Chinese flora, learn about how plants are named and classified in science,

© Alan Hamilton and Pei Shengji 2024. *History and Future of Plants, Planet and People* (Alan Hamilton and Pei Shengji)
DOI: 10.1079/9781789248944.0012

and, through staying in their villages, coming to appreciate the great stores of knowledge about plants held by the ethnic minorities of Yunnan.

As mentioned in Section 6.2.1, the 20th century was a turbulent time in Chinese history. When the Communist Party came to power in 1949, China became viewed with hostility by the United States, which imposed on China an embargo on the sale of military equipment and related materiel, including rubber. At the time, China lacked a domestic source of rubber and, since rubber was deemed an essential commodity, the Chinese Government had little option but to adopt a policy of self-reliance. A two-pronged scientific research programme was launched to see whether any solutions could be found.

The first prong in the research was to investigate whether there were any suitable, latex-producing, species of plants available locally. Led by Tsai Xitao, the research on this theme eventually resulted in the shortlisting of about ten locally available species that might be suitable. All of them, except *Hevea* (which is in the *Euphorbiaceae*), were in the botanical families *Apocynaceae* and *Moraceae*. The shortlisted *Apocynaceae* included three species of indigenous *Urceola*, plus the exotics, *Funtumia elastica* (earlier introduced from Africa) and *Cryptostegia grandiflora* (introduced from Madagascar). Those in the *Moraceae* included the indigenous *Bleekrodea tonkinensis*, a small indigenous tree found in rocky karstic regions, and *Ficus elastica*, which had been introduced from India.

Prior to 1949, there had been only three plantings of the South American rubber tree, *Hevea*, in China, all based on planting stock imported from Singapore by overseas Chinese – stock that had, most likely, originated from the *Hevea* plants that the British had brought earlier to the Malay Peninsula. The first of these plantings was in 1901 on the island of Hainan, the second in 1904 in Xishuangbanna Prefecture in South Yunnan and the third in 1945, also in Xishuangbanna. The last of these was a small (<1 ha) plantation that was unsuccessful. However, the conclusion eventually reached by Tsai Xitao and his team, based on the evidence that they had gathered, was that only with *Hevea* was knowledge sufficiently well advanced for it to be worthwhile advancing the research to the mass-production level.

The second prong in the Chinese Government's search for a solution to its rubber problem was to request the Chinese Academy of Sciences (CAS) to undertake a vegetation survey to see if there was anywhere in the country that was suitable for the commercial planting of rubber. Because latex-producing plants are primarily tropical, the survey concentrated on the southern provinces of Fujian, Guangdong, Guangxi, Hainan and Yunnan. Eventually, after six years of fieldwork (1952–1958), the conclusion was reached that the most suitable place for commercial rubber planting was the land lying below an altitude of 900 m in Xishuangbanna Prefecture. Of the three counties in Xishuangbanna, most of the land lying below 900 m is situated in Jinghong County, with an extension into Mengla County to its east. This area, which borders on Laos and Myanmar, lies only just within the Tropic of Cancer, but, nevertheless, has a climate typical of that associated with lowland tropical rainforest elsewhere in the world, thanks to a favourable regional pattern of atmospheric circulation.

Launching commercial production of a species of plant that has only recently been domesticated is not necessarily a simple process, especially when the planting is to be in a place well outside its natural range. Because there was so much urgency about making progress, in 1958 the Provincial Government of Yunnan opened a special institute, the Yunnan Provincial Research Institute of Tropical Crops, to undertake research into all aspects of establishing a domestic rubber industry, using *Hevea* as its resource base. *H. brasiliensis* is a notoriously difficult species to propagate by vegetative means (Muzik and Cruzado, 1958), but, after ten years of hard work, the scientists and technicians at the research institute managed to develop a new variety (TJ-one) that was capable of being propagated by bud grafting. Other experimental work undertaken at the institute revealed that planting at a density of about 500 trees per hectare resulted in flows of latex comparable in volume to those that were being obtained from existing rubber plantations in Southeast Asia.

Under the leadership of Tsai Xitao, the scientists involved in the rubber research suggested to the government that it would be advantageous for the country to establish a permanent botanical institute within the part of Xishuangbanna which has a lowland rainforest climate in order that research into tropical plant resources could continue on a permanent basis. The proposal was accepted and, in 1959, the construction of Xishuangbanna Tropical Botanical Garden (XTBG) in Mengla County began. Today, XTBG is one of the few botanical gardens in China that are constituent parts of the CAS. A statue of Tsai Xitao with two of his students has been erected in the garden (Fig. 12.1).

The orderly development of XTBG was severely disrupted when the Cultural Revolution erupted in 1966. Contacts with foreign scientists were broken off, about one-third of the garden's plant collections were destroyed and, for a number of years, Tsai Xitao was made to live in a cowshed charged with being an intellectual who was opposed to the academic authorities. Pei and his colleague, Xu Zaifu, were put under pressure to denounce Tsai. Starved of resources for their work, Pei and Xu only managed to emerge from this ordeal because they had learned basic survival skills from the local ethnic minorities. In the words inscribed on the garden's website, as of 2021:

> Through hunting and gathering excursions with farmers, these botanists began to discover that the best way to find out about plants and trees was to talk to local farmers who had names for them all. As Pei Shengji described it, 'If we had not learned about the plants in the forest from the local people we would have died of hunger and disease.'

Later, Pei and Xu became successive directors of XTBG, Pei in 1978–1986 and Xu in 1987–2001.

By the late 1960s, sufficient was known about *Hevea* and its potential for cultivation in China to allow large-scale planting to begin. Factories for the primary processing of the latex were constructed close to the new plantations. Planting was initially on state-owned farms, becoming supplemented by private plantings once this had become politically possible. Today, a large percentage of all land lying below an altitude of 900 m in Xishuangbanna Prefecture is

Fig. 12.1. Statue of Professor Tsai Xitao and two of his students. Photo: P.S. (2021).

covered by plantations of rubber or other tropical crops. Figure 12.2, which was taken from an aeroplane near the Mekong River in Xishuangbanna, illustrates the extent of rubber planting (dark green striped areas).

The effects of so much landscape transformation have been dramatic. In the early 1950s, lowland tropical rainforest covered about 70% of Xishuangbanna lying below an altitude of 900 m; today, the equivalent figure is about 30% and much of the remaining forest is degraded, altered in species composition and impoverished in biodiversity. The agricultural climate has deteriorated, and water supplies are now less reliable (Hu Huabin *et al.*, 2008; Liu Wenjie *et al.*, 2011). The local society has been transformed from a traditional agricultural type that has existed in this area for over 2000 years into one that is largely industrial. Many traditional farming villages have disappeared, and former swidden farmers are now industrial workers.

In its earlier years as a socialist state, China adopted a model for its economic development based on the one which was then current in the Soviet Union. This involved strong central planning and economic production based on communes. As illustrated by the example of cotton and the Aral Sea described in Section 5.4.2, this model of socio-economic development could be insensitive to the particularities of local environments and ignore traditional botanical and ecological knowledge. At one time, the relentless logic of central planning, combined with a keenness by the authorities to adopt the most modern technologies, resulted in the entire lowland tropical rainforest zone of Xishuangbanna being earmarked for replacement by plantations of rubber and

Fig. 12.2. Rubber plantations in Xishuangbanna, China. Photo: P.S. (2021).

other tropical crops. Fortunately, the excesses of pursuing economic development in this single-minded ruthless way, untempered by other considerations, began to fade towards the end of the 1970s, when a more nuanced approach became adopted.

A seminal event was the Second National Symposium on Rubber and Tropical Economic Crops held at Jinghong, the capital city of Xishuangbanna Prefecture, in 1979. At it, Pei Shengji and his colleagues in XTBG managed to argue successfully that the conservation of traditional ecological knowledge in Xishuangbanna would be advantageous both for the country as a whole and for local agriculture. What had happened with local agriculture was that, with the arrival of the socialist state and of 'scientifically-based production', the authorities had decided to replace the traditional irrigation system of the Dai people with concrete structures, concrete being deemed to be a 'scientifically progressive material' (Gao Lishi, 1998). In fact, as the scientists pointed out at the symposium, the traditional irrigation system of the Dai is a much more flexible and efficient way to irrigate the paddy fields than depending on hard engineering. The traditional system relied on wooden dams, bamboo dykes, earth-banked channels and long-standing understandings between the farmers on the sharing of the water, backed up by religious beliefs that gave protection to the forests in the catchments (see Chapter 14).

Additional to the above economic reasons for conserving traditional botanical and ecological knowledge in Xishuangbanna, the XTBG scientists further argued that it would be advantageous for China to re-establish scientific links with the outside world. A specific example of the merit of doing so involved research on the taxonomy of *Gymnosporia* (*Maytenus*) *senegalensis*, a small African tree in the botanical family *Celastraceae*. This research had eventually resulted in the development of a new anticancer drug from *Maytenus hookeri* and *Maytenus austroyunnanensis*, woody plants native to Yunnan and nearby parts of Southeast Asia. Both of these species have been used traditionally in Xishuangbanna for treating human tumours.

In 1972, it had been reported in the international scientific literature that, in a programme supported by the National Cancer Institute (NCI) in the United States, random screening of extracts of plants to test for their biological activity had identified an African species of *Maytenus* as a promising lead for anticancer drugs (Kupchan et al., 1972a,b). At that time, no species of *Maytenus* were known from China, but a subsequent taxonomic study by Pei and Li Yanhui revealed that all of the Chinese species that had previously been assigned to the genus *Gymnosporia* should be transferred to *Maytenus* (Pei Shengji and Li Yanhui, 1981). Then, further research into the properties of the Chinese species of *Maytenus* revealed that *M. hookeri* was the species with the highest quantity of maitansine, the active principle. Subsequent research resulted in the approval of a standardised crude drug known as Maytenus no. 1 Tablet, a medicine that soon became famous in China because it had been requested to be sent to Beijing by the doctors who were treating Zhou Enlai, the first premier of the People's Republic of China (in office, 1949–1976).

13 Commercialisation of the Hani's Knowledge of Fleagrass – A Case Study in Access and Benefit-Sharing (ABS) in China

Abstract

The third part of the Convention on Biological Diversity (CBD) requires countries that develop new commercial products based on biodiversity to share the benefits that they receive with the countries that were the sources of the biodiversity. Attempts to implement this agreement have run into many practical problems. The case described in this chapter, together with that described in Chapter 14, have been used by China and the United Nations Development Programme (UNDP) to demonstrate what has been achieved with implementing the third part of the CBD in China. This one concerns the development of new products to repel mosquitoes and treat mosquito bites, based on the indigenous knowledge of fleagrass (*Adenosma buchneroides*) of the Hani people of Xishuangbanna.

13.1 The Historical Context

Countries that have ratified the Convention on Biological Diversity (CBD) have agreed to conserve the biodiversity that lies within their borders, to use their biological resources sustainably and to ensure that there is an equitable sharing of benefits when access to their biodiversity is granted that leads to the development of new commercial products (see Section 6.11). The phrase 'access and benefit-sharing' (ABS) in this context refers to the terms of agreements reached between those national organisations that are responsible for safeguarding their countries' biodiversity and those commercial companies that are interested in exploiting it to develop new commercial products. It was envisaged when the United Nations established the CBD in 1992 that ABS agreements would be between biodiversity-rich countries in the Global South, which were seen as needing assistance with the safeguarding of their biodiversity, and businesses in the Global North, which were seen as better-equipped to develop new commercial products. As mentioned in Section 6.11, countries have struggled to turn ABS into an effective conservation tool.

The Kunming Institute of Botany (KIB) and its sister organisation, Xishuangbanna Tropical Botanical Garden (XTGB), have been leaders in China in trying to find ways to make ABS work. KIB and XTGB's approach in this matter has been informed by their experiences in helping to create an indigenous rubber industry in China and in helping to save Yunnan's tropical forests (see Chapter 12). Essentially, KIB and XTGB have acted as brokers between the interests of local communities, businesses and themselves, with themselves taking the role of custodians of biodiversity. This approach involves some benefits from the commercialisation of products to be returned to KIB and XTGB, to support their continuing existence. According to one international expert on ABS, acting as brokers in ABS arrangements can be a valuable role for non-governmental organisations and national scientific institutes, such as KIB and XTGB, to take (Laird, 2002). However, it carries the risk of being seen as stooges for industry (MacDonald, 2010).

KIB's involvement with ABS has taken place within the context of a world witnessing a spectacular growth in business sectors interested in medicinal and aromatic plants (MAPs) (see Section 6.3.3). Calculated globally, it has been estimated that the annual imports of MAPs into all countries rose by 100% between 1991 and 1997 (Lange, 2000). Since then, imports have continued to skyrocket (Barata et al., 2016). In China's case, the great expansion of its industrial capacity since 1980 has resulted in a big expansion in its middle class, which has produced an upsurge in demand for natural products.

One consequence of the increased demand for natural products outside and inside China is that a number of companies engaged in MAPs-related business sectors have approached KIB to seek its collaboration with the identification of Chinese plant species and associated indigenous knowledge that might be worth investigating for the development of new commercial products. The director of KIB has tended to pass on such enquires to Pei Shengji, because of his historical position as the institutional leader on these topics. The international companies that have approached KIB to seek its collaboration have included France-based L'Oréal (the world's largest cosmetic company), Chanel (a high-end, French-owned, fashion house), dōTerra (a US-based marketing company selling essential oils and related products) and the Japanese chemical and cosmetics company, Kao Corporation. In response to this upsurge in interest, KIB and XTBG have created a 15-person team of ethnobotanists to investigate the potential in China for the development of new natural products.

On the domestic front, a company with which KIB has become closely involved is Dr Plant, a Beijing-based cosmetics group. In 2015, at the initial stages of collaboration between KIB and Dr Plant, Pei Shengji had recommended the use of the orchid Dendrobium officinale in skincare products, based on information obtained from a combination of ethnobotanical and laboratory research. Since then, Dr Plant has become heavily invested in products containing D. officinale. At the same time, it has provided support for the development of an orchid cultivation industry in Pu'er City Prefecture in Yunnan, which has removed some of the pressure on wild Dendrobium orchids. It has also provided funds to KIB and XTBG to support their community-based conservation work. In China, D. officinale and similar species have been used in traditional medicine to promote the secretion of body fluids. For instance, the drug *Feng-dou* has traditionally been used by Chinese opera singers to help them sustain their voices. Part of the upsurge in demand for orchid products had come from officials requiring throat medications to make it easier for them to deliver long drawn-out orations at gatherings of the people.

KIB's efforts to develop ABS arrangements in China have been recognised by the United Nations Development Programme (UNDP), which has provided financial assistance through the Biodiversity Finance Initiative (BIOFIN – a partnership between UNDP and the European Union to promote good financial management alongside conservation). UNDP's institutional partner in China is the Ministry of Ecology and Environment (formerly, the State Environmental Protection Agency – SEPA). The case study described below (on fleagrass) and that described in Chapter 14 ('Revitalising the Traditional Conservation Practices of the Dai') were chosen by UNDP/SEPA to demonstrate what has been achieved with implementing ABS in China at side events at the Conferences of the Parties to the CBD at Sharm El-Sheikh, Egypt in 2018 (COP14) and at Montreal, Canada, in 2022 (COP15).

13.2 ABS and Fleagrass

In 1975, when Pei Shengji was visiting a house of a member of the Hani ethnic minority in Xishuangbanna Dai Autonomous Prefecture, Yunnan, he discovered that a fragrant herb, of a type that he had not previously encountered, had been placed under the mattresses of the beds. Intrigued, he asked a villager to take him to the place where the plants had been growing. Later, in the laboratory, he identified the plant as *Adenosma buchneroides*, a blue-flowered member of the plant family *Plantaginaceae* that had not previously been recorded in China.

A. buchneroides is a species of uncertain evolutionary origin. It is possible that it has only ever existed as a cultivated domesticated plant, having evolved from a wild species through human selection. Its putative ancestor has been suggested as *Adenosma indiana*, a species which is widely distributed in Southern China and occurs in Southeast Asia. In China, *A. indiana* is used medicinally to treat colds, dermatitis and other ailments.

In 1976, Pei returned once again to the Hani area of Xishuangbanna and discovered that the reason why it was being placed in bedding was because the people knew that it killed and repelled haematophagous (blood-sucking) insects, such as bedbugs, fleas, lice and mosquitoes. He also learnt that it was being strewn on the floors of houses for the same reason and was an ingredient in herbal medicines used for the treatment of insect bites and subsequent swellings (Shen Peiqiong *et al.*, 1990; Gou Yi *et al.*, 2018). Other uses included as a personal adornment and as a source of perfume. The English-language name that became assigned to the species was fleagrass. Phytochemical analyses carried out at KIB revealed that *A. buchneroides* contained an essential oil with at least 39 chemical constituents, the bioactive properties of which included insect repellence and anti-inflammatory and antibacterial activity.

Worldwide, it is common for traditional communities to know of species of plants useful for repelling or killing haematophagous insects. The strewing of insect-repelling plants on floors, in the same way as the Hani, was once a common practice in Britain, where the two species commonly used were mugwort (*Artemisia vulgaris*) and tansy (*Tanacetum vulgare*), both in the family *Asteraceae* (Allen and Hatfield, 2004). In Buganda, the species that was traditionally strewn on floors was *etteete* (*Cymbopogon nardus*), a lemon-scented relative of lemon grass (*Cymbopogon citratus*) (Roscoe, 1921; Hamilton, 2020).

At the time of Pei's initial visits, the Hani of Xishuangbanna Dai Autonomous Prefecture were swidden farmers who used forested land for rotational agriculture. Their system of rotation involved, first, cutting down patches of forest and burning the trash, then planting dryland rice in the first year, followed by maize in the second, after which the land was left alone for the forest to recuperate. Then, after 13–15 years, the forest was cut again and the cycle repeated. Pei discovered that the way that the fleagrass was being grown was by interplanting it with dryland rice in the first year of cultivation. The rice, which grows faster than fleagrass, reached maturity first and was then harvested, after which the fleagrass, now liberated from the shade of the overtopping rice, was able to grow freely and mature, at which stage it, too, was harvested. In this way, the farmers were obtaining two harvests of useful crops from the same patch of land in one season. (It should be noted that the agriculture of the Hani living in Xishuangbanna Prefecture differs from that of the Hani living in adjacent Honghe Prefecture, where the people grow wetland (rather than dryland) rice in hillside paddies – see Section 3.3.2.)

Additional to the crops that they cultivated in their swidden fields, the Hani of Xishuangbanna have traditionally obtained many types of plant and animal products from the forests. One of them has been the leaves of the tea tree, *Camellia sinensis*, which they have tended in semi-wild stands in the forests (*not* in parallel rows of trimmed hedges, as in tea estates). As mentioned in Section 3.5.2, it has been a long-standing custom of the Hani to carry bundles of tea leaves harvested from tea trees down to the villages of the Dai and the Han Chinese in the lowlands and exchange them for commodities to take back to their communities.

By the 2010s, the traditional way of life of the Hani in Xishuangbanna had almost completely disappeared, thanks to population growth, the widespread replacement of traditional agriculture by plantations of crops, including of rubber and tea, and government policies that discouraged swidden farming (Gou Yi *et al*., 2018). Industrially produced products had extensively replaced traditional herbal remedies and many younger Hani in Xishuangbanna knew next to nothing about fleagrass. Yet, just across the border in Laos, the Hani who lived there were still growing fleagrass in their traditional way, this part of Laos not having experienced the dizzying pace of economic development and cultural transformation that has characterised China. Elsewhere, fleagrass remains a widely distributed species in Vietnam, the country where the original collection of *A. buchneroides* for scientific study was made in the early 20th century.

Fig. 13.1. Pei Shengji with Hani villagers in a field of fleagrass, China. Photo: Yang Zhiwei (2021).

Stimulated by the new interest in natural products that had developed in China, in the early 2010s members of a team of KIB and XTGB ethnobotanists returned to the area where the Hani farmers had formerly planted fleagrass to see if any remaining plants could be found. They were fortunate in being able to discover a few plants of fleagrass that had managed to persist after the abandonment of the swidden fields, collected seed and took it back to XTGB to sow in its experimental beds. Once the seed had been bulked up, they returned to the Hani communities with the seed and then experimented with the Hani on ways to cultivate it in open places in managed tea forests. Figure 13.1 shows Pei Shengji (with hat) in a field of fleagrass with Mr Qian Zhou, Mrs Piao Mei and Mr Yue Piao, who are Hani from Daka Village in Xishuangbanna. They are accompanied by publicists working for Dr Plant.

There have been several beneficiaries from this project. The Hani have benefited from receiving training in the planting and cultivation of fleagrass, such that they are now able to produce a product of sufficiently high quality to be usable by industry. This has given them a new, long-term, source of income. This demonstration of the continuing value of their traditional knowledge should have helped the Hani realise the usefulness of retaining it and passing it on to future generations. The creation of an additional use for patches of semi-wild tea forest should have given the Hani more incentive to maintain them, which is useful for the conservation of the global genetic diversity of *C. sinensis*, which is a major world crop. On the commercial front, continuing laboratory research by scientists in KIB and XTBG on the essential oils in fleagrass resulted, in 2021, in the launch of two new commercial products. One is a mosquito repellent and the other a soothing agent to be applied to mosquito bites. Both have met the required regulatory standards and now their industrial production is being taken forward by Huang-Ya, a local pharmaceutical group.

14 Revitalising the Traditional Conservation Practices of the Dai (China)

Abstract

The Dai are a Chinese minority people who live in the lowland tropical forest zone of Xishuangbanna. Traditionally Buddhist, their nature-friendly lifestyles, backed by their religious beliefs, have helped to conserve an exceptionally rich diversity of species. The banning of religion in China during the Cultural Revolution of 1966–1976 led to the undermining of the traditional conservation practices of the Dai and a great loss of biodiversity. The ban on religion was lifted in 1982. This case study describes the work led by ethnobotanists from the Kunming Institute of Botany and Xishuangbanna Tropical Botanical Garden since 2014 to revitalise the traditional conservation practices of the Dai. It has been recognised by UNESCO as a good example of the usefulness of the concept of Sacred Natural Sites for conserving linked cultural and natural heritage.

14.1 The Dai of Xishuangbanna and Their Traditional Buddhist religion

Along with the case of commercialisation of the Hani's knowledge of fleagrass described in Chapter 13, the project described in this chapter was chosen by the Chinese Ministry of Ecology and Environment (formerly, the State Environmental Protection Agency – SEPA) and the United Nations Development Programme (UNDP) to demonstrate at the Conferences of the Parties (COP) of the Convention on Biodiversity (CBD) in 2018 and 2022 what has been achieved with implementing access and benefit-sharing (ABS) in China. The project is concerned with the traditional plant-resource system of the Dai and the steps taken to revitalise it since 2014. The Dai, like the Hani, are a Chinese minority ethnic group living in Yunnan. Traditionally, Dai and Han Chinese communities living in the lowlands of Xishuangbanna have exchanged commodities with the Hani and other ethnic groups living in the uplands, among them rice for tea (Fig. 3.4).

Figure 14.1 illustrates the plant-resource system of the Dai. It shows a compact village with its temple beside a river meandering through paddy fields. The paddies are used to grow wetland rice, the Dai's staple food. The fuelwood forest shown in the foreground is of *Senna siamea* (*Fabaceae*), a species introduced into Xishuangbanna from Thailand at least 600 years ago and which is planted on marginal lands. Not marked on the sketch are the Dai's

Fig. 14.1. Traditional plant-resource system of the Dai, China. Reproduced from: Pei Shengji (1985).

home gardens, which are species-rich and commonly contain 40–60 species of useful plants, among them fruit trees, vegetables, spices, bamboos and plants providing fibre. Historically, the Dai's home gardens are believed to have been the sites of domestication of many East Asian tropical crops. The hills seen in the background of the sketch are covered by lowland tropical forest, which provides the Dai with a variety of plant products.

Traditionally, the Dai are Buddhists of the Theravada tradition (see Map 4.1). Buddhism, which is a particularly nature-friendly religion, views compassion for the natural world and human development as being inextricably linked. This has resulted in the territory of the Dai in Xishuangbanna being excep-tionally biologically diverse (Liu Hongmao et al., 2002; Pei Shengji, 2010). One aspect of the respect that the Dai accord to nature is the safeguarding of certain hills, known as Nong Holy Hills, one of which is depicted on the left side of the river in Fig. 14.1. Ethnobotanical surveys have revealed that, until recently, there were 1200 forest-covered Nong Holy Hills in Xishuangbanna, the total area of forest associated with them being 80,000 ha (about 2% of the prefecture's area).

The way that the Dai have traditionally regarded their holy hills has been described as follows (Pei Shengji, 2010):

> The sacred forest or Holy Hill is known as 'Nong' in the Dai language. Traditionally, each village had its own Nong forested hill, where in the traditional concepts of the Dai, the gods reside. All plants and animals that inhabit the sacred forest are considered to be companions of the god and are 'sacred beings' living within the god's 'garden'. In addition, the Dai believe that the spirits of their ancestors and great and revered chieftains go to the sacred forest to live following their departure from the world of the living. Any violence to or disturbance of plants and animals in the forest will be punished by the gods. Therefore, hunting, gathering and cutting are strictly prohibited.

In Section 21.3, we mention the holy mountain of Mt Kawagarbo in the Meili Snow Mountain Range in Northwest Yunnan and how it attracts many Buddhist pilgrims each year. The most famous sacred mountain in Tibetan Buddhism is Mt Kailash in the Western Himalayas, which, as well as being sacred to Buddhists, is sacred to Hindus, Jains and adherents of Bon. According to the traditional ecological wisdom of the Dai, their own existence is sustained in their settlement areas by maintaining harmonious relationships between forest, water, paddy rice fields and themselves.

In 1966, Mao Zedong launched the Cultural Revolution with the aim of sweeping away old ways and superstitions (see Section 6.2.1). This unleashed a concerted attack on the religious belief-system of the Dai, which, up to then, had helped to maintain the forests, the biodiversity and the agriculture of low-land Xishuangbanna. Fortunately, from a conservation perspective, this highly destructive episode in Chinese history began to ease towards the end of the 1970s. In 1982, China adopted a new article (no. 36) in its constitution, which granted Chinese citizens freedom of religion. Since then, being a Buddhist is no longer a crime and an increasing number of Chinese people are considering themselves to be Buddhists (see Section 6.9) (Madsen, 2011). However, the brief period of attack on religion had a devastating effect on the conservation of biodiversity in Xishuangbanna. By the year 2000, only 200 Nong Holy Hill forests remained, covering, in total, an area of only 2000 ha.

14.2   Project Strategy and Initiation

Since 2014, an ethnobotanical team drawn from members of the Kunming Institute of Botany (KIB) and Xishuangbanna Tropical Botanical Garden (XTBG) guided by Professor Pei Shengji has been engaged in a range of activities intended to revitalise the traditional conservation practices of the Dai. Six villages in three counties of Xishuangbanna Prefecture were selected for project implementation on the basis of previous acquaintance. The populations of these villages range from 220 to 820 accommodated in 44–134 households.

One village, Manyuan, was selected to be the pilot on the basis of being one of the few Dai villages which still possessed a well-preserved forest on its Nong Holy Hill (even though reduced in area) and also because traditional cultural rites associated with the hill were still being performed. Manyuan, which has a population of about 440 people accommodated in 81 households, is situated at an altitude of 450 m on the banks of the Lancang (Mekong) River. It is a middling village financially by Xishuangbanna standards. The average annual income per capita (as of 2021) was 21,000 RMB (US$3000). Like other Dai villages, Manyuan was greatly affected when the Chinese economy was liberalised in 1980 – in its case, particularly by the spread of commercial plantations of rubber, fruit trees and vegetables. One consequence of the opening-up was to create volatility in the incomes of farmers, who now became exposed directly to the volatility of the free market.

The strategy adopted by the project was to start by using the Nong Holy Hill at Manyuan as the focus of attention and then to expand the work, step-by-step, to embrace other aspects of conservation at Manyuan and extend it to the other five project villages. The project team considered that building internal capacity within the villages was just as important as the direct interventions that it could make itself (for example, through influencing government policy). The thinking was that this would help ensure that the ideas and activities that had been introduced by the project would continue to exist beyond the project's lifetime. In total since 2016, more than 25 training events have been held in the six project villages, in total involving about 450 villagers.

After initial introductions to the Manyuan community, the next step taken by the team was to invite the community to propose the names of people for a Manyuan Village Conservation Group. In response, the community put forward the names of five people – the village head, an herbal doctor, the traditional keeper of the Nong Holy Hill, a village elder and a lady to represent the women. The involvement of this group, right from the start, has ensured that the villagers have been active participants in all the project's work, from the development of an initial workplan to the implementation of each of the project's specific activities. It has helped ensure that the achievements of the project have benefited the whole community, not just a few individuals. Later on, similar Village Conservation Groups were established in the other five project villages.

Project interventions and related training activities at Manyuan have had three aims. One is the conservation and restoration of the forest on the Nong Holy Hill, including the restoration of associated land that had become eroded and degraded after deforestation. Training sessions connected with this aim have covered the selection of species to plant for afforestation and their propagation, transplantation and post-planting care. The other two project aims are conservation of traditional medical knowledge and the revitalisation of home gardens. Training sessions to back up these two aims have covered the local names of plant species, their scientific identification, local distributions, traditional uses, and planting and horticultural techniques.

To promote policy development, the project has organised annual meetings with representatives of a number of line agencies, including SEPA, Public Health and a Dai Medicine hospital. These meetings have provided opportunities for the Village Conservation Groups and members of households to demonstrate their home gardens to government officials and outside experts. In that way, the communities and the line agencies have been able to engage in direct face-to-face exchanges with one another, allowing both sides space to explain their ideas and their practical experiences. In 2018, the Xishuangbanna Government agreed to the registration of a Xishuangbanna Ethnomedicine Association, which, as of 2023, has 260 members. Pei Shengji is its Honorary President. One of the activities in which the association has been engaged is to establish an apprenticeship scheme, under which students spend three years studying under experienced traditional doctors based on the master/pupil system of learning that is traditional in China and Buddhism. More than 20 students graduated in 2023.

14.3 Project Activities and the Nong Holy Hills

A survey of the Nong Holy Hill at Manyuan revealed that it contains 1.5 ha of well-preserved forest and a large surrounding area in which the forest had been degraded. Four of the 36 species of trees identified in the forest were on the Chinese Endangered Species Red List, among them *Antiaris toxicaria* and *Mangifera siamensis*. Following this inventory, a tree-planting programme was initiated on 2 ha of the degraded land associated with the Nong Holy Hill and, as of 2023, this had resulted in the planting of over 8000 seedlings belonging to 82 species. Among the species planted have been *Dipterocarpus turbinatus*, *Gmelina arborea*, *Horsfieldia amygdalina*, *Knema furfuracea*, *Mesua ferrea*, *Myristica yunnanensis*, *Parashorea chinensis* and *S. siamea*. The planting has been undertaken by members of the Manyuan Village Conservation Group, with the assistance of about 80 other villagers. The whole Nong Holy Hill, whether carrying well-preserved forest or intended for restoration, has been declared a Medicinal Plant Conservation Area (MPCA), along the lines of the ones that have been established at Ludian and Yongzhi further north in Yunnan (see Chapter 21).

On the basis of what has been achieved at Manyuan, a proposal was made in 2022 to Mr Dao Haiqing, the mayor of Jinghong County City (the administrative area in which Manyuan lies), to have a 'Nong Forest Day' annually, when all Dai villages in the county will be encouraged to help restore their Nong Holy Hill forests. The proposal was well received and, last year (2023), the Xishuangbanna Bureau of Forestry and Pasture declared that, on 22 May each year, all the Dai villages in the prefecture will be encouraged to look after their Nong Holy Hill forests and plant trees.

14.4   The Conservation of Medicinal Plants and Associated Traditional Knowledge

Dai Medicine is one of the four types of minority traditional medicine that are supported by the Chinese Government (see Section 7.3.3). It is an ancient tradition with a history traceable back over 2000 years. A considerable number of the 1260 plant species reported to be used are rainforest plants. A long-standing feature of the Dai's traditional medical culture has been the planting of medicinal plants in the precincts of Buddhist temples. Since 2018, the project has supported the rehabilitation of the temple courtyard in Manyuan by enriching it with medicinal plants. As at the beginning of 2023, 2000 m² of the precinct had been planted up with a total of 9600 plants belonging to 186 species, among them *Arundina graminifolia*, *Clinacanthus nutans*, *Dendrobium nobile*, *Dendrobium officinale*, *Dracaena cambodiana*, *Maytenus austroyunnanensis* and *Maytenus hookeri*. Included in the workplan for 2023 was encouragement to be given to the villagers to grow *D. nobile* in the open upper balconies of their traditionally constructed Dai houses. This is intended to serve as a complement at community level to the development of industrial plantations of *D. officinale* that has been ongoing in Yunnan (see Section 13.1). Figure 14.2 is a photograph taken from the air of Manyuan Village. The temple is the red-roofed building towards the top left. The Nong Goly Hill with it forest remnant is in the foreground. Note the rows of plantations crops in the fields.

At Manyuan the project has encouraged the construction of a traditional Dai herbal-steaming centre close to the Buddhist temple. Herbal-steaming is a long-standing curative practice in Dai Medicine. This facility is intended to be both a source of income for the community and to contribute towards the health of its members. As of now, the centre is open to treat 40–50 people a day, using a traditional local herbal formula containing 19 species of plants. The treatment is free-of-charge for villagers. Visitors are charged 40 RMB (US$5) for a 50-minute session. Traditionally, this treatment has been prescribed for the curing of colds, rheumatism, back and leg pain, and women's illnesses.

14.5   Revitalisation of Traditional Home Gardens

Project activities to revitalise the home gardens began in 2018 when a decision was made to concentrate on encouraging the growing of green vegetables. This

Fig. 14.2. Manyuan Village, Xishuangbanna, China. Photo: Mr Liu Guangyu (2020).

choice was based on the health-giving benefits that their eating will bring to the villagers and because it creates for them a new source of income – the latter related to the new markets that have opened up in Xishuangbanna through the growth in the numbers of domestic tourists coming to the prefecture to feel close to nature and through curiosity about the colourful ethnic minorities who live there and their traditional religious practices. The first practical step taken to launch the project was to start a demonstration garden in Manyuan. From this modest beginning, this facet of the project has spread to other garden-owners in Manyuan and to all the other five villages included in the project's programme. As of 2023, a total of 20,000 individual plants of over 36 species had been planted in 15 demonstration gardens. The species planted include *Elsholtzia kachinensis*, *Gymnema inodorum*, *Momordica charantia*, *Oenanthe javanica*, *Oroxylum indicum*, *Perilla frutescens*, *Solanum nigrum*, *Solanum spirale* and *Solanum torvum*. The tourism side of the project is continuing to develop. As of March 2024, a large parking area for cars has been constructed at Manyuan, houses have been opened for homestays and 35 villagers have set up stalls to sell fruit and vegetables.

15 Research into the Environmental History of Tropical Africa

WRITTEN IN THE FIRST PERSON BY ALAN HAMILTON

Abstract

This case study describes A.H.'s contributions towards knowledge of the climatic, vegetational and agricultural history of tropical Africa and its implications for conservation of biodiversity. His main research methods have been pollen analysis of sediments of post-40,000 BP age and biogeographical studies of the distributions of forest plants and animals. His work helped to overturn the Pluvial Theory, the accepted theory of Quaternary climatic and vegetational change up to the 1960s. After 1989, he became involved in practical conservation efforts at Bwindi Forest, Uganda, and on the East Usambara Mountains, Tanzania, two of the places which his work on environmental history had revealed to be prime sites in tropical Africa for the conservation of forest biodiversity.

15.1 How This Research Began

My interests in the geology of the Quaternary Period and tropical rainforest began, respectively, when I was about 10 years old and when I was an undergraduate studying Botany at the University of Cambridge (see Preface). Because of these interests, I was very happy to be offered an opportunity in 1966 to undertake research on the Quaternary history of tropical Africa, while working as a demonstrator in the Department of Botany at Makerere University College in Uganda. My research on Africa's Quaternary history since then has focused on the last 40,000 years (the Upper Quaternary), which is the period datable by radiocarbon. The Quaternary Period, which began 2.6 million ya (years ago), has seen the rise of the human species to a position of ecological domination on the Earth and has been marked by a series of global ice ages. The last global ice age lasted from about 115,000 ya to 10,000 BCE.

Looking back from the present, it is astonishing to realise how much has changed in scientific knowledge of the history of the environment since the 1960s. At that time, there was no particular concern about the destruction of tropical forest and climate change was not a political issue. These topics were not mentioned by the academic staff who taught the undergraduate course in natural sciences that I took at Cambridge University in 1963–1966. The word 'biodiversity' had not been invented. Only one lecturer in the Department of Geology at Cambridge (John Dewey) was undertaking research on plate tectonics, a theory that at least one of his colleagues openly derided as absurd. Plate tectonics has subsequently revolutionised geology.

Three things that *were* known in 1966 about the Quaternary environmental history of the Earth were: (i) that there had been a series of ice ages at North Temperate latitudes; (ii) that there had also been major changes in climate in tropical Africa; and (iii) that fossils, thought to be the ancestors or close relatives of our own species, could be discovered in Quaternary sediments in Africa. As for the number of glaciations, the current thinking was that there had been four glaciations in North Temperate latitudes and that these were correlated in tropical Africa with four glacial advances on the high mountains and with four periods of high lake levels at lower altitudes. The theory of Quaternary climatic change in tropical Africa then in vogue was one known as the Pluvial Theory, which postulated that each ice age at northern latitudes (a glacial) was marked by a *cool, wet climate* in tropical Africa (a pluvial) and that each intervening warmer period in the north (an interglacial) was marked by a *warmer, drier climate* in tropical Africa (an interpluvial) (Wayland, 1934, 1952).

15.2   Pollen-Analytical Research in East Africa

Pollen analysis has been the principal tool that I have used to investigate the environmental history of East Africa, pollen being the tiny (mostly 15–60 μm) particles produced by stamens (parts of the male organs of plants) which, after making contact with stigmas (parts of the female organs of plants), cause fertilisation and the formation of seed. Insects and the wind are the principal vectors that transfer pollen from stamens to stigmas. Pollen grains have decay-resistant coats and can be very abundant in the atmosphere. In consequence, after falling to the ground, they can become abundant in the sediments that form under lakes, swamps and bogs. The pollen contained in these sediments provides records of the types of plants that have existed in the past and, by inference, the environmental conditions under which they were living.

The principal way that pollen analysis has been used to reconstruct the Upper Quaternary environmental history of East Africa involves choosing sites containing sediments of Upper Quaternary age, collecting cores of sediments from the sites and then transferring them to a laboratory to study the types of pollen that they contain. The results are usually expressed in the form of diagrams showing how the abundances of different types of pollen change with depth (see Fig. 15.2 for an example). Next, the diagrams are classified into zones, each characterised by a particular assemblage of pollen types. Then, for each zone, the type of vegetation present in the past is inferred from the assemblage of pollen types present and subsequently, from this reconstructed vegetation, the characteristics of the environment that existed at the time.

The principal environmental factors considered in these assessments are temperature, climatic moistness and human influence. Normally, the age of sediments increases with depth in cores. Radiocarbon dating is the principal technique used for determining the absolute ages of sediments of Upper Quaternary age.

Some pollen grains extracted from sediments of Upper Quaternary age in East Africa are shown in Fig. 15.1. Photographs 1 and 2 are of the same grain but taken at different focal depths. Similarly, photographs 3 and 4 are of the same grains, but at different depths of focus. The pollen grain shown in photos 1 and 2 is a 28,000-year-old grain in a core of sediment extracted from Kamiranzovu Swamp in Nyungwe Forest, Rwanda. It belongs to the wet-loving genus *Platycaulos* in the *Restionaceae*, a family of grass-like plants that is no longer found in Rwanda. The two types of pollen in photos 3 and 4 are 13,000-year-old grains found in a core of sediment extracted from Kuwasenkoko Swamp, also in Nyungwe Forest. They are (larger) *Dendrosenecio*-type pollen and (smaller) *Helichrysum*-type pollen, both of which are produced by species belonging to the family *Asteraceae*. The *Dendrosenecio*-type pollen probably comes from the genus *Dendrosenecio* (the giant groundsels), which are characteristic plants of the high-altitude Afroalpine Belt on the East African mountains (see Fig. 15.4). The finding of

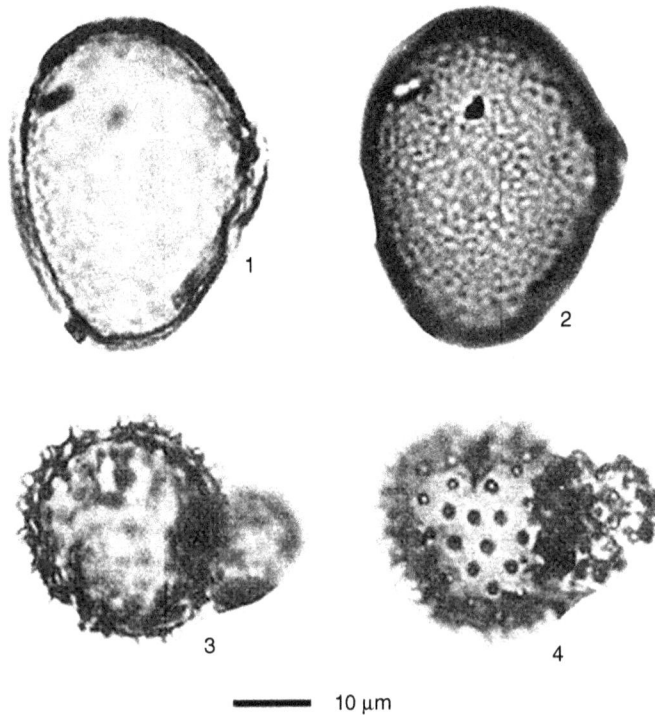

——— 10 μm

Fig. 15.1. Fossil pollen grains from Rwanda. Photo: A.H.

the *Platycaulos* and *Dendrosenecio*-type pollen grains in Quaternary sediments in Nyungwe Forest, which is situated well within the Montane Forest Belt, shows how greatly the flora and vegetation of the locality have changed during the relatively recent geological past.

Under the guidance of my research supervisor, Michael Morrison, I concentrated in my doctoral studies on working out how to interpret those pollen diagrams that already existed for East Africa or were about to be completed, rather than collecting new cores of sediment myself and constructing new pollen diagrams. The sites to which these existing or soon-to-be finished pollen diagrams relate are shown on Map 15.1. They are for sites: (i) on Mt Kenya and the Cherangani Hills in Kenya (Coetzee, 1967); (ii) on the Rwenzori Range in Uganda (Livingstone, 1967); (iii) in the Rukiga Highlands in Southwest Uganda (Morrison, 1968; Morrison and Hamilton, 1974); and (iv) at localities in the northern part of Lake Victoria (Kendall, 1969). Tragically, Michael Morrison passed away prematurely in 1970 and the final stages of the work, as described in his 1974 paper on Southwest Uganda, were undertaken by me.

The research which I undertook during my postgraduate studies in 1966–1970 had three components (Hamilton, 1972). One was a study of the morphology of the pollen of selected East African species, the purpose being to be able to identify the grains when I encountered them either in the modern pollen rain or in the fossil state (Hamilton, 1976a). The second component was an investigation of the characteristics of the modern pollen rain in Uganda, the purpose being to be better able to reconstruct past vegetation from the assemblages of pollen found in pollen diagrams. The third component was to add to

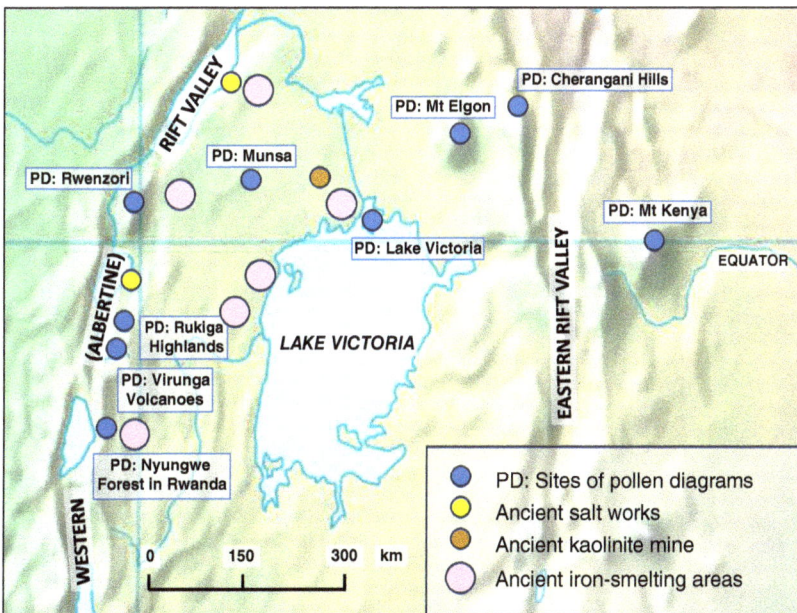

Map 15.1. Archaeological sites and sites of pollen diagrams in the greater Lake Victoria region.

the existing knowledge of the ecology of East African plants (see Section 15.5). The general principle followed in these studies in relation to reconstructing environmental history was the one that underlies much of geological science, namely that 'the present is the key to the past' (Lyell, 1830–1833).

In 1976, I received a grant from the British Government enabling me to continue with my investigations into the East African past. Together with Alan Perrott, a research assistant, I spent seven months doing fieldwork in East Africa, mostly in Kenya and Rwanda. One part of the work was a study of the characteristics of the modern pollen rain along two transects running up the Kenyan side of Mt Elgon (Hamilton and Perrott, 1980). The transects extended altitudinally from the agricultural/savannah zone below 2000 m, up through various types of montane forest (2000–3200 m) and into the Ericaceous and Afroalpine Belts at higher altitudes (see Fig. 15.4). One of the transects was situated on a drier aspect of the mountain and the other on a wetter. This study included the use of pollen traps to determine the absolute numbers of pollen grains being deposited from the atmosphere per unit surface area per unit time. This work confirmed that the montane forest on Mt Elgon is a much higher producer of pollen than the vegetation found at higher and lower altitudes. It showed that there are some pollen types, for example *Podocarpus*-type (produced by two species of conifers on Mt Elgon) and *Urticaceae* (produced by species in the stinging nettle family), can be abundant in the pollen rain, even when the plants producing the pollen are growing a long distance away.

15.3 Reinterpretation of Pollen Diagrams from East Africa and a New Hypothesis about Quaternary Climatic Change in Tropical Africa

Towards the end of my doctoral studies, I turned my attention to reinterpreting those pollen diagrams for East Africa that had already been published. When I did so, I was amazed to find that the Last Glacial Maximum (LGM – c.24,500–17,000 BCE), which was the last time that glaciers globally had been at their maximum extent, had been climatically cool and *dry* and that the postglacial period that followed had been warmer and *wetter*. This result was not anticipated, being at complete variance with the Pluvial Theory, which postulated that the ice ages were marked by a cool, *wet* climate in East Africa and that the interglacials (such as the present postglacial) were warmer and *drier*.

An example of one of the pollen diagrams that I reinterpreted is shown in Fig. 15.2. It is a redrawn version of a diagram that Dan Livingstone of Duke University in North Carolina (USA) constructed, based on his counts of pollen in samples of sediment taken from a 6-m-long core from Lake Mahoma. This is a high-altitude lake situated at 2960 m in the upper part of the Montane Forest Belt of the Rwenzori Range, which lies on the border between Uganda and DR Congo. The lake occupies a kettle hole formed behind a terminal moraine dating to the LGM. Three levels of the core have been dated by radiocarbon dating, as indicated.

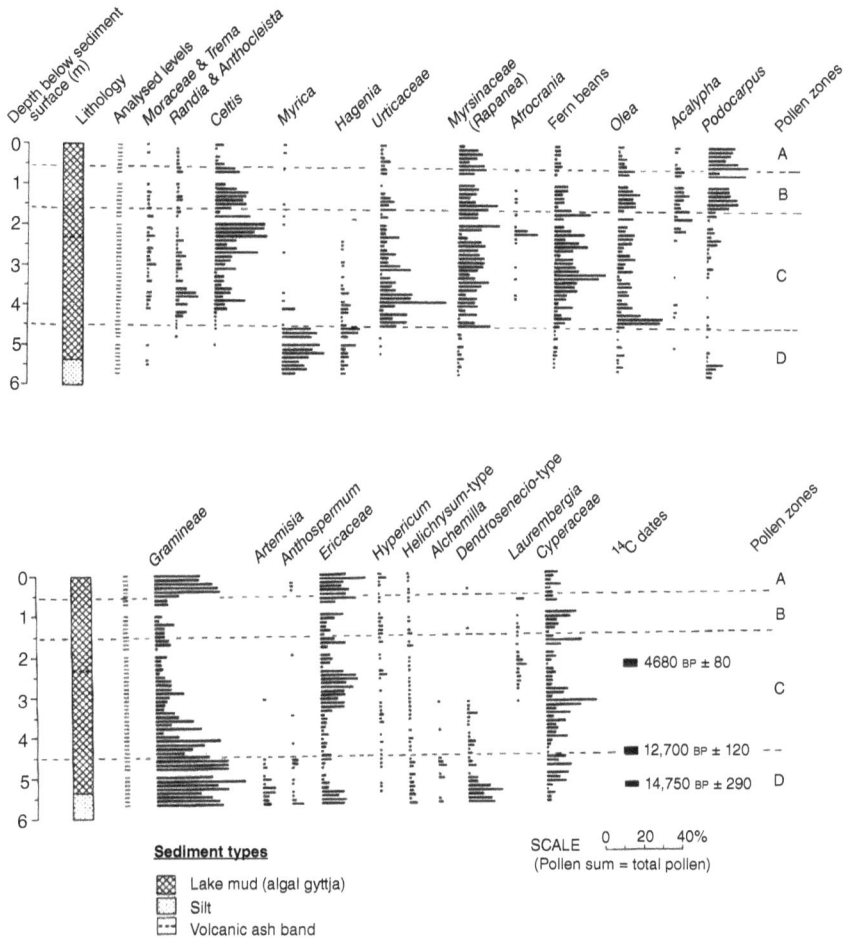

Fig. 15.2. Pollen diagram for Lake Mahoma, Rwenzori, Uganda.

The pollen diagram is classified into four zones, A–D. According to my reinterpretation, the climate during Zone D times, which dates to the last part of the last ice age, was cold and dry. Among other indications, coldness is suggested by the high values of *Dendrosenecio*-type pollen, and dryness by the high values of *Myrica* and grass (*Gramineae*) pollen, and also by the low values of *Urticaceae*. *Dendrosenecio* (the giant groundsel genus) is characteristic of the high-altitude Afroalpine Belt (Fig. 15.4). The *Myrica* pollen is thought to have come from the tree *Morella salicifolia*, which is characteristic of drier types of montane forest and montane woodland. The abundant grass pollen has probably come mostly from low-altitude savannah and been blown up in the wind to the altitude of the lake.

A warmer and wetter climate in Zone C times is suggested by the high values of *Rapanea*, a montane forest tree, fern spores (beans) and *Urticaceae*, which is a family of herbs, shrubs and climbers that is particularly common in wetter types of montane forest (Hamilton, 1976a). It is also indicated by the high values of *Celtis* and *Moraceae*, a genus and family of plants that can be well-represented among the trees in lowland African rainforests. *Celtis* and *Moraceae* are known to produce large quantities of well-dispersed pollen and much of their pollen in the lake sediments at Mahoma is believed to have been blown up from the lowlands.

The rise in *Podocarpus* pollen at the Zone C/B boundary is paralleled in many other pollen diagrams for East Africa. It is believed to have been caused by a major transition from a wetter to a drier climate at *c*.3000–2500 BCE that was experienced across the whole of tropical Africa. The rise of grass pollen and fall in *Celtis* pollen at the Zone B/A boundary are believed to have been due to the extensive replacement of lowland tropical forest by grass-rich vegetation, following more intensified agriculture and pastoralism at *c*.1000 CE (see Section 3.3.1).

My reinterpretation of the Lake Mahoma and other pollen diagrams available for East Africa in 1968 indicated that East Africa was subject to a cool and *dry* climate during the LGM and that the climate of the subsequent postglacial had been warmer and *wetter*. This was so radically different from the established Pluvial Theory that I had some concerns about whether it was really true. Therefore, it was a relief for me, soon after reaching my conclusion, to receive a scientific paper from R.L. Kendall, one of Dan Livingstone's students. It contained the results of his research into the Upper Quaternary history of Lake Victoria and showed that he had independently come to the same conclusion (Kendall, 1969).

Lake Victoria is a large, low-altitude lake (1135 m) straddling the Equator and supporting a small area with a lowland tropical forest climate in its north-western hinterland (the homeland of the Baganda). Kendall's pollen diagrams, along with chemical and physical analyses of the lake's sediments, showed unequivocally that the level of the lake was much lower than now during the LGM, that it had filled up during the transition from the LGM to the postglacial, and that it had finally started to spill over its brim and flow down the course of the river Nile at about 12,000 BCE. There was very little forest in the proximity of the lake during the LGM, forest was at its maximum extent at 9000–1800 BCE and extensive reduction in the extent of forest, presumably to make way for agriculture, occurred at *c*.1000 CE.

After completing my doctoral research, I took time out from pollen analysis to write a field guide to the forest trees of Uganda (Hamilton, 1981b), begin a career as a university lecturer in Northern Ireland and undertake desktop research into the modern distributions of plants and animals in African tropical forests (see Section 15.5). Returning to a second phase of pollen analysis in 1976, I set myself the task of testing a hypothesis that had originated from a combination of my previous pollen work and my desktop research into the distributions of plants and animals in Africa's tropical forests. The hypothesis was that, during each world glaciation, the climate in tropical Africa had generally been cool and dry, with the extent of

forest much restricted, and that, during the interglacials (including the present postglacial), the climate had been warmer and wetter, with forest more
extensive. From my research into the modern distributions of forest species
of plants and animals in African forests, I had identified certain localities
that were predicted to have been places where forest survived during the
glacial maxima. The nearest of these refugia to Uganda was postulated to
have been situated in Kivu Province in Eastern DR Congo, possibly extending slightly into Southwest Uganda. I had further postulated that it was from
this refugium that forest had spread out northwards and eastwards across
Uganda when the climate became wetter and more suitable for forest at the
beginning of the postglacial.

My plan was to test this hypothesis of forest spread by undertaking pollen
analysis on cores of sediment collected from two localities – one that was predicted to have remained relatively moist during the LGM, with at least some
forest remaining, and the other from a locality that was predicted to have
been relatively dry, with little, if any, forest present. Originally, the plan was
to site the first locality in Southwest Uganda, close to where it abuts on Kivu
Province in the Congo. However, by then, Idi Amin had become President
of Uganda and I thought that he might not take kindly to two foreigners
working in remote places near Uganda's borders carrying equipment (for
boring into bogs) whose parts superficially resembled gun barrels. Therefore,
I decided to change the sample site to nearby Rwanda and, specifically, to
Nyungwe (Rugege) Forest, which is close to Kivu. The second sample site,
where the climate was predicted to have been arid during the LGM, was the
eastern side of Mt Elgon, which is in Kenya. In the event, Alan Perrott, my
research assistant, and I did manage to collect cores of sediment from two
sites in Nyungwe Forest and from three from the Kenyan side of Mt Elgon,
as well as from a site on the Aberdares Range in Kenya. On analysis in the
lab, the results proved broadly consistent with the initial hypothesis that 'glacial in tropical Africa = cool and dry' and 'interglacial = warmer and wetter'
(Hamilton, 1982).

Other pollen-analytical work that I have undertaken in tropical Africa
included pollen analysis of two cores of peat collected by Alayne Street-Perrott
in Ethiopia (Hamilton, 1982) and, together with David Taylor, pollen analysis
of cores of sediment collected by one or both of us in the Rukiga Highlands
of Southwest Uganda (Hamilton, 1982; Taylor, 1990). More recently, David
Taylor and I have reinterpreted environmental change in three previously
published pollen diagrams for widely separated sites situated in the highlands
along the eastern rim of the Albertine Rift Valley. From this reinterpretation,
we concluded that there was a major climatic event at 40,000 BCE, probably related to a large-scale shift of the Earth's tectonic plates (see Section
15.4.1). David Taylor, Rob Marchant and their associates have advanced the
work further, including concerning the histories of pastoralism and large-scale
cereal-growing in Uganda (see Section 15.4.2) (Marchant et al., 1997, 2006;
Marchant and Taylor, 1998). Figure 15.3 shows Adrian Mwesigye, a former
student of Botany at Makerere University, working with me in 1984 to collect
a core of sediment for pollen analysis from Muchoya Swamp in the Rukiga
Highlands.

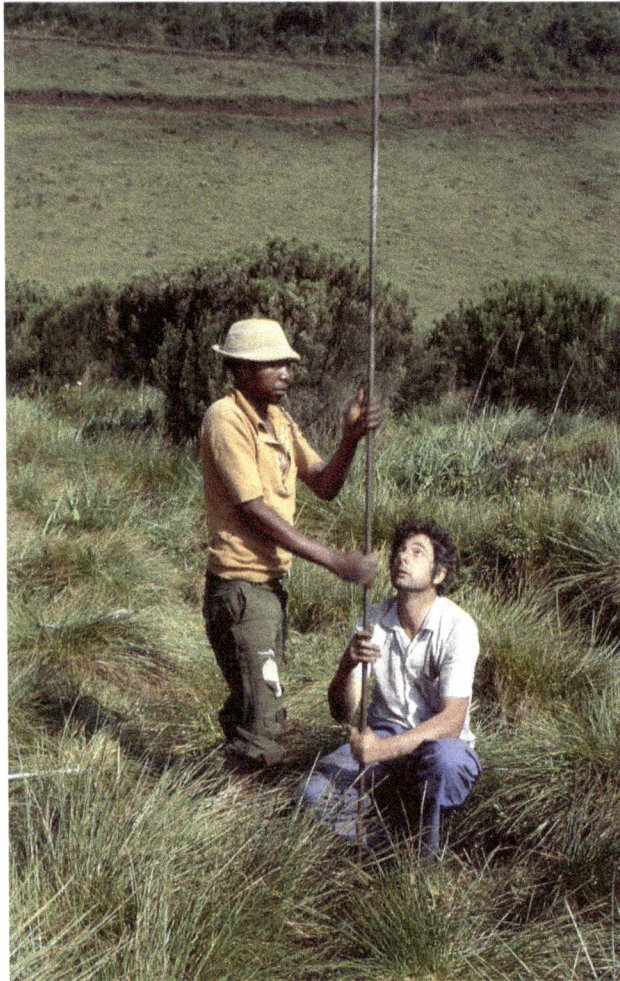

Fig. 15.3. Collecting a peat core from Muchoya Swamp, Uganda. Photo: David Taylor (1984).

15.4 Synthesis of Findings about the Post-40,000 BP Environmental History of Tropical Africa (especially East Africa)

This section contains highlights of the environmental history of tropical Africa (especially East Africa) during the last *c*.40,000 years, based on the syntheses that I have occasionally attempted (Hamilton, 1982, 1992; Hamilton and Taylor, 1991; Hamilton *et al*., 2016). Contemporary written records for inland East Africa start to become available from the mid-19th century CE. For earlier times, I have tried to make use of all the information available that tells something about what happened in the past, regardless of the academic discipline. Among the disciplines considered have been archaeology, biogeography, geomorphology, glaciology, historical linguistics, oral history and palaeontology. Doubtless, there are many gaps in my scanning of the literature.

15.4.1 The tropical African environment during the last ice age and up to a mid-postglacial transition towards a drier climate

In general, the climate of tropical Africa during the last ice age, which extended from about 115,000 ya to *c*.10,000 BCE, was colder and drier than during the succeeding postglacial (10,000 BCE to the present). However, in detail, the climate of the last ice age was far from uniform. For brief periods, it was just as warm and wet as during the postglacial. One such brief period at *c*.40,000 BCE terminated abruptly when it became suddenly much colder and drier (Hamilton and Taylor, 2018). Contemporaneous with this climatic downturn, there were other major geological and geographical events. Volcanoes were particularly active along the Western (Albertine) Rift Valley, there was a massive volcanic eruption in Italy (the Campanian Ignimbrite super-eruption), there was a short reversal of the Earth's magnetic field (the Laschamp Event) and large armadas of icebergs invaded the North Atlantic. As in tropical Africa, the climate of Western Eurasia similarly transitioned abruptly to becoming much colder and drier, which is believed to have contributed to the final demise of the Neanderthals and the complete takeover of Western Eurasia by *Homo sapiens* (Sepulchre *et al.*, 2007; Hoffecker *et al.*, 2008). It can be hypothesised that the root cause of these events was a major shift of the Earth's tectonic plates.

The last time during the Quaternary when there was a maximum ice cover on the Earth is known as the LGM, dating to *c*.24,500–17,000 BCE (Clark *et al.*, 2009). In tropical Africa, temperatures were at a minimum, aridity at a maximum and forest cover was at its smallest extent. As everywhere else in the world, people in Africa lived by hunting and gathering. Agriculture had not yet been invented. The opposite ice age environmental extreme – when temperatures, climatic moistness and forest cover were at their maxima in tropical Africa – came to an end at about 3000–2500 BCE.

A good indicator of the end of this warmest, wettest part of the postglacial (prior to when humans started to have much influence on the climate) is a rise in values of *Podocarpus*-type pollen in pollen diagrams, as in the diagram for Lake Mahoma (Fig. 15.2). This rise can be seen even in pollen diagrams for places far away from where *Afrocarpus* and *Podocarpus*, the two genera that produce this pollen type, are likely to have been growing. Its strength as a maker of time is due to the vast amounts of pollen that these two genera of conifers produce and the exceptional abilities of their pollen to be carried long distances by the wind. As mentioned in Section 3.2, it seems likely that there was a causal relationship between this climatic event and the onset of farming in sub-Saharan Africa and, as mentioned in Section 3.3.1, the initiation of the expansion of Bantu-speaking people in sub-Saharan Africa.

In between the environmental extremes of the full glacial and the full interglacial climatic modes, the tropical African environment experienced a number of major events. One was an apparent 'switching on' in East Africa of the Southeast Monsoon at *c*.12,500 BCE. Nowadays, this is the moister of the two monsoons that bring rain-bearing clouds to East Africa, but the orientation of the moraines that formed on the mountains in Eastern Africa that were glaciated during the LGM show that, at that time, the Northeast Monsoon was the moister of the two (Hamilton and Perrott, 1979; Hamilton, 1982). There was a dramatic environmental event at $c.11,000$ BCE in East Africa, when soils were destabilised and thick layers of inorganic sediment accumulated rapidly in sedimentary basins. Tectonic movements may have been the cause (Thompson and Hamilton, 1983).

As mentioned in Section 3.2, North Africa had a much wetter climate than later between $c.9000$ and 3000 BCE and, during the latter part of this period, herders were pasturing their livestock in extensive parts of what is now the Sahara Desert. The earlier part of this moist climatic phase, between $c.9000$ and 6000 BCE, was particularly wet and there were connections between the faunas of what are now widely separated river systems and water bodies in the Sahel of North Africa. For example, there were interchanges of fish between Lake Chad and the far-away river Nile (Drake *et al.*, 2010).

15.4.2 Early agriculture within a tropical rainforest setting in East Africa

Agriculture within a tropical rainforest environment has been extensively practised in the most westerly part of East Africa, including in Uganda, from the first centuries BCE (Taylor, 1990). Since there is linguistic evidence that Bantu-speaking people, with their forest-orientated agriculture, settled in the Great Lakes Region of East Africa at 1000–500 BCE (see Section 3.3.1), it is likely that they were responsible for at least some of this agriculture.

There is some evidence of earlier agriculture within a rainforest environment, albeit of a localised nature. A pollen diagram for one site in extreme Southwestern Uganda suggests that there was a long phase of local forest clearance dating to 1700–750 BCE (Hamilton *et al.*, 2016). From nearby Rwanda, there is archaeological evidence that sorghum and finger millet were being cultivated during the first millennium BCE and that the people associated with it were smelting iron (see Section 3.3.1). Since there is linguistic evidence that the Bantu-speaking people who settled in the Great Lakes Region of East Africa acquired the knowledge of how to grow cereals and work iron from Nilo-Saharan speakers (Ehret, 1982, 1998), it is likely that it was Nilo-Saharan, rather than Bantu, speakers, who were the first people to practise agriculture within a tropical rainforest setting in East Africa.

Palynological and other sediment analyses, backed up by archaeological research, show that there was a major reduction in the extent of lowland forest in Western Uganda and in the northern hinterland of Lake Victoria at $c.1000$ CE (Taylor *et al.*, 1999; Ssemanda *et al.*, 2005; Ryves *et al.*, 2011; McGlynn *et al.*, 2013). At the archaeological site of Munsa in the savannah zone of Western Uganda, there was a contemporaneous establishment of an economy based on cereal-growing and the large-scale keeping of cattle (Map 15.1) (Lejju *et al.*, 2005, 2006). This intensification of food production was accompanied by an increase in soil erosion and the siltation of a swamp that, in consequence, became covered with the giant sedge *Papyrus*. As mentioned in Section 3.3.1, the major reduction in forest in Uganda at $c.1000$ CE can be connected through oral history with the origins of hierarchical societies in the Great Lakes Region and, specifically, with the origins of the Kingdom of Bunyoro-Kitara in the savannah zone of Western Uganda and the Kingdom of Buganda in the rainforest zone near Lake Victoria.

15.5 Ecological and Biogeographical Studies

15.5.1. Vertical (altitudinal) distribution of plant species in East Africa

Palynologists (people who study pollen) have used the past upwards and downwards altitudinal movements of plant species in East Africa, as apparent from pollen diagrams, as a surrogate way to measure how temperatures have changed in the past. This method is based on the observation that all species of East African plants have upper and lower altitudinal limits to their ranges and the meteorological knowledge that temperatures in East Africa decline, on average, at a rate of about 6°C per 1000 m (the lapse rate). As a consequence of the association of plant species with altitude, it is possible to represent the natural vegetation types present on the East African mountains as a diagram showing their distribution in relation to altitude (Fig. 15.4).

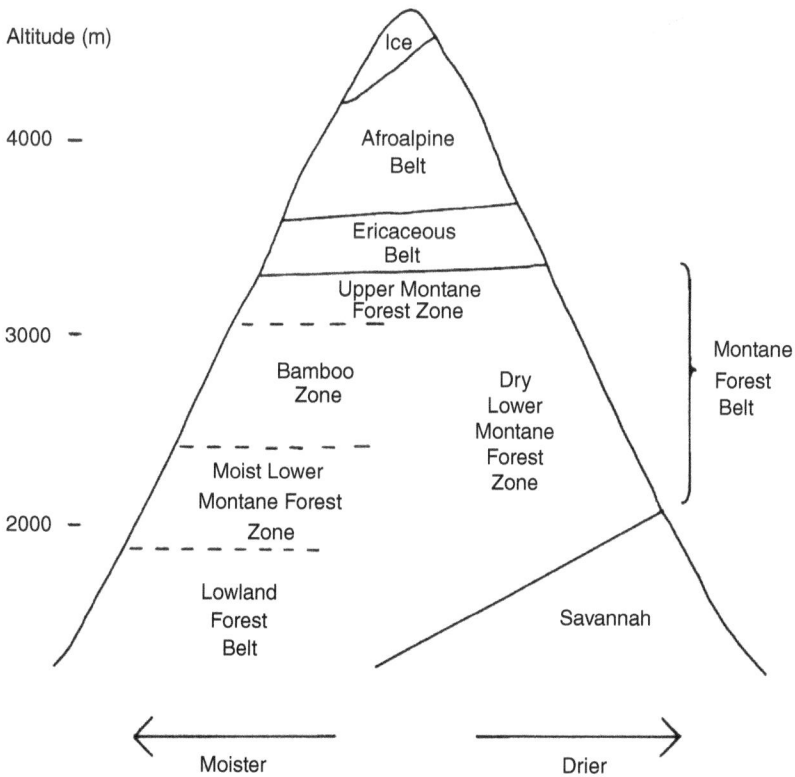

Fig. 15.4. Belts and zones of vegetation on the East African mountains.

Observations that I made in the field during the course of my doctoral studies indicate that it is particularly the *upper* altitudinal limits of their ranges that are most limited by temperature (Hamilton, 1972). One of these observations is the fact that several species of plants reach their highest known altitudinal limits close to a hot spring on Mt Elgon, as was first noted by Hedberg (1951). Another is the observation that the vegetation belts are altitudinally inverted in places where the shape of the land prevents cold air draining away to lower altitudes at night. For example, the Ericaceous Belt occurs altitudinally *below* (rather than above) the Montane Forest Belt in the virtually enclosed valley surrounding Kuwasenkoko Swamp in Nyungwe Forest in Rwanda (Hamilton, 1982). A third observation is that, although species tend to ascend to roughly the same altitudes on all the mountains on which they occur, this altitude is actually displaced somewhat downwards in climatically moister areas, which is where daytime cloud is most persistent and which, accordingly, have lower mean maximum temperatures. In wider geographical perspective, this lowering of vegetation zones is an example of a well-known ecological phenomenon known as the *massenerhebung* effect. This is the lowering of tropical vegetation zones in climatically wetter places, on small, isolated peaks and on mountains close to the sea (Richards, 1964; Grubb, 1971).

After completing my doctoral studies, I followed up my research on the altitudinal distribution of plant species in East Africa in three ways. One was an investigation of the altitudinal ranges of species of forest trees in Uganda, based on the altitudes at which specimens in herbaria had been collected or their presence had been noted in the Working Plans of the Forest Department (Hamilton, 1975a). Another was an investigation of changes in forest vegetation with altitude on the East Usambara Mountains of Tanzania (see Section 16.3). Both of these studies showed that, when the whole of Uganda or all the East Usambara Mountains are considered as single entities, there are no 'critical altitudes' at which the floristic composition of the forest changes abruptly. Rather, the types of tree species present in the forests change steadily as the altitude increases.

The third follow-up study was a survey of altitudinal changes in the floristic composition of forest vegetation on Mt Elgon in Kenya (Hamilton and Perrott, 1981). This study differed from the two mentioned above in that it included the entire higher plant flora (except epiphytes) – that is, trees, lianas, shrubs and herbs. It was based on the field-recording of species encountered in sample plots positioned at altitudinal intervals along two transects, one on a wetter aspect of the mountain and the other on a drier. These were the same transects as were used for studying the modern pollen rain, which meant that quantitative comparisons could be made between the abundance of pollen types in the pollen rain and the abundance of their parent species in the vegetation (see Section 15.2). One of the conclusions of this study was that both the tree/liana flora and the herb/shrub flora become increasingly impoverished in species as the altitude increases. Another was that the forest vegetation on the wetter and drier slopes become increasingly similar floristically as the altitude increases, a finding in line with the observations of others (Hedberg, 1951; Langdale-Brown et al., 1964).

15.5.2 Horizontal (geographical) distribution of forest plant and animal species in tropical Africa

A field guide to the tropical forest trees of Uganda that I wrote in 1971 included information on the distributions of the species in the country (Hamilton, 1981b). On examining the distributions, I discovered that the patterns shown by many species could not be explained solely with reference to their ecological requirements (Hamilton, 1974). Underlying the ecological factors were gradients of decreasing numbers of species away from the southwest of the country towards the north and the east. To explain these gradients, I postulated that they had originated through the differing abilities of species to spread from a forest refugium that had existed in Kivu Province in DR Congo during the arid LGM (labelled 'East Congo core area' on Map 15.2). I thought that this refuge may have possibly extended slightly into extreme Southwest Uganda. Forest tree species able to grow at *very* high altitudes were an exception to the pattern. They tended to be widely distributed, occurring on all the mountains that reach to a sufficiently high altitude. I postulated that this was because they had been able to persist during the arid LGM, being under reduced evaporative stress under the cold temperatures that are found at higher altitudes.

Later, I followed up this hypothesis about forest spread with a study of the relationships between the distributions of forest tree species in Uganda and the ways that their fruits or seeds are dispersed, as judged by their morphological characteristics (Hamilton, 1975b). This showed that species with large fruits or seeds that appear to be animal-dispersed, but are too heavy to be carried by those that can fly, are concentrated in their distributions to the extreme southwest. In contrast, the species that have fruits or seeds adapted for dispersal by birds or fruit bats tend to widely distributed, as do those with very lightweight seeds with hair-like plumage that enables them to be wafted along easily in the wind. This finding was taken as evidence backing up my hypothesis about post-LGM forest spread.

Just as the conclusions from my pollen work about the climate of the last ice age in tropical Africa had conflicted with the established Pluvial Theory, so too did my hypothesis about the distribution of forest being restricted in Uganda during the LGM conflict with the currently accepted scientific theory. This theory, which was associated particularly with the ornithologist Reggie Moreau (1897–1970), was that forest, especially montane forest, was much more extensive in tropical Africa during the Quaternary glacial periods (including the LGM) than it is now and that, during the interglacials (including the postglacial), its areal extent was restricted (Moreau, 1966). Moreau was convinced of this theory because he had noted that the avifaunas of the montane forests on the scattered high mountains of tropical Africa have many similarities with one another and thought that this could only have been achieved if montane forest had been much more extensive during the glacial periods than it is now, making it relatively easy for the birds to fly from one high mountain to another.

After I had published my hypothesis arguing for '*glacial* = *forest retraction*; interglacial = forest expansion' in Uganda, Professor van Zinderen Bakker, a pioneer researcher into the Quaternary environmental history of Africa, wrote to me to suggest that I undertake a general review of the historical significance of the modern patterns of distribution of plants and animals in African tropical forests. I agreed and, on doing so, discovered that there was other biogeographical evidence supporting my hypothesis of a forest refugium in Kivu Province during the LGM. Furthermore, the sites of several other former forest refugia in tropical Africa could be identified (Hamilton, 1976b). The gradients of decreasing numbers of forest tree species northwards and eastwards in Uganda were paralleled in other taxonomic groups, such as birds and mammals, and there were similar gradients associated with the other former refugia (Map 15.2).

I followed up this general review of African forest biogeography with more detailed studies of the distributions of passerine birds and forest monkeys. Both supported and helped to refine the 'glacial = forest retraction; interglacial = forest expansion' hypothesis (Diamond and Hamilton, 1980; Hamilton, 1988b). A dynamic image of the history of African tropical forests during the Quaternary emerged – one of forests expanding and contracting with the rhythm of the global ice ages, with the sites of the forest refugia tending to form at the same places during each of the contractions. The pulsating forest had acted as a pump for the creation of new species. During episodes of extensive forest cover, those species that were more mobile had been able to spread from their existing refuge areas and reach others, where some of them managed to form sufficiently robust populations to be able to survive during future periods of ice age tropical aridity. During times of forest retraction, the isolated populations of species in the different refugia had little or no genetic contact with one another and started to evolve in their own separate ways. The end result was to present to the modern-day taxonomist a portfolio of related taxa occupying different refuge areas, some seemingly identical to one another

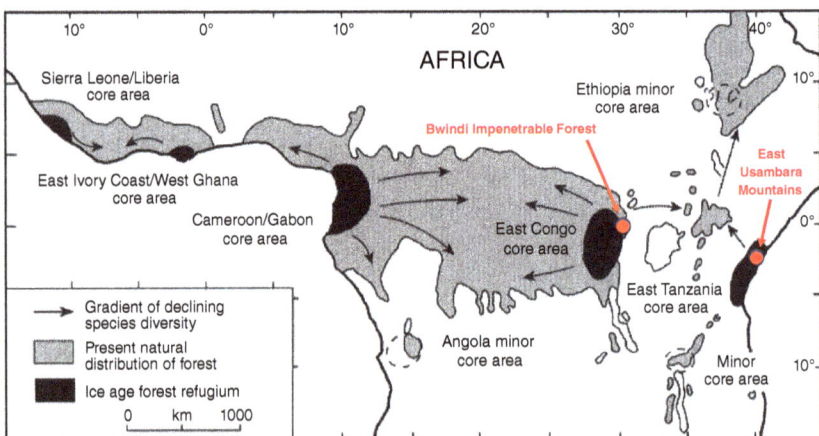

Map 15.2. Equatorial Africa showing the present natural distribution of tropical forest, Quaternary forest refugia and gradients of declining species diversity.

and others differing slightly, at subspecific level, at specific level or even determined to belong to different genera.

I concluded from the results of my research into the history of forest in tropical Africa that the sites of the former Quaternary refugia should be high priorities for the conservation of biodiversity (Hamilton, 1981a; Hamilton et al., 2001). This was because these are the places with the greatest concentrations of plant and animal species, where their populations are likely to be most genetically diverse and where they have the greatest chances of surviving future arid climatic episodes. In one publication, I singled out two places in East Africa that, I thought, were particularly high priorities on this account – the forests of the East Usambara Mountains in Tanzania and Bwindi Impenetrable Forest in Uganda (Hamilton, 1982). As it turned out, I later became engaged at both of them in practical conservation efforts intended to increase the chances of survival of their forest species (see Chapters 16 and 18).

Moreau, the person most closely associated with the overturned 'glacial = forest expansion; interglacial = forest retraction' hypothesis, was a British colonial scientific civil servant who worked during 1928–1946 at a scientific research station at Amani on the East Usambaras. I lived at the same station myself during 1986–1987 (see Chapter 16). During his time at Amani, Moreau had noticed that the upper altitudinal limits of tree species in the East Usambara forests are depressed in altitude compared to further inland in Africa, including in Uganda. I was able to confirm that this is, indeed, the case (see Section 16.3). Moreau attributed the depression in altitude of forest trees on the East Usambaras to the persistent presence of low-lying cloud hanging around the upper slopes of the mountains, reducing daytime temperatures. Since Moreau's time, it has become known through pollen-analytical and other research that forest did, indeed, persist on the East Usambaras during the Quaternary and also that the climate was exceptionally misty (Lovett and Wasser, 1993; Fjeldså and Lovett, 1997; Marchant et al., 2006; Mumbi et al., 2008; Finch et al., 2014).

There was one other study of the horizontal (geographical) distribution of species in African tropical forests in which I have been involved. This was on a much smaller scale and started from curiosity about why individual specimens of trees within a forest are rooted at precisely the places where they are. What I did, starting in 1968, was to select a small area (80 m × 80 m) of forest in Mpanga Forest Reserve near Kampala, identify and map the trees, and then follow this up by resurveying the same trees and noting any deaths and newcomers at approximately 10-year intervals for about 50 years. On more recent occasions, David Taylor has worked with me to inventory the plots (Taylor et al., 2008). The total number of individual trees in the plot was originally about 450, belonging to 55 species. The most recent survey was in 2016. Because I have repeated measurements on the same trees, these data have proved useful to add to others for big-data studies of the ecology of tropical forests and how they have responded and contributed to recent anthropogenic climate change (Lewis et al., 2009; Hubau et al., 2020; Sullivan, 2020; Cooper et al., 2024).

16 A Conservation Crisis Caused by Destructive Logging in the East Usambara Mountains, Tanzania (1986–1987), and a Sequel in Mexico (1988)

WRITTEN IN THE FIRST PERSON BY ALAN HAMILTON

Abstract

The forests of the East Usambara Mountains in Tanzania are small in total area, exceptionally rich in biodiversity and, as catchments, vital for providing water supplies to the people living in the surrounding lowlands. In 1986, logging supported by the Finnish aid agency FINNIDA had felled almost all of the forests that were accessible to heavy-duty logging machinery. Written from a personal perspective by A.H., this chapter describes how an IUCN scientific team assisted the Tanzanian Government to halt the logging and establish Amani Forest Nature Reserve, the first of several such Forest Nature Reserves now existing in the country. This work led to the follow-on engagement of A.H. to prepare an environmental impact assessment of a planned aid-supported forestry project in Mexico, where there was an additional problem of poor forest management connected with the presence of drug cartels.

16.1 Invitation and the Conservation Challenge

While working in my office in the University of Ulster in Northern Ireland in 1986, I received a telephone call from the International Union for Conservation of Nature and Natural Resources (IUCN) asking if I was available for a year to assist with the conservation of the forests of the East Usambara Mountains in Tanzania (see Map 15.2). They wanted me to head up a scientific team to gather information for a new management plan that was to be prepared for the forests. They explained that there was an urgency about starting the job, because the forests were being logged destructively and only a small area of 'accessible' unlogged forest remained ('accessible' referring to being accessible to heavy-duty logging machinery).

I was immediately interested, because I knew from my own research that the forests of the East Usambaras are exceptionally rich in their numbers of

© Alan Hamilton and Pei Shengji 2024. *History and Future of Plants, Planet and People*
(Alan Hamilton and Pei Shengji)
DOI: 10.1079/9781789248944.0016

species of trees, birds, amphibia and other taxonomic groups that are found in no or few other places in the world (see Section 15.5.2) (Hamilton, 1981a, 1982). Loss of the forests would result in an irretrievable reduction in the total biological wealth of the Earth. I felt that having urged for extra effort to be put into saving the forests, I should respond positively to IUCN's request, now that I was being presented with an opportunity to do something practical. Accordingly, I requested the University of Ulster for a 1-year leave of absence so that I could accept IUCN's offer. At first, the university flatly refused, but, on appeal, the university authorities relented and agreed to let me go, but only on condition that I return for a 3-month period during the year to cram in my essential teaching.

Geographically, the East Usambaras Mountains form one of a group of isolated mountain ranges and individual summits, known as the Eastern Arc Mountains (Lovett and Wasser, 1993). Situated on the eastern side of East Africa, mostly in Tanzania, the Eastern Arc Mountains differ from many of the high mountains found further west in tropical Africa in being composed of ancient hard Precambrian rocks, rather than much more recent volcanics. The East Usambaras are the closest to the sea of any of the Eastern Arc Mountains, being visible from the port of Tanga on the shore of the Indian Ocean, which is situated only 40 km to the east.

The East Usambaras consist of a main range and three isolated mountains on its seaward side, Mts Mlinga (1069 m), Mhinduro (1033 m) and Mtai (1060 m). The main range, which levels off on to a dissected plateau at an altitude of 900–1050 m, is surrounded on all sides by a steep escarpment and has a number of rocky marginal peaks, the highest being Nilo (1506 m). There are several tea estates on the plateau, as well as some villages, each with its own agricultural area. On the plain below the escarpment, at an altitude of about 300 m, there are further villages, some extensive sisal estates and, within Longuza Forest Reserve, a large plantation of teak (*Tectona grandis*). Teak is a valuable timber tree of Asian origin. The teak plantation at Longuza had been started in 1954 and much expanded after 1961.

The industrial logging on the East Usambaras in 1986 was being carried out by Sikh Saw Mills (SSM), a state-owned company that was being assisted in its efforts by the Finnish Agency for International Development (FINNIDA) via the intermediary of Jaakko Pöyry, a Finnish forestry consultancy company. A number of Finnish forestry experts had been employed to work in the saw-mill and help in the forest, including to assist a team from Tanzania's Forest Division to inventory the quantity of timber present in the forests. A considerable amount of equipment had been supplied by Finland, including milling machinery, chainsaws, bulldozers and large trucks to transport the logs from the forests to the mill. Figure 16.1 is a photograph of a bulldozer extracting timber from Kwamkoro Forest Reserve on the plateau of the main range.

IUCN had been alerted to the seriousness of the environmental degradation that was happening on the East Usambaras by a scientific paper that reviewed the literature on the biodiversity of the forests and the threats that it was facing (Rodgers and Homewood, 1982). One threat that was emphasised in the review paper was the illegal underplanting of some of the forests on the

Fig. 16.1. Extracting timber in Kwamkoro Forest Reserve, East Usambara Mts, Tanzania. Photo: A.H. (1986).

plateau with cardamon (*Elettaria cardamomum*), the pods (fruits) of which are highly valued as a spice. Reportedly, the pods were being smuggled over the border for sale in Kenya, where they fetched a better price. The underplanting was said to have begun in about 1960, but no supporting evidence was given for this date, and it is possible that the planting of cardamom had begun earlier (see Section 3.5.4) (Iversen, 1991; Dhakal *et al.*, 2012). Underplanting with cardamon poses a serious threat to the survival of the forests, because it involves cutting down the smaller trees and the undergrowth and hoeing the soil, activities which inhibit the regeneration of the trees.

Two inventories of the stocks of timber available in the forests had been commissioned prior to the 1986 inventory, one in 1977 that covered all the accessible forests and the other in 1985 that was restricted to Kwamkoro Forest Reserve. The results were used by FINNIDA to demonstrate that there was still plenty of standing timber available in the forests, but this failed to mollify the conservationists. To answer the continuing criticism, FINNIDA decided to modify the methodology of the third inventory (that underway in 1986) so that *all* species of trees encountered along the inventory transects were included, not just the timber species (as before). IUCN was requested to assist in the identification of the trees and to report, more widely, on the environmental values of the forests. An excellent Tanzanian field taxonomist, Christopher Ruffo, had been engaged by IUCN to help with the identification of the trees and a knowledgeable field assistant, Mr R. Abdallah, to collect biological specimens for scientific identification (in practice, the field team collected hardly any). IUCN had also engaged three members of the staff of the University of Dar es Salaam as local consultants. They were Kim Howell (zoologist), Mike Bruen (catchment specialist) and Idris Kikula (remote sensing analyst). The whole exercise was labelled the Amani Forest Inventory and Management Planning Project (AFIMP).

16.2 Historical Background

The moist climate of the East Usambara Mountains makes it certain that they would have been almost entirely covered by forest before it started to be cleared for agriculture and the same is true of the West Usambaras, a sister range to the East Usambaras separated from it by the deep Lwengera Valley. Research by archaeologists has demonstrated that agriculture was first introduced into this area by Bantu-speaking people about 2000 years ago (see Section 3.3.1) (Soper, 1967). During the course of our own research, we dug deep soil pits in six patches of tall forest chosen for their long-established, old-grove, appearance and discovered that all of them contained evidence of past human activity. Two contained Iron Age pottery, one passed through a house foundation and all of them contained abundant charcoal at depth (Hamilton and Bensted-Smith, 1989). These discoveries suggest that little, if any, forest on the East Usambaras has been uninfluenced by the past activities of people. It adds to the body of evidence showing that this is the case for more or less all tropical forests everywhere (see Section 3.2).

The coastal strip of East Africa came under Arab influence during a phase of rapid geographical expansion of Islam that followed the death in 632 CE of the Prophet Muhammad (see Section 3.5.4). Through their influence, Swahili, a Bantu language with a large infusion of Arabic, emerged during subsequent centuries as a lingua franca spoken along the coastal strip of Africa. From the early 16th century, the Portuguese and, later, other European powers began trading along the eastern coast of Africa, but they did not venture far inland before the mid-19th century. Meanwhile, from about 1700, Omani Arabs were expanding their influence from their homeland at the head of the Persian Gulf, decisive events in their advance including the capture of Mombasa from the Portuguese in 1697 and the decision of the Omani ruler, Said bin Sultan, to move his capital from Muscat to the island of Zanzibar in 1840. Bin Sultan made his move so that he could better oversee the valuable plantations of cloves (*Syzygium aromaticum*) on Zanzibar and the lucrative trade in African slaves.

In the late 18th century, the Washambaa, the indigenous people who lived (and live) on the Usambaras, came under the rule of the Kilindi dynasty. The Kilindi established their capital at Vugha on the southern edge of the West Usambaras, where it was in a good position to control a caravan route that passed close to its base. Feuding within the Kilindi dynasty began to weaken its grip on the Washambaa in the mid-19th century, a reduction in authority that coincided with the time of arrival in the area of the European powers of Britain and Germany, both of whom had developed an interest in acquiring territory in

East Africa. The first contact between Europeans and the Kilindi Kingdom was made by a British expedition in 1824 – the year after the one when the British had signed a series of treaties with the Sultan of Zanzibar that were intended to curb the Arab slave trade. The first Christian mission station on the Usambaras was Lutheran, established by the Germans in 1852. The first Anglican mission was opened by the British in 1867. The attempts by the British to curb the Arab slave trade proved unsuccessful at first. The last permanent slave market in Zanzibar did not close until 1876 and owning slaves in Zanzibar was not made illegal until 1897.

As mentioned in Section 5.3.5, Germany was awarded with the mainland of what is today the country of Tanzania at the Berlin Conference of 1884–1885. While diplomats in Europe were discussing the fate of East Africa, the Carl Peters' Society for German Colonisation and the British East African Company were competing with one another to create political realities on the ground by signing treaties with local rulers (see Section 5.4.3). Carl Jühlke, representing the Carl Peters' Society, made a far-reaching treaty with a Washambaa chief living in the village of Mgambo on the plateau of the East Usambaras in 1885. The treaty stated that virtually the whole of the East Usambaras was to be handed over to the Germans for all time, with just one small area (at Mzirai) being set aside for the local people as a 'native reserve'. It is doubtful whether the Washambaa chief at Mgambo had much comprehension of the European concept of exclusive sole private ownership of land, which, from the European point of view, was the underlying principle that allowed landowners to do more or less anything they liked on their legally held land.

Soon after the conclusion of the Berlin Conference, German colonists began to establish plantations on the East Usambaras and the German colonial administration made a start on gazetting Forest Reserves. However, quite soon thereafter, a public outcry erupted in Germany over the harsh way that the Carl Peters' Society was treating Africans, which bore similarities to the brutality that King Leopold II had unleashed on the Congo (see Section 5.3.5). The response in Belgium was similar to that in Germany in that the government decided to take over control of the territory and manage it itself, which it did in 1912. The various treaties made by the Germans during their time of occupation of the East Usambaras still form the legal basis of land ownership on the mountains.

Various plantation crops were tried by the Germans on the East Usambaras and in its surroundings. The favourites at first were coffee on the plateau of the main range and sisal in the lowlands. In 1902, the government opened a scientific station in the village of Amani on the southeastern edge of the plateau and began creating a huge (300 ha) botanical garden extending over the extraordinary wide altitudinal range of 400 to 1100 m. Already by 1907, 650 species of plants had been planted in the garden to test their suitability for the local conditions.

One of the scientific achievements of the Amani scientific station when it was run by the Germans was to debunk the assumption that was held by the

German colonial administration that the presence of magnificent tall evergreen forest on the plateau area of the East Usambaras signals inherently fertile soils (Iversen, 1991). In fact, as the scientists demonstrated, the soils of the plateau are old, highly leached and nutrient-poor, and the only reason why the luxuriant forest is able to exist is because of very tight nutrient cycling between the soil, the trees and the litter. Anyway, reality soon intervened, because the yields from the coffee estates were declining continually. The response of the planters was to switch from coffee to tea, which is more tolerant of acidic, nutrient-poor, soils. A similar transition from coffee to tea happened in the central highlands of Ceylon (Sri Lanka) for the same reason (Wenzlhuemer, 2010).

After Germany's defeat in the First World War, the responsibility for the governance of Germany's colonies in Africa passed to the League of Nations (see Section 6.10). In turn, the league divided German East Africa into two small parts on the west, which were handed to the Belgians and became Rwanda and Burundi, and a larger part on the east, which was handed to the British and became Tanganyika. In 1921, the British created a Forest Department in Tanganyika and reproclaimed the Forest Reserves that the Germans had already established on the East Usambaras. More Forest Reserves were added during the next 20 years. A forest policy along similar lines to Uganda's was formulated for Tanganyika in 1929 (see Section 5.6). This gave a high priority to the protection of natural forests for environmental reasons and divided the Forest Reserves into Central Forest Reserves (CFRs) under the central government and Local Forest Reserves (LFRs) under local administrations.

Agricultural research restarted at Amani in 1926, but, in 1949, the Amani research station became repurposed for the study of malaria and other vector-borne diseases. A new institution, the East African Agricultural and Forestry Research Organisation (EAAFRO), was created to serve the whole of East Africa and a headquarters was created for it at Maguga near Nairobi in Kenya. Some of the scientists at Amani moved to Maguga. These developments were part of a general reorganisation of scientific research in East Africa by the British related to their intention to form an East African Federation covering Kenya, Tanzania and Uganda. The creation of the East African Fisheries Research Organisation (EAFRO) and the Lake Victoria Fisheries Service (LVFS), mentioned in Section 2.3.2, were part of the same reorganisation.

In 1961, Tanganyika achieved political independence and, in 1964, it was united with Zanzibar and became Tanzania after an insurrection on the island had resulted in the deposition of the sultan and his Arab-dominated government. In 1967, Julius Nyerere (1922–1999), the first president of Tanzania, issued the Arusha Declaration, which outlined his political ideology of *ujamaa*, a word borrowed from the Arabic *jamā'a*, which means 'a group [of people]'. Nyerere's vision for Tanzania was of a country advancing through cooperation and collectivism (rather than competitive capitalism and individualism), a worldview which, as explained in Section 4.4, resonates with traditional ways of thinking in sub-Saharan Africa. (The Swahili *ujamaa* is similar in meaning to the Bantu *obuntu*.) For Nyerere, the pursuit of *ujamaa* in economic affairs required the nationalisation of larger industries and the establishment of agricultural communes. This followed the politico-economic model that then existed in the left-leaning Soviet Union.

Among the businesses on or near the East Usambaras that were nationalised were the sisal and tea estates that were owned by Europeans. Marvera Tea Plantation, which was owned by an Indian company, remained in private hands. Another Indian-owned estate, Bulwa Tea Plantation, was bought from its owner, Indar Singh Gill, along with a small sawmill that was later moved to Tanga and became SSM. Once SSM was under government ownership, the inexperienced Tanzanian managers found it difficult to operate it profitably and it fell into a state of disrepair, which is why the Tanzanian Government had requested FINNIDA to step in and turn it into a viable enterprise. Indar Singh Gill (1903–1993) was an enterprising businessman, who had been born in the Punjab, emigrated to East Africa at the age of 21 to work on the railways, established a sugarcane plantation in Uganda and opened sawmills in Uganda, Kenya and Tanzania. Indar Singh Gill's misfortune in losing his sawmilling business in Tanzania followed the loss of his businesses in Uganda in 1972, when Idi Amin expelled all the Asians from the country and expropriated their property (see Section 5.7).

In 1972, the Tanzanian Government decided for political reasons to decentralise nearly all the CFRs and place them under local government control. The effect of this was environmentally disastrous, because the local governments, seeking cash for their own purposes, proved too eager to issue felling licences for timber and failed to ensure that the forests retained their long-term values. Large numbers of trees were felled in the Forest Reserves and there was an inrush of farmers seeking to expand their farms. Another (or additional) explanation has been offered for these destructive activities. It has been suggested that this was a (delayed) response to the resentment that had been built up when the Forest Department was being run by the British, who tended to be strict about enforcing forestry rules (Kessy, 1998; Mgaya, 2016). Figure 16.2 is a photograph of recently cleared submontane forest on public land near Amani planted with bananas.

The speed with which the degradation of the forests was happening in Tanzania alarmed those scientifically informed people who knew the importance of the forests for securing water supplies to the lowlands, including for the towns and cities. In response, in 1976, the Tanzanian Government decided to restore central government control over those Forest Reserves that were deemed to be most critical for maintaining supplies of water. Renamed Catchment Forest Reserves, they were, in principle, to be managed to serve long-term, far-sighted, interests. All the Forest Reserves on the East Usambaras that were not already under special protection were turned into Catchment Forest Reserves. This was on the basis of their importance for maintaining flows in the Sigi River, water from which was extracted to supply the piped water system of Tanga Town. The Forest Reserves that already received special protection were a few plantations of exotic trees, the largest of them Longuza Forest Reserve.

Fig. 16.2. Recently cleared forest planted with bananas, East Usambara Mts. Photo: A.H. (1986).

16.3   My Involvement and Observations on the Forest Types, Soils and Climate

At the start of my work in the East Usambaras, Robert Bensted-Smith, my IUCN supervisor based in Nairobi, drove me to Tanga and introduced me to government officials, the Jaakko Pöyry foresters and the staff of the medical research centre at Amani, where I initially found lodging in its guesthouse. Among the government officials were Israel Mwasha of the Forest Department, who was my designated counterpart, and Mr Rajabu, the Natural Resources Officer for Tanga Region. The idea of Western experts and local counterparts learning from them was one fashionable in development aid projects at the time (see Section 5.7). Mr Rajabu's brief was huge, being responsible for all of the natural resources in the region, which has a land area about the size of Belgium. These resources included the forests, the water supplies and the marine fisheries. He mentioned to me once that he knew that Japanese fishing fleets were operating illegally in Tanzania's territorial waters just beyond the area visible from the land, but had no means of checking directly what was happening because he lacked a patrol boat. My brief introduction to the AFIMP project concluded with the presentation of a Land Rover, courtesy of FINNIDA. Robert then returned to Nairobi to continue with his office duties.

My first impressions of Tanzania, a country that I had only visited briefly before, was of a welcoming people, a run-down economy and a country bursting with aid projects. Israel Mwasha introduced me to how the various sectors of the economy had been divvied up among foreign aid agencies and non-governmental organisations (NGOs). For example, the Finns were assisting forestry, the British were involved in a cement factory and the Germans had the railways and water supplies. This was novel for me. Aid agencies were not conspicuous in either Kenya or Uganda at the time. In Tanzania, a remarkably large percentage of the few vehicles on the roads had doors marked with the logos of development aid agencies and foreign NGOs.

The depressed state of the economy was obvious. The nationalised sisal estates were derelict, the tea estates degraded and there was very little to buy in the shops. I never saw bottled beer openly available for sale, but, like many other commodities, it was available on the black market smuggled in from Kenya. However, on the positive side, there was a feeling of national unity that contrasted with the divisive ethnic politics of Kenya and Uganda. The system of electing government leaders at village level seemed to be working well. This system involved an arrangement whereby the Village Chiefs were elected from among the village elders and the Village Secretaries from among the younger members of the community who had received some level of education. I noticed that the decision by the government to adopt Swahili as the national language was appreciated at Amani.

On arrival at the East Usambaras, I found four inventory teams from the Tanzanian Forestry Department, each accompanied by a Finnish forester, hard at work on the new AFIMP inventory. The procedure involved members of the inventory teams identifying, measuring and recording the trees as they advanced at breakneck speed along carefully positioned sample transects running through the forests. The rate of daily advance of each team was much too fast for Christopher Ruffo or me to be able to confirm the identities of all the trees, as their names were shouted out by the tree-identifiers for noting down by the data-logger. In any case, some of the tree species were unfamiliar to me, because they did not grow in Uganda, where I had a better knowledge of the trees (see Preface). I did not believe that much of use for conservation and sustainable development would emerge from the inventory. So, instead, I decided to concentrate the research which I undertook myself on trying to discover more about the full range of forest types on the mountain (not just in 'accessible' areas) and about how the forests functioned ecologically.

In practice for the inventory teams, the 'accessible' forests on the East Usambaras consisted of those on the plateau area of the main range and those on leveller ground below the escarpment. In order to discover the full range of forest types that were present on the mountains, I ran three transects up the total range of altitudes at which forest occurred, which was 290–1220 m. Two of the transects were positioned on different aspects of the main range and one was on Mt Mtai to its east. Surveying the transects was time-consuming, which limited the number of transects that we were able to do. The survey method used involved placing three sample plots at every 100 m of altitude, one each on ridges, slopes and valleys. This was an adaptation of a method previously used by Jon Lovett to study the forest tree floras of three other Eastern Arc Mountains (Lovett, 1996). Jon spent a few days with me at the start of my work to demonstrate the technique in the field. In addition to recording the types and sizes of trees present in the plots, we also examined the characteristics of the herbaceous vegetation and of the upper layers of the soil.

Analysis of the data collected along the sample transects revealed that the representation of species of trees changes continuously with altitude, without any altitudes at which there are abrupt floristic changes (see Section 15.5.1). In spite of this, the forests are clearly divisible into two main altitudinal types, below and above an altitude of about 850–900 m, as had been recognised earlier by others (Moreau, 1935; Pócs, 1976). The two types can conveniently be called lowland and submontane forests, because of their obvious floristic and physiognomic similarities to such forest types further inland in Africa, including in Uganda. The lowland forest is semi-deciduous and has an abundance of tree species in the families *Moraceae*, *Sapotaceae* and *Ulmaceae*. The submontane forest is evergreen and includes trees, such as *Allanblackia*, *Drypetes gerrardii*, *Parinari excelsa* and *Strombosia scheffleri*, which, in inland Africa, are characteristic of the Moist Lower Montane Forest Zone (see Fig. 15.4). However, the boundary between the two forest types on the East Usambaras is at about 850–900 m, which is about 500–550 m lower than in inland Africa.

The exceptionally low altitudes of species' occurrences on the East Usambaras, compared with inland Africa, is dramatically shown by the types of trees present on the isolated rocky summits that fringe the plateau of the main range and on those of the three isolated mountains to its east. Examples of montane forest tree species present at exceptionally low altitudes at these places include *Rapanea melanophloeos* on Nilo, *Agauria salicifolia* on Mtai and *Podocarpus latifolius* on Mlinga. There are two species of tree heathers (*Erica* spp.) close to the summit of Nilo (1506 m). Tree heathers are typical of the Ericaceous Belt, which, on inland mountains in tropical Afria, typically has a lower altitudinal limit of about 3200 m (Fig. 15.4).

The upper layers of the soil profiles under submontane forest have two visual features that are absent from the soils under lowland forest. One is the presence of an extremely dense mat of intertangled roots close to the soil surface, many of the roots having thick stubby ends arranged in short compact sprays. The other is a layer, up to 10 cm thick, of dark-coloured mor humus immediately above the root-rich horizon. Measurements made in the field of the acidity of the upper 10 cm of the soil profile (taken immediately under the litter) revealed that it rose steadily from pH 7 at 300 m to pH 6.5 at 850 m, then rose steeply to an average of just under pH 5 at 900 m and finally to about pH 4 at 1050 m. The altitude of ~850 m at which the acidity starts to rise steeply is the same as that of the boundary between the lowland and submontane forest. Our observations and measurements on the soils confirmed the finding of the scientists who worked earlier at the German scientific station at Amani that the soils of the plateau area of the main range are inherently infertile.

Moreau, who had earlier noted the altitudinal depression of vegetation zones on the East Usambaras, had attributed it to the presence of persistent low daytime cloud hanging around the upper slopes of the mountains, especially during the rainy seasons (Moreau, 1935). He debated the extent to which the vegetation depression was caused by lower daytime temperatures, blocking direct solar radiation, or by occult precipitation, augmenting the amount of water available to the plants. In wider geographical context, the altitudinal depression on the East Usambaras is an example of a widespread phenomenon known as the *massenerhebung* effect (see Section 15.5.1) (Richards, 1964). It was first discovered by Alexander von Humboldt (see Section 6.5.1).

An unusually large number of meteorological stations have recorded data on daily temperatures and rainfall totals on and near the East Usambaras, thanks to the presence of a number of scientific research stations and the former European-owned estates. When I analysed these data, along with those for some other meteorological stations in Tanzania, I found that mean maximum temperatures (i.e. daytime temperatures) were depressed on and near the East Usambaras by about 4–5°C, compared with those further inland (Hamilton, 1998). Mean minimum temperatures (i.e. night-time temperatures) were essentially the same. This supports the 'persistent low daytime cloud' hypothesis of Moreau. When I was living at Amani, it was easy to observe from the lowlands the presence of this persistent cloud cover, especially during the rainy seasons.

I found evidence that the climate of the East Usambaras has been changing during recent decades and, in apparent response, so too the vegetation and flora (Hamilton and Macfadyen, 1989). The meteorological data showed that rainfall had become less predictable since about 1960. I detected some temperature anomalies in the records of some of the meteorological stations and decided to back up this research by recording statements about local climate change from a few of the long-term residents at Amani. Records of what they said can be seen in a book that I later wrote with Robert Bensted-Smith (Hamilton and Bensted-Smith, 1989). All of those interviewed agreed that the climate has changed during recent decades. They had noticed a greater frequency of torrential downpours and reduced mist, and the temperature has felt warmer. Biological changes noted included upslope movements of certain crops (e.g. coconut, citrus fruits and mangoes) and the presence of endemic malaria on the plateau of the main range, where formerly it had been absent. Tamás Pócs, a Hungarian scientist who was familiar with the mountains, had noticed a decrease in the luxuriousness of epiphytes. The East Usambaras are not the only place in the tropics and subtropics where malaria has moved upslope during recent decades. It is also known from the Rukiga Highlands in Southwest Uganda, the Bamenda Highlands in Cameroon and Mengla County in Yunnan (China) (Pascual *et al.*, 2006; Xiang *et al.*, 2018; Yufenyuy and Nguetsop, 2020).

16.4 *Maesopsis eminii* and Other Invasive Species

It was already known in the 1930s that some of the species of plants that had been planted in Amani Botanical Garden had spread into the indigenous forests and had become invasive. In our work, studying the forests at scattered places in some detail, we encountered 12 invasive tree species, plus one common invasive shrub (*Clidemia hirta*). The three tree species that seemed to be of the greatest conservation concern were *Maesopsis eminii* (mainly in submontane forest), *Azadirachta indica* (in lowland forest) and *Millettia dura* (in submontane forest). According to my previous observations in Ugandan forests, the forests of the East Usambaras were exceptionally strongly infested with invasives. It is suspected that the vulnerability of the East Usambara's forests to invasives is because these forests, along with the others on the Eastern Arc Mountains, are, in effect, 'continental islands' (Kingdon, 1990). As mentioned in Section 2.3, isolated oceanic islands are particularly susceptible to biological invasions and so, it seems, are isolated continental islands.

M. eminii is a fast-growing, light-demanding, tree which grows to a height of about 25 m (exceptionally 40 m) and has a straight cylindrical bole. It is native to lowland tropical forests in West and Central Africa, as far east as extreme Western Kenya. Although the wood provides only relatively low-grade timber, the fast growth rate of the tree, together with its straight cylindrical bole, has made it a favourite of foresters in Uganda since the 1930s to 'enrich' natural forests after they had been opened up by selective logging (Philip, 1962). There is a small plantation of *Maesopsis* within the grounds of the research station at Amani. It has been claimed that the species was first introduced to the mountains only in 1962 (Iversen, 1991). However, local residents at Amani told me that the plantation at Amani had existed for much longer. *Maesopsis* is not invasive in its Ugandan homeland.

Regardless of when and how *Maesopsis* was introduced, it has certainly spread rapidly through the submontane forests without direct human assistance. The black-and-white casqued hornbill, a large frugivorous bird, has collaborated in its spread. Every morning during the *Maesopsis*-fruiting season, flocks of this hornbill fly across the Lwengera Valley from roosts on the West Usambaras to gorge on the fruits of *Maesopsis*. It is easy to imagine the birds dropping *Maesopsis* seeds contained in their hard, beak-proof, shells on to the forest floor after consuming the fleshy outer parts of the fruits.

Pierre Binggeli, a Swiss postgraduate student studying at the University of Ulster, joined the IUCN field team for three months to study the ecology of *Maesopsis* (Binggeli, 1989). His studies included research on the seedbanks present in the forest soils. He discovered that viable *Maesopsis* fruits are common in the soils beneath submontane forests, regardless of whether or not there are *Maesopsis* trees in the vicinity. The seeds germinate when moist and exposed to direct sunlight and are capable of very rapid growth. It is common to observe *Maesopsis* seedlings and young trees growing in the submontane forest where there have been large treefalls opening up the forest canopy and allowing light to stream directly on to the forest floor. We never saw *Maesopsis* seedlings and small trees in places with a dense canopy of well-grown trees.

Industrial logging, as was being practised on the East Usambaras in 1986–1987, was creating very favourable conditions for the spread of *Maesopsis*. It resulted in the extensive opening-up of the forests through the making of wide tracks to allow the big log-transporting trucks to enter the forest, large loading sites where the logs were hoisted on to the trucks and bulldozed tracks to each of the chain-sawed trees to pull out the logs. Large numbers of small trees that were in the way were being knocked over in the process. It was not immediately clear to me which native species, if any, had been ecologically displaced by the arrival of *Maesopsis*. *Cylicomorpha parviflora*, a wild relative of the cultivated pawpaw, is a candidate, since it is a fast-growing, light-demanding, species found in large gaps created by treefalls. We encountered it very rarely.

The IUCN team compared the properties of the soils under the dense canopies of well-grown submontane forest with those found in the gaps created by large natural treefalls and with those found in places where *Maesopsis* trees are abundant. We discovered that, both in the large treefall gaps and where *Maesopsis* trees are abundant, the dense root-mat and the layer of mor humus, as found under well-grown submontane forest, are absent. The topsoil under a *Maesopsis* canopy is reduced in acidity by about one pH point and the litter is much thinner. Red, clay-rich, subsoil can be seen at, or very close to, the soil surface and there are signs of active erosion. I concluded that, where *Maesopsis* is dominant in the canopy, the soil is a wasting asset. It is vulnerable to being reduced in volume, now that it is no longer receiving protection from the elements through a thick covering of litter and by the presence of a dense binding root-mat.

16.5   Conservation Conclusions and Project Result

After considering the various findings of the IUCN research team, I came to the conclusion that continuing with the mechanical logging of the forests would yield immediate economic benefits but was unjustifiable considering the longer term. The bulldozing of passages through the forest to every tree earmarked for felling and the construction of tracks for lorries to reach the loading sites of the logs were obviously causing a huge amount of disruption to the forest ecosystem. A large number of small trees were being pushed over in the process and extensive areas of red subsoil were becoming exposed to the elements. Compounded by the negative influences of *Maesopsis*, it was concluded that the catchment quality of the submontane forests was becoming seriously downgraded.

Even from the perspective of seeing the forest only as a source of commercial timber (lacking any other values), it seemed unlikely that the forests would be able to return to a loggable state in just 35 years, as had been assumed for the cutting cycle in the management plan for Kwamkoro Forest Reserve (Iversen, 1991). It seemed a shame that *Cephalosphaera usambarensis*, one of the main species being felled in the submontane forests, was being harvested to make plywood for tea chests, a low-grade use for such a magnificent and near-endemic tree. One specimen of *Cephalosphaera* encountered during the forest inventory was measured by the Finnish foresters as being over 200 ft (61 m) tall and, so far as they knew, was the tallest tree that had ever been discovered in an African tropical forest.

After a few months on the East Usambaras, I returned to Northern Ireland for three months to teach my courses, as my university required. By then, I was physically and mentally run-down, having contracted malaria, developed jungle sores and was not eating well. I was dejected, because it seemed to me that it was impossible to stop the destructive logging.

However, soon after resuming my work at Amani, I had a stroke of luck. Word reached me that SSM was clear-felling some lowland forest in the southeast of the main range to prepare the ground for an expansion of the teak plantation at Longuza. This felt like a critical moment to make a stand. My understanding of Tanzanian law at the time was that clear-felling of indigenous forest was not permitted. What I did was to drive immediately to Tanga to inform Mr Rajabu, who agreed to return with me to the forest and see for himself what was happening. A manager from SSM met us on-site and, after some discussions, Mr Rajabu ordered the clear-felling to stop, pending a permanent decision at higher governmental level.

One of my major tasks during my final months living at Amani was to collate the information gathered by the IUCN team through field research and consultancy reports. Later, after receiving an analysis of the data collected by the AFIMP inventory, Robert Bensted-Smith and I worked up this material into a book (Hamilton and Bensted-Smith, 1989). However, my immediate purpose in spending time at Amani thinking about the information and writing was to extract key points of relevance to the new management plan, in particular what to do about the mechanical logging. By then, it was clear that the richest forests biologically on the main range of the mountains were those on its south- to southeast-facing aspects (Hamilton, 1988a). Uniquely, this was also the only place where intact forest still existed extending in a continuous altitudinal belt from the lowlands through the submontane forest up to the rocky summits. This was ecologically significant, since it had been discovered that several species of birds migrate during the course of the year between the lowland and submontane forests (Stuart, 1989). Furthermore, this part of the main range has the highest rainfall and is the most critical for maintaining water supplies to the lowlands, including Tanga Town. Figures 16.3 and 16.4 respectively are photographs illustrating the difference in sediment load between a river flowing down from a largely forested catchment and one flowing from an area of the forest that is being actively logged.

There was a decisive meeting of the Supervisory Board of AFIMP in Dar es Salaam in June 1987, which was towards the end of my period of contract to IUCN. At this meeting, Mike Bruen (catchment specialist) and myself summarised the available biological, ecological and hydrological information and its implications for conservation and sustainable development. We pointed out the damage that was being caused by mechanical logging and the threats posed by *Maesopsis* to the ecological integrity and catchment quality of the forests. We recommended the establishment of a Nature Reserve on the southeast corner of the main range. At the conclusion of the meeting, the Supervisory Board of AFIMP decided to place a permanent ban on mechanical logging in the forests. Although not stated directly, it was my impression that it was more the hydrological than the biodiversity argument that had won the day. In 1997, the government declared a Forest Nature Reserve, as the IUCN team had recommended. This turned out to be the first of a network of Forest Nature Reserves that have since been created in Tanzania (Ract *et al.*, 2024). Several of the new Forest Nature Reserves are on other Eastern Arc Mountains.

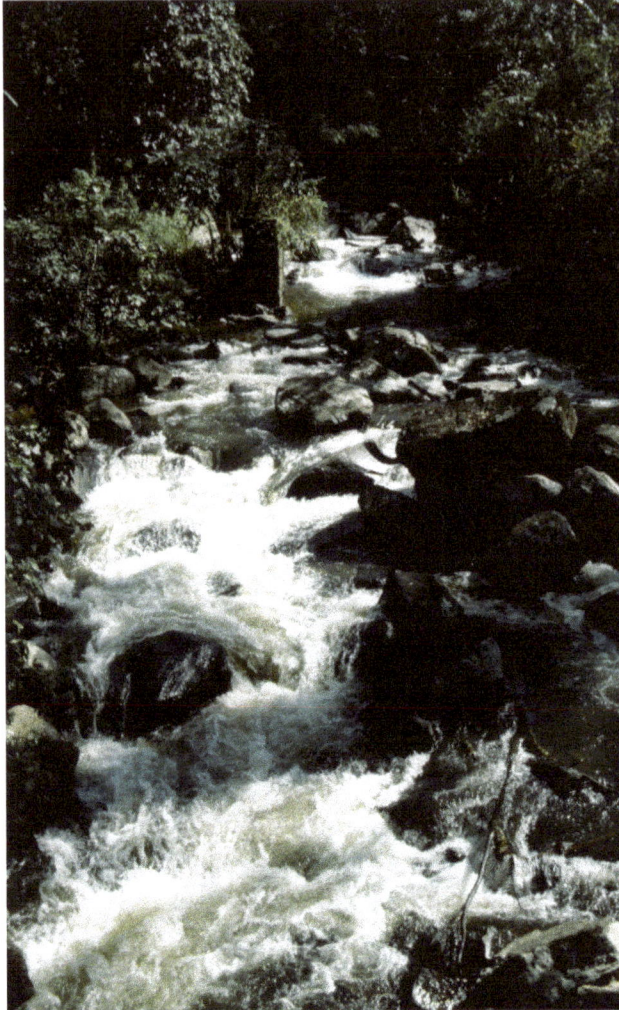

Fig. 16.3. Sigi River flowing from a largely forested catchment, East Usambara Mts. Photo: A.H. (1987).

16.6   A Sequel in Mexico

One of the countries which FINNIDA was assisting with its forestry in the 1980s was Mexico, where it was working with the Mexican Forest Department to plan a large forestry project in the states of Oaxaca and Guerrero (Hamilton and Taylor, 1988). Funding for the project was to be through a loan to Mexico from the Inter-American Development Bank. As in Tanzania, Jaakko Pöyry was delivering the Finnish forestry aid on the ground.

Under Mexican law, forestry developments of this type had to be backed by an environmental impact assessment (EIA), which had to be written by a non-Mexican consultant. Presumably because of my work in Tanzania, I was

Fig. 16.4. Sediment-laden stream flowing from a forest being actively logged, East Usambara Mts. Photo: A.H. (1987).

contacted by phone in my office at the University of Ulster in 1988 asking if I was available for a month to undertake the EIA. I agreed to do the job, since it was only for one month and I could find a month to spare when I was free of teaching obligations. However, I made one condition, which was that money be made available to support David Taylor, one of my research students, so that he could undertake some background research into forests and forestry in Mexico and identify suitable people for me to consult in the universities.

Strangely, at least in my previous experience, I was subsequently phoned up twice by two different men, one ringing from Finland and the other from Mexico, informing me that on no account during the EIA was I to make contact with anyone in a Mexican university. However, this was a request with which

I could not comply, since, in my situation, the Mexican universities were, realistically, the only places where I stood a chance of finding people with the knowledge that I needed. In the event, David did manage to find two suitably qualified academics, one to accompany me to each of the two states. As I learned more about Mexico, I started to understand why government departments in Mexico, such as those involved with forestry, might be wary of the universities. Ever since the Mexican Revolution of 1910–1920, only one political party, the Partido Revolucionario Institucional (PRI – Institutional Revolutionary Party) had been in power and the resulting void in political opposition had been filled by the universities. Student demonstrations were common and sometimes ruthlessly suppressed (Ordorika, 2021).

It became obvious to me, soon after arriving in Mexico, that the project had substantial inadequacies. The emphasis was all on constructing sawmills and building roads into previously hard-to-access forests to extract timber. Hardly any consideration was being given to the abilities of the forests to regenerate or to ensuring the proper engagement of the communities. The omission of the latter seemed strange, because, under the law, forests in Mexico belong to the communities – either to *communidades*, if the local communities are indigenous, or to *ejidos*, if composed of *mestizos* (people of mixed heritage). Mexico differed from Britain and its former colonial territories (such as Tanzania) in lacking Forest Reserves – one of the consequences of the Mexican Revolution (see Section 6.7). My assessment was that, if the forestry project went ahead, it would result in denuded hillsides and in communities having even smaller bases of local natural resources than before. Since the forestry project was to be financed by a loan, Mexico would be saddled with more national debt and, it could be foreseen, even more Mexicans would feel the urge to migrate to the Global North, as represented by the neighbouring United States.

There was a revealing incident in Guerrero when I went to the office of the chief of the Forest Department to make my introductions. According to my contract, he was required to provide me with transport so that I could visit at least a few forests and communities. However, instead of providing me with transport, he flatly refused and, at a meeting, in front of his senior staff, ranted at me for what seemed an eternity, shouting abuse that I lacked the competency to do the job. Fortunately, we were sitting on opposite sides of a huge round table made of high-grade hardwood, which created at least some distance between us. This was an experience remarkably similar to one that I had had with a senior official in Finland after I had finished my work on the East Usambaras – similar right down to the existence of a magnificent huge round hardwood table separating the chief of FINNIDA's forestry bureau and myself seated on opposite sides. I learnt later that the forestry chief of Guerrero had formerly been the forestry chief of Oaxaca but had lost his position following charges of corruption.

Feeling depressed from this experience, I returned to my hotel for a rest, but was roused later by a gentle knocking on the door. Standing outside was one of the forestry officials who had attended the meeting in the forestry office, at which he had not uttered a word. Mysteriously, he began by saying 'I am not here', then added that, if I visited the office of a certain senior member of the department in the evening, I might learn something to my advantage. The staff member in question was the one in charge of the water department, which, according to the institutional setup in Mexico, was a junior section of the Forest Department (Departmento de Forestales y Aguas). I had actually visited the same person earlier in the day and asked if there was any information available on the relationships between the forests and water issues useful for the EIA, but he told me that no such information existed.

It was the custom for senior government officials in Mexico to return to their offices in the evenings after the junior staff had left, so that they could quietly get on with their work without fear of interruption. On returning to the Departmento de Forestales y Aguas in the evening, I found the official in question seated at his desk. Without a word, he signalled that I should look at some documents that he had placed on a table open ready for my inspection. Contrary to what he had told me earlier, a lot of information about the relationships between the forests and water existed. It showed that inept logging had resulted in many problems of erosion, siltation of dams and declines in the quality of catchments. After an hour or so, he broke his self-imposed silence and we started to converse freely about issues of forest management and hydrology in Guerrero.

I was incensed by what I had learnt in Oaxaca and Guerrero and, on return to Mexico City, was determined to write the EIA before I left Mexico and deliver it in person to the responsible Mexican civil servant. Earlier, when I had visited her in the Ministry of Ecology and Urban Development (SEDUE) to make my formal introduction, she had given me little time and had acted dismissively. This time it was different. She welcomed my report and told me that EIA reports for development aid projects in Mexico were normally uncritical and gave carte blanche to the developers to do as they pleased. She offered me the use of her official car to see the famous monarch butterflies at a roosting site on a nearby sierra, but I was too exhausted to do so and took a quiet look around a craft market instead.

I suspected that the problems that forestry was experiencing in Guerrero were connected in some way with the illicit drug trade. After the head of forestry had refused to provide me with transport, the local office of SEDUE stepped in and offered me the use of a vehicle so that I could have a quick look at a small part of the Sierra Madre del Sur associated with four or five *ejidos*. What I saw was a landscape that had been devastated by extensive tree-cutting, followed by heavy livestock-grazing and burning. There was virtually no regeneration of trees. I was surprised to see a helipad high on the sierra and was told that this was used by a US drug-control agency to fly helicopters in pairs, one to spray fields of narcotic plants with herbicide and the other to keep a watch over the first, with a marksman squatting in its open door to shoot at anyone on the ground who decided to take a potshot at the helicopters. I have since learnt that this part of Mexico is notorious as a hotbed for the linked evils of organised crime, the illegal growing of opium poppies and massive deforestation (see Section 6.3.3) (Grandmaison et al., 2018; García-Jiménez and Vargas-Rodriguez, 2021).

This experience, combined with the earlier one on the East Usambaras, strengthened my growing conviction of the usefulness of scientists involved in biodiversity conservation working with hydrologists for mutual benefit (see Section 9.2). It also led me to thinking about how the relationship between the pure and applied sciences might be improved for the sake of sustainable development. In Guerrero, a lot of useful scientific information on the relationships between the management of the forests and the management of water existed, but it was not being put to practical use. A somewhat similar situation existed in Tanga Region, where an impressive amount of research had been carried out by the German Agency for Technical Cooperation (GTZ) for a Tanga Water Master Plan (1976), but there seemed little connection between this and FINNIDA's contributions to forestry.

I once asked Mr Rajabu, as the government official responsible for the management of natural resources in Tanga Region, what he did with the consultancy reports that he received from aid agencies. He opened the cupboard behind his desk, pointed to a shelf of consultancy reports gathering dust inside and said, 'I put them in here'. And that, apparently, was where they ended their days. Later, when I became Plants Conservation Officer for WWF-International, I found that a similar disconnect exists in the 'standard' approach to plant conservation between the people who make Red Lists of endangered species and those who work in the field to try and save them (see Section 8.6). These experiences made me receptive to the values of participatory action research and participatory adaptive management for achieving plant conservation, when I came to learn of their existence (see Section 9.7).

17 Conservation and Community Knowledge, Mt Kinabalu, Malaysia (1992–1998)

WRITTEN IN THE FIRST PERSON BY ALAN HAMILTON

Abstract

At 4094 m, Mt Kinabalu in Sabah (Malaysia) on the island of Borneo is the highest mountain situated between the Himalayas and the highlands of New Guinea. An ethnobotanical survey carried out during 1992–1998 by scientists working with villagers living around the foot of the mountain greatly increased scientific knowledge about its flora. For instance, the number of species and infraspecific taxa of palms known to occur increased from 48 to 74. Much more information became available about the distributions of the palms on the mountain and about their conservation statuses, methods of management and uses. Contrary to existing Red List assessments, the survey found that few of the species of palms were in imminent danger of extinction, at least so far as their presence on the mountain was concerned.

One of the jobs with which I was entrusted when I joined WWF-International as its Plants Conservation Officer in 1989 was to see through the publication of a book, *Palms for Human Needs in Asia* (Johnson, 1991). The palm family (*Arecaceae*) is one of exceptional usefulness to people, providing edible nuts, beverages, construction materials, basketry materials, thatch, and more. Concentrated in the wet tropics, there are many fewer species of palms in tropical Africa than in either of the Americas or Southeast Asia, an imbalance in numbers which is believed to have been caused by the trend towards a drier climate that Africa has been experiencing during the last 23 million years (see Section 2.1) (Baker and Couvreur, 2012).

Palms for Human Needs in Asia includes assessments of the conservation statuses of species of palms in India, Indonesia, Malaysia and the Philippines, each compiled by a taxonomic expert. I was aware at the time when I joined World Wildlife Fund (WWF – now World Wide Fund for Nature) that there are many scientists who propose places where conservation efforts should be concentrated based on their academic research – as I had done myself earlier in

© Alan Hamilton and Pei Shengji 2024. *History and Future of Plants, Planet and People* (Alan Hamilton and Pei Shengji)
DOI: 10.1079/9781789248944.0017

my career (see Section 15.5.2). I seemed to me that there was a gap in the standard approach to plant conservation, as described in Section 8.6, in turning recommendations for *in situ* conservation based on Red Listing into conservation realities on the ground. I thought that it would be interesting to see how far it might be possible to do so in the case of *Palms for Human Needs in Asia*.

So far as I knew, WWF-Malaysia was the only WWF National Organisation in the early 1990s to employ an officer dedicated full-time to plant conservation. His name was Balu Perumal and, in the interest of building partnerships for plant conservation within the WWF family, I decided to visit WWF-Malaysia to discuss with Balu and the other staff of WWF-Malaysia about how the conservation of the endangered palms of Malaysia might actually be achieved.

At the time, the rainforests of Malaysia were being subjected to rapid loss and degradation through the logging of trees for timber and wood pulp, and through their replacement by plantations of oil palm (*Elaeis guineensis*), eucalyptus and black wattle (*Acacia mangium*). Malaysia consists of three parts, one Peninsular Malaysia and the other two, Sarawak and Sabah, which are on the island of Borneo. The following figures for Sabah show just how massive has been the loss of tropical forest: the cover of primary rainforest in Sabah was reduced from 2.8 million to 0.3 million ha between 1975 and 1995, and the cover of old grove forest in its Commercial Forests from 98 to 15% between 1970 and 1996 (Vaz, 2006). The Malaysian Government was sensitive to criticisms about its development policies in relation to rainforest, whether these came from within the state or from outside. This presented WWF-Malaysia with a strategic challenge. What WWF-Malaysia had decided to do was to adopt a strategy of working closely with the Malaysian Government to locate the ideal places from a scientific perspective to site protected areas, lobby for them to be created and thereafter assist, such as it was able, to see that they were properly managed.

Balu and the other staff of WWF-Malaysia advised me that a good place to mount a palm conservation project would be Mt Kinabalu (4094 m), which is the highest mountain situated between the Himalayas and New Guinea. Mt Kinabalu is possibly the most biodiversity-rich place for its size in the world, containing an estimated 5000–6000 species of vascular plants in an area of less than 2000 km². As an example of its botanical wealth, over 720 species of orchids have been recorded, which is more than there are in either the whole of Europe or the whole of East Africa, which are much larger areas (Wood *et al.*, 1993). Despite Mt Kinabalu being a magnet for botanists since 1851, the palm flora of the mountain – and, indeed, the whole of Sabah – was still poorly known in the early 1990s (Dransfield and Johnson, 1991). This was due, in part, to the difficulties of collecting specimens of palm, which can have huge leaves and inflorescences concentrated at the top of tall trunks lacking any branches. Climbing palms to collect specimens is a daunting prospect for the ordinary field botanist.

My visit to discuss a possible palm conservation project in Malaysia coincided with the time that Tony Cunningham, Gary Martin and I were drawing up plans for the People and Plants Initiative (PPI), an international programme designed to increase global capacity in applied ethnobotany (see Chapter 10).

Gary and I discussed the possible usefulness of ethnobotanical research being used as an approach to finding ways to save Mt Kinabalu's palms. When I suggested this idea to WWF-Malaysia, they responded that it would be worth trying, partly because they thought that an ethnobotanical project might be a good way to improve relations between the local people and Kinabalu Park, a protected area covering Mt Kinabalu declared in 1964. Gary then visited Malaysia to discuss the possibility of an ethnobotanical project on Mt Kinabalu with WWF-Malaysia and others in detail. Among those he consulted were Sabah Parks, the agency responsible for managing Kinabalu Park, and academics at Universiti Kebangsaan Malaysia, a national university situated in Kota Kinabalu, which is the capital city of Sabah. WWF-Malaysia made two of its trainee conservation officers, Lin Idrus and Agnes Lee Agama, available sequentially to assist with the work.

Mt Kinabalu, which is mainly composed of granite, is largely forest-covered below 3700 m, above which the landscape becomes rocky and open, the soil having been stripped away during Quaternary glaciations (Fig. 17.1). The vegetation changes floristically with altitude and can be classified into lowland forest, montane forest and, above the forest limit in places where there are pockets of soil, ericaceous and alpine scrub. Locally on Mt Kinabalu, there are areas of serpentine rocks carrying distinctive types of vegetation and plant species, including a number of species of endemic orchids. Outside the park, the lower slopes of Mt Kinabalu are covered with a complex patchwork of agriculture and forest under various conditions of occupancy and tenure. The indigenous people are the Dusun, many of whom, as of 1992, were following a traditional way of life that involved growing dryland rice in slash-and-burn

Fig. 17.1. Mt Kinabalu, Sabah, Malaysia. Photo: A.H. (1993).

clearings created in the forest, tending agroforestry plots, and gathering wild plant and animal resources in the forest. Some Dusun were engaged in monetary sectors of the economy, such as tourism and horticulture.

Projek Etnobotani Kinabalu (PEK), the ethnobotanical project at Mt Kinabalu led by Gary Martin between 1992 and 1998, produced a much clearer picture of the types of palms growing on the mountain than had existed before (Martin *et al.*, 2002). PEK began with an ethnobotanical inventory of the whole flora, not just concentrating on the palms. The intention of widening the survey beyond palms was to gain an overall understanding of the botanical knowledge of the Dusun and their dependencies on plants, and gain some insight into the political, socio-economic and cultural contexts surrounding their interactions with plants.

Nine communities situated in a ring around the foot of the mountain were involved in the project, a geographical positioning that enabled all aspects of the mountain to be covered. Two people within each community were engaged for the survey, one an older person who was well-acquainted with the forest and the other a younger literate person, to be responsible for collecting plant specimens for scientific identification and to record information. Figure 17.2 is a photograph of one of the younger community ethnobotanists who were engaged in the survey. The types of standard information to be collected with each plant specimen were discussed with the communities and an herbarium label devised accordingly. Among the categories of information recorded for each specimen were the names of the localities where they were collected, their altitudes and their environmental characteristics; also, the local names of the plants and local knowledge about their distributions, their uses and their methods of management. Because of the special importance of Mt Kinabalu for orchids and their vulnerability to over-collection, collecting orchids was discouraged.

The achievements of PEK over six years (1992–1998) by community ethnobotanists working part-time can be compared with those of the more than 200 scientific botanists who have visited Mt Kinabalu to collect plants over a 147-year period (1851–1998), their lengths of stay ranging from several days to several years. Altogether, PEK greatly increased the number of types of plants known from the mountain, for example increasing the number of new taxa in a sample of dicotyledon families by 16%. Considering just the palms, the figures for the community ethnobotanists were as follows: 404 numbered plant collections, comprising 19 genera and 74 species and infraspecific taxa. The equivalent figures for the scientific botanists (collections made up to 1992 only) were: 372 collections, ten genera, and 48 species and infraspecific taxa. Presenting these figures another way, there were 41 species and infraspecific taxa that were collected by both the community ethnobotanists and the scientific botanists, 33 by the community ethnobotanists only and seven only by the scientific botanists.

Apart from numbers of species, much more information became available through PEK on the local distributions and conservation statuses of the palms than had existed before; also, about their local names, uses, ecologies and methods of management. For example, the scientific botanists who had studied

Fig. 17.2. Community ethnobotanist and wild durian fruits, Mt Kinabalu. Photo: A.H. (1998).

palms earlier on Kinabalu had, in combination, recorded the local names and uses for just two species (4.2% of the total), even though it is likely that they could have obtained this information from the Dusun helpers who most likely assisted them as guides or porters. In *Palms for Human Needs in Asia*, it is mentioned that the conservation statuses of 112 (85%) out of the 131 species and subspecific categories of palms listed for Sabah are unknown (Dransfield and Johnson, 1991). With the data from the community inventory now available, it was thought that few of the 81 species and infraspecific taxa of palms known to grow on Mt Kinabalu were in imminent danger of extinction, at least so far as their presence on the mountain was concerned. (Note on the calculation: the figure of 81 includes seven species that had been found by the scientific botanists but were not encountered in PEK.)

   Several factors are believed to be responsible for the greater productivity of the PEK collectors, compared with the scientific botanists (Martin *et al.*, 2002).

Positive ones for the PEK collecting teams were that the elders were familiar with the forest; they collected specimens throughout the year; they tended to collect at lower altitudes, which are floristically the richest; and they collected in all sorts of habitats, from the most closely tended to the very wild. In contrast, the scientific botanists, most of whom have been from Europe, the United States or Peninsular Malaysia, have tended to know little about the ecologies of the plants and the details of their distributions on Mt Kinabalu prior to their visits. They have tended to avoid those parts of the mountain where people live, concentrate on wilder habitats, use the same trails as one another and collect at higher altitudes. For the Dusun, the top of the mountain carries a special meaning. It is a sacred place, where the spirits of their ancestors reside (see Section 8.3) (Verschuuren *et al.*, 2010).

PEK illustrates the advantages of including local people as integral members of teams formed to identify and tackle plant-related issues of conservation and sustainable development. It turned out that, for the Dusun, the conservation of the palms was not a pressing issue. More urgent for them was conservation of medicinal plants and associated traditional knowledge. The elders were concerned that members of the younger generation were not acquiring the traditional knowledge of their people. Three of the causes of cultural erosion were identified as: the young spending less time with the elders; being taught in schools in Bahasa Malaysia, which is the official national language (rather than in Dusun, the language of the Dusun); and the influence of newly introduced evangelical Christian missions, which discourage traditional medicine (see Sections 7.5.3 and 8.1).

A number of development sub-projects were spun off from the PEK floristic inventory, mostly aimed at encouraging conservation of the Dusun's traditional knowledge. One was a series of collaborative initiatives with a local Dusun non-governmental organisation to develop a curriculum in the Dusun language for use in voluntary pre-schools. Another was the creation of nature trails in the forest to demonstrate the traditional uses of plants. A third was the building of resource centres in villages, these being buildings constructed in the traditional style containing information on traditional knowledge, for example through including displays of local crafts such as baskets.

The strong interest in conservation of knowledge of medicinal plants expressed by the communities was followed up by supporting the development of a booklet (two versions – Bahasa Malaysia and English) describing some of the most commonly used medicinal plants, together with their uses and the ways to prepare medicines that were safe to use in home settings (Agama, 2002). Care was taken not to disclose to a wider public knowledge about the plants that the Dusun wished to keep secret (see Section 6.11). The communities were involved in all stages of its preparation – information-gathering, data-sharing, compilation, editing and deciding on how the booklet was to be to be accessed and distributed.

18 Conservation on the Frontline: Bwindi Impenetrable Forest, Uganda (1966–2004)

WRITTEN IN THE FIRST PERSON BY ALAN HAMILTON

Abstract

In the 1970s, the Forest Department in Uganda lost control of Bwindi Forest Reserve, the forest with the highest overall biodiversity score in the country and the home of half of the world's mountain gorillas. There was widespread illegal tree-felling, hunting and alluvial gold-mining, and it was feared that all the gorillas would be lost. In 1990, the Ugandan Government commissioned an Establishment Plan to turn the forest into a National Park to be managed by the better-resourced Uganda National Parks. The communities living near the park resented the restrictions placed on their activities, began to light fires inside the forest and threatened to kill all the gorillas. A turnaround for the park and the people came after an ethnobotanical survey revealed the types of natural resources within the forest that interested the communities. With the development agency CARE acting as a facilitator, agreements were reached between the park and the people allowing the harvesting of certain natural resources within defined areas inside the park in a controlled way and, in cases where this was deemed to be impossible, the development of alternatives to these resources outside the park's boundary.

18.1 Bwindi Forest and Growing Conservation Concern

Bwindi Impenetrable (henceforth, 'Bwindi') is a small forest of 321 km² situated in the Rukiga Highlands of Southwest Uganda. Its western border abuts on Kivu Province in DR Congo and, on its Ugandan sides, it abuts on some of the most densely populated parts of rural Africa. It was estimated in 1994 that 100,000 people lived in Uganda within 5 km of its boundary, most of them being smallholding farmers (IUCN/WCMC, 1994). The forest, which extends in altitude from 1190 to 2607 m, is unique in Uganda for straddling the transition between lowland and montane forest. Its uppermost part just extends into the zone of mountain bamboo (*Yushania alpina*) (see Fig. 15.4). The topography is steep and there are marked differences in the types of trees and other plants found on ridges, slopes and in valleys (Hamilton, 1969). Figure 18.1 is a photograph

© Alan Hamilton and Pei Shengji 2024. *History and Future of Plants, Planet and People* (Alan Hamilton and Pei Shengji)
DOI: 10.1079/9781789248944.0018

Fig. 18.1. Farmland on the margin of Bwindi Forest, Uganda. Photo: A.H. (1997).

taken of a smallholding farm on the margin of a relatively low-lying part of Bwindi Forest (seen in the background). The crops present include sorghum and banana.

Bwindi lay on the eastward margin of a major refugium in East Congo for forest survival during the Last Glacial Maximum (LGM – *c.*24,500–17,000 BCE) (see Section 15.5.2 and Map 15.2). Judging by an exceptionally high diversity of species of plants in the low-lying northern part of Bwindi, where the Ishasha Gorge is situated, the East Congo refugium may have extended a little into Uganda in this area. Among the trees found near Ishasha Gorge and nowhere else in Uganda, there is one species, *Leplaea mayombensis*, and one genus, *Allanblackia*, that have exceptionally massive seeds that have obviously evolved to be dispersed by animals but are far too large to be carried by those that can fly. Having such fruits may have inhibited their abilities to spread further into Uganda when the climate moderated after the LGM (Hamilton, 1975b). Elsewhere in tropical Africa, *L. mayombensis* shares with the gorilla a disjunct distribution across the Congolese Basin, being found on both of its two sides, but not in the vast area in between. *Allanblackia* reappears on the Eastern Arc Mountains near the Indian Ocean coast in East Africa, but it is represented there by a different species (see Section 16.3).

Bwindi is a frontline locality for the conservation of global biodiversity (WWF and IUCN, 1994–1997). It was one of two prime sites in East Africa for the conservation of forest biodiversity that I mentioned in a synthesis of the environmental history of East Africa, the other being the East Usambaras (see Chapter 16) (Hamilton, 1982). It has the highest overall biodiversity score of any forest in Uganda (Howard, 1991). Internationally, Bwindi is famous for being the home of half of the world's mountain gorillas. Hydrologically, it lies very close to the ultimate source of the river Nile. Two streams, emerging from either end of Ahakagyezi, a swamp situated about 5 km from the forest's southeastern corner, flow in diametrically opposite northwestern and southeastern directions, initially into the Kiriruma and Ishasha rivers. The Kiriruma flows into the Kagera River, Lake Victoria, the Nile and Lake Albert, while the Ishasha flows more directly into Lake Albert via the lowlands to the west of the Rwenzori Range. Once reunited in Lake Albert, the water from the two ends of Ahakagyezi Swamp flow mixed together down the White Nile towards Egypt. The United Nations Educational, Scientific and Cultural Organization (UNESCO) listed Bwindi as a World Heritage Site in 1994 for its importance for the conservation of natural heritage.

Pollen analysis and archaeological research show that agriculture began to be practised extensively in Uganda at c.350–1 BCE, with an indication from one site close to Bwindi (Ahakagyezi) of localised forest clearance as early as 1700 BCE (see Section 15.4.2). However, pollen analysis carried out on sediment collected from a swamp situated actually *within* Bwindi has indicated that the surrounding area of the swamp continued to be covered by forest, even while it was being cleared away elsewhere (Marchant and Taylor, 1998). The dauntingly steep slopes of Bwindi and the exceptionally high acidity of some of its soils may have helped keep at least some of its forest intact (Chenery, 1951).

Bwindi was declared a Central Forest Reserve by the British colonial administration in 1932 and as an Animal Sanctuary by the newly independent Government of Uganda in 1964. When I first visited the forest, which was in 1966, it was being managed by the Forest Department in its normal way. There was none of the paraphernalia of modern tourism – no entry fees, professional trekking guides and tourist lodges. Gorilla-viewing had not become an industry. A little legal pit-sawing, mainly of *Podocarpus latifolius*, was taking place. I was not aware of any infringements of Forest Department regulations.

The name of the Forest Reserve was changed officially from 'Impenetrable' to 'Bwindi' during the 1960s by a Norwegian forester working on an aid project. However, the name of the National Park created in 1991 incorporates both the old and new names. The name 'Impenetrable' is apt. It refers to the dense tangly vegetation that smothers its more open steep slopes and can make movement through the forest away from established trails extremely arduous. The denseness of the undergrowth can be such as to make it sometimes difficult to spot nearby gorillas and even the elephants which, in 1966, were still living in the forest.

As mentioned in Section 5.6, forestry and forests started to undergo radical changes from 1966. The accession of Idi Amin to power in 1971 initiated a phase of widespread forest loss and degradation, which is still continuing. When I joined WWF-International in 1989 as a conservation officer, Tom Butynski, an American biologist attached to the Department of Zoology at Makerere University, was living at Bwindi, carrying out conservation-orientated research and making recommendations to the authorities in Uganda and to international conservation agencies of ways to improve the forest's management (Butynski, 1984, 1986). Butynski estimated that, by the mid-1980s, 61% of the forest had been heavily logged illegally by pit-sawing, that the best hard-woods had been removed from an additional 29% and that only 10% of the forest remained intact. Illegal hunting and alluvial gold-mining were rampant. Only about 30 elephants survived in the forest.

Butynski's perseverance in his research and lobbying started to pay off in 1986, when an Impenetrable Forest Conservation Project was launched with the support of the New York Zoological Society and World Wildlife Fund (WWF – now World Wide Fund for Nature), and greater control over illegal activities started to be exerted (IUCN/WCMC, 1994). Support was provided to the local Game Department, systematic monitoring of the forest was expanded and outreach educational programmes were initiated with the communities. CARE, an international humanitarian agency, launched a community devel-opment project in 1989 with the support of WWF and the US Agency for International Development (USAID). The name selected for this programme was 'Development Through Conservation' (DTC), which was chosen to empha-sise how the continuing presence of the forest could bring lasting benefits to the people. The Global Environment Fund (GEF) launched a trust fund to back community development projects (Victurine and Oryema Lalobo, 2001).

Gorilla-based tourism at Bwindi got underway in the 1990s. The financial benefits for the government agency responsible for managing the National Park have been substantial. Initially, this was Uganda National Parks (UNP) and, later, Uganda Wildlife Authority (UWA). Today (2024), a large proportion of the total income of UWA comes from the sale of gorilla trekking permits, the cost of which now stands at US$700 per person for foreign non-resi-dents, US$600 per person for foreign residents and 250,000 Uganda Shillings (US$70) for East African citizens.

The wisdom of allowing mass gorilla tourism (for rich people) is debatable. Butynski had recommended that gorilla-viewing should only be allowed if trek-king rules are in place prohibiting tourists from getting too close to the gorillas and are rigidly enforced (Butynski, 1984). He believed that occasional laxness in this respect posed an acute risk of transmitting diseases from the tourists to the gorillas. He also pointed out that habituated gorillas are an easy prey for trappers. As a result of such considerations, strict rules have been put in place limiting the ways that tourists are allowed to approach the gorillas, but a study has shown that they are routinely disregarded (Weber et al., 2020). Warning of the possibility of disease transmission between humans and goril-las is not scaremongering. Forty thousand to sixty thousand years ago, the pathogen that causes falciparum malaria jumped from the gorilla to the human (see Section 2.3). Malaria is a serious disease in Uganda, where there are an estimated 13 million cases every year (see Section 5.2).

18.2   The Establishment Plan and the Creation of Bwindi Impenetrable National Park

My own direct involvement in efforts to strengthen the management of Bwindi began in 1990, when I happened to be in Uganda on other WWF business and by chance encountered Eric Edroma, the Director of UNP, in a doctor's surgery in Kampala. Dr Edroma asked me if I was available immediately to head up a small team to prepare an Establishment Plan for a new National Park at Bwindi, his expectation being that UNP would soon be taking over the responsibility for managing the forest from the Forest Department. He added that there was an urgency about having the plan available, due to a debate on Bwindi's future being scheduled in the Ugandan parliament and having an Establishment Plan in hand was a legal requirement for a new National Park to be created. The preparation of an Establishment Plan for a new National Park required those who were contracted to prepare it to consult with researchers (such as Butynski), local government officials and local communities to find out what they thought about having a National Park and elicit their recommendations about how it should be managed. In view of the urgency, I changed my schedule and accepted Dr Edroma's request.

The three-person Establishment Plan team which went to Bwindi for the consultations consisted of Jonathan Baranga, the Deputy Director of UNP, Justice Tindigarukayo, the Principal Game Biologist in the Game Department, and me. A representative of the so-called relinquishing agency, the Forest Department, should have been with us, but no one had been made available. Some years later I learnt from the staff of the Forest Department that they knew that their management of Bwindi (and other Forest Reserves) had had many deficiencies since the early 1970s and that UNP's management of the National Parks had been better, but they thought it unfair to justify the transfer of Bwindi from the Forest Department to UNP solely on this account. They pointed out that UNP had been receiving generous funding from foreign agencies (such as WWF), but the Forest Department had received next to nothing. An illustration of the desperate lack of resources that the Forest Department was facing is shown by the startling statistic that, in 1988, the salaries in the Forest Department stood in real terms at just 0.4% of their 1962 values (Howard, 1991). For my part, when I joined WWF in 1989, I discovered that its thinking in terms of protected areas was orientated strongly towards National Parks and little consideration had been given to the value of Forest Reserves for conserving nature.

Actually, the Forest Department had not been complacent in thinking about how to improve the management of the Forest Reserves. Field investigations of the distributions of species of trees, primates, birds and butterflies in Uganda's 12 principal Forest Reserves were used to devise a new strategy for the conservation of Uganda's forest biodiversity (Howard, 1991). This study had identified Bwindi as the forest with the highest biodiversity score in the country and the Forest Department was floating the idea of creating a new category of Forest Reserve, called the Forest Park, especially for Bwindi. As mentioned in Section 16.5, the concept of Forest Nature Reserves was also on the agenda in Tanzania, where the first of several, Amani Nature Reserve, was established on the East Usambaras in 1997.

When we in the Establishment Team consulted local government officials in Kigezi to discover their views, they told us that they had always been sympathetic to Butynski's proposals to strengthen the management of Bwindi Forest, but added that Butynski would have been more effective in his lobbying if he had visited their offices more often to explain his concerns in person, mentioning that this is the way that Africans prefer to do business. Lengthy reports prepared by researchers packed with data and analyses were certainly considered useful, but they needed to be backed up with verbal communications if they were to have much impact. This resonated with the way that Mr Rajabu, the Regional Natural Resources Officer in Tanga, Tanzania, had responded when I asked him what he did with the data-packed technical reports that he received from development aid agencies. As mentioned in Section 16.6, he told me that he put them in his cupboard and there, it seemed, they went to sleep.

The local government officials whom we consulted in Kigezi were professionals employed on a national basis and were not necessarily from the local area. In contrast, the three packed meetings held by the Establishment Team with local communities heard directly from the people who would be most directly impacted by the creation of the park. Figure 18.2 is a photograph showing one of the consultations. The Establishment Team is seated in front of the house, with me writing down the proceedings. Not a single voice at any of the community meetings was raised in favour of the proposed park, although, more positively, no one actually called for Bwindi to be degazetted and the forest to be converted to farmland. One reason for this might have been that the reserve had existed for sufficiently long for the fact of its existence to have become embedded as a given in the local psyche.

Some people at the community meetings expressed the view that they had a natural right to enter the forest and collect any natural resources that they needed. There was a plea for UNP not to introduce dangerous wild animals,

Fig. 18.2. Establishment Team community meeting, Bwindi Forest (1991).

such as lions, into the new National Park, a fear possibly grounded in the observable fact that all the existing National Parks in Uganda were in savannah areas (see Section 8.2.3). The idea of a forest-covered National Park may have been inconceivable. However, at the heart of the communities' problems, as was forcefully expressed by some, was the interest of White People in viewing gorillas. It was suggested that this interest would be better served if the gorillas were removed and taken to zoos where they could be more easily examined at close quarters. As one person put it: 'Why don't you take away the gorillas and leave us in peace?'

While I was completing the editing of the final version of the Establishment Plan, I received an oral submission from the Batwa ('pygmies'), transcribed and presented by a third party (Elena Kingdon). This read:

> This is a strong request that some land (outside the Forest Reserve or proposed national park) be given to them for farming and settlement. The area of land required is not large. They state that their traditional way of life is no longer viable and that their future depends on having permanent title to agricultural land.

This petition was included in the Establishment Plan as finally submitted (Hamilton *et al.*, 1990). In 1990, the Batwa at Bwindi numbered about 500, a tiny percentage of the total number of people living in the area. In Section 3.3.1, we mention the depth of knowledge that the Batwa at Bwindi had of the forest. Jacob Bandusya, one of the Batwa, was a key member of the team that undertook the ethnobotanical survey described in Section 18.3. According to the Acknowledgements given in the survey report, he made major contributions to the work through his good humour and skills as a field observer (Cunningham, 1996).

Research at Bwindi undertaken by Dominic Byarugaba, one of the postgraduate students who benefited from the People and Plants Initiative (see Chapter 10), revealed something of the traditional conservation practices of the Batwa (Byarugaba, 2008). It showed, for instance, that the Batwa recognise many types of *Dioscorea* yams and that one of their traditional conservation practices has been to leave parts of the tubers behind in the ground when they harvest them to eat, so giving the plants a chance to recover and grow. Bwindi, as of 1990, was a classic example of the 'tragedy of the commons' – the loss of natural resources that can happen in the absence of sufficient controls to safeguard against their unsustainable harvesting (Hardin, 1968). For the Batwa to continue to harvest produce from wild plants in the forest in modern times, they would have to have support from the law. Figure 18.3 is a photograph of Dominic Byarugaba in Bwindi Forest with the tuber of a yam known locally as *ebihama*, which is eaten by the Batwa. It requires detoxification before eating.

In view of the need for a strengthening of the law at Bwindi to prevent the complete loss or degradation of the forest, the Establishment Team emphasised that its management must be strengthened as a matter of urgency, but did not specifically state whether this should be through the Forest Department or UNP. Anyway, in 1991, the Ugandan parliament voted to create the National Park and the forest passed into the stricter hands of UNP. No one knows how Bwindi would have fared if it had become a Forest Park under the Forest

Fig. 18.3. Tuber of edible yam from Bwindi Forest. Photo: A.H. (2000).

Department and if the Forest Department had become a more effective organ-isation. However, given what has happened to Forest Reserves in Uganda since 1991, as described in Section 5.6, most likely all the gorillas would now be gone, just as have all the elephant.

18.3 Threats to the National Park and the Value of Applied Ethnobotany

There was little local community support for the National Park when it first was created, attributable to the rushed manner in which it had been established and the meagre level of consultation that there had been with the communities

(Wild and Mutebi, 1996). Resentment at the existence of the park was quickly made apparent by 16 fires breaking out in the forest during the dry season that followed the Park's establishment. Five per cent of the forest was burned and threats were made to kill the gorillas. Under the forest's new status, the local people enjoyed no rights at all in the Park – not even, for instance, the right to collect the bark from the medicinal tree *nyakibazi* (*Rytigynia* sp.), without which, they said, 'they would die' (Cunningham, 1996). *Nyakibazi* was being used as the key ingredient in local medicines to treat intestinal parasites, which were a big medical problem locally. According to one survey carried out at the time, 89% of people living near Bwindi were infested with whip-worm (*Trichuris trichiura*) and 34% with roundworm (*Ascaris lumbricoides*) (Ashford et al., 1990). Combined with the strict policing now in place, this was a recipe for conflict, because, being caught collecting *nyakibazi* in the forest could result in a fine or imprisonment, yet having access to the bark was con-sidered by the people to be a survival necessity.

Realising that it could not hold the conservation line without support from the local people, UNP took the decision in 1991 to explore the possibilities of allowing limited and controlled harvesting of some of the Park's natural resources (Wild and Mutebi, 1996). I was contacted to ask if I knew of someone who could carry out a baseline ethnobotanical survey and recommended Tony Cunningham who, at the time, was writing a report for WWF on medicinal plants, conservation and healthcare (see Chapter 10). Tony agreed to take on the job and, together with DTC, put together a survey team of local peo-ple knowledgeable about the forest's natural resources. One of its members was Jacob Bandusya. The baseline survey had to be completed in just a few months, in view of the urgency of finding a solution. Later, it was followed up at a more leisurely pace by postgraduate students of Makerere University car-rying out detailed studies on particular plant resources, such as that of Dominic Byarugaba on yams and one by Maud Kamatenesi on *nyakibazi* (Kamatenesi et al., 2014).

Two major recommendations emerged from the baseline survey – one applying to those natural resources that, it was thought, could potentially be harvested sustainably from the forest and the other to those resources that could not (Cunningham, 1996). For the former, it was recommended that multiple-use zones be established inside the Park close to its boundary, within which harvesting was to be permitted once management systems to ensure that the harvesting was being undertaken sustainably had been put in place. For those natural resources that could not be harvested on a sustainable basis from the forest, it was recommended that alternative ways of providing the resources required be found instead. This could be, for instance, by substituting one plant resource with another (e.g. replacing timber extracted from the forest with timber grown in plantations) or it could be by substituting a resource with something quite different (e.g. substituting wood fuel with kerosene).

In 1993, permission was given by UNP headquarters to start the process of establishing multiple-use zones inside Bwindi, beginning with pilots in three of the 18 parishes that bordered the Park. The staff of DTC facilitated the process of bringing together representatives of the communities and the park, the intention initially being for them to gain a better understanding of each other's positions (Wild and Mutebi, 1996, 1997). Eventually, the communities revealed more about the types of forest resources in which they were interested and who, especially, within the communities was particularly interested in them. Having this information made it easier for the two sides to agree on where the boundaries of the multiple-use zones inside the Park should be sited. This then led on to the drawing up of Memoranda of Understanding (MOUs) between the park and each of the pilot communities. These set out who, within each community, carried the responsibility for ensuring that the community as a whole complied with its terms. This was to help ensure equitability and to lower the risk of disgruntled people breaking the rules. A new cadre of Community Conservation Rangers was established to facilitate future communication between the communities and the park. At first, the communities were suspicious of UNP's true intentions and only really began to believe that the agreements were genuine when they began to collect the materials from the park and were not harassed by the Park's guards.

The process of establishing the MOUs included the formation of user-groups within each parish for each category of forest resources and also, for each parish, an umbrella group, *Ekibiina Kya'beihamba* (Forest Society), to represent all categories of resources. Discussions between the communities and DTC, in its role as a facilitator, then moved on to consider how the Forest Societies should be constituted and their officials selected, which then led on to actual elections. It turned out that most of those elected were the leaders of the specialist user-groups or their nominees. However, also chosen were some people who had no direct connections with the forest, such as chiefs, religious leaders, and even some staff of UNP and DTC who had managed to gain the respect of the communities.

This community development work by DTC built upon its understanding of how the communities were organised. It was found that there were three pre-existing community groups or organisations that were recognised as having local authority. These were the *abataka* (respected long-term residents), *Ebibiina Bya'ngozi* (Stretcher-bearer Societies) and RC1s. None of them had direct connections with the forest. The *abataka* ('elders') form one of the traditional cornerstones of these societies. The *Ebibiina Bya'ngozi* were new organisations formed during the 1980s to carry the sick on stretchers to clinics. In this region, travelling to clinics can require transporting the sick on foot for distances of up to 50 km. The RC1s were the basal level of a new official system of local government that had been established after President Yoweri Museveni assumed political power in 1986.

The first agreements reached were about beekeeping. These agreements allowed a limited number of people with a special interest in this activity to place a stipulated number of beehives within the multiple-use zones of the Park. This resulted, eventually, in an average of about 1000 hives being installed within each of the multiple-use zones of the three pilot parishes. Beekeeping at Bwindi is mainly an older-man activity and some of the beekeepers will have been *abataka*. In these societies, elders are respected, and it is the tradition for hive-keepers to share their honey with their families and friends – cultural features that help bind the communities together. Allowing beekeeping would have generated some general goodwill towards the park. Those involved in developing the MOUs at Bwindi considered beekeeping a benign activity with regard to its impact on the forest's ecology. So too, it was judged, was the collection of most medicinal plants, including *nyakibazi*. The approach for regulating the collection of most medicinal species within the multiple-use zones was to make species-specific agreements on maximum allowable harvests.

The high-use, high-impact, forest resources for which alternatives outside the park were deemed necessary included wood (used for construction and fuel), young tree stems (which were being harvested in large numbers to provide stakes for climbing beans) and some of the species that provided materials for basketry. For wood, the approach was to encourage the establishment of tree nurseries within the farmland surrounding the Park, an activity which proved to be popular. It was estimated that, during the first six months of 1994, 357 tree nurseries were in production within the communities. *Sesbania sesban*, a small, fast-growing tree capable of being coppiced, was promoted for the production of bean stakes. One type of basket that was in demand locally was the tea-basket, large numbers of which were being made for use on local tea estates and by local people who had planted tea on their farms in outgrower schemes. DTC made representations to the estates requesting them to find alternatives, such as providing collecting nets or plastic buckets.

Bwindi lies in a region of endemic political instability (Hamilton *et al.*, 2000). Political conflicts claimed 600,000 lives in Uganda between 1971 and 1986 (Doornbos, 1987) and, in Rwanda, which is only a short distance to the south, 800,000 people were massacred in a genocidal slaughter in 1993. Kivu Province, abutting Bwindi to the west, has been one of the politically most disturbed parts of the world for years, tying down one of the United Nations' largest peacekeeping missions. Bwindi itself was in the international news in March 1999, when Paul Wagaba, a community ranger, and eight tourists were murdered, possibly by Rwandese rebels who had entered from Kivu. There was a prediction in a British newspaper that the deaths would lead to the destruction of the Park, as tourists kept away, revenues dried up and local support evaporated (Millar and Woodward, 1999). But local support for the Park did not evaporate, which suggests that, by then, relationships between the park and the communities had resulted in the park's management system becoming sufficiently robust to cope with such shocks.

19 Resource Conflicts at Ayubia National Park, Pakistan (1996–2004)

WRITTEN IN THE FIRST PERSON BY ALAN HAMILTON

Abstract

Forests on the foothills of the Hindu Kush and Himalayan ranges in Northern Pakistan and neighbouring parts of Afghanistan and India help to protect the catchment of the river Indus, the lifeblood of the Pakistani economy. One of the remaining forests in Pakistan is contained mostly within Ayubia National Park, where an ethnobotany-based project was mounted in 1996–2004 to investigate the relationships between the park and the communities and to see whether ways could be found to prevent further degradation of the forest. It was found that the illegal harvesting of fodder and fuelwood in the park by women living in nearby villages was hampering the forest's ability to regenerate. Based on the information collected, a plan was drawn up by the ethnobotanical team to promote greater sustainability in the use of plant resources. This involved regularising the harvesting of fodder and fuelwood inside the park by incorporating the women's collecting groups into the management system of the park and promoting the provision of alternative resources outside the park. The plan was rejected on the basis that it would cause too much disruption to existing socio-economic arrangements and could have unpredictable results. Subsequent events in the nearby Swat Valley confirmed that, for the use of natural resources to be made ecologically sustainable in this part of Pakistan, a stronger voice for women is needed in decision-making circles.

19.1 The Choice of Ayubia as a Project Site

One of my tasks when I became Plants Conservation Officer of WWF-International in 1989 was to review those requests for funding received by WWF-International that related to plants. One such request arrived from the Pakistan Forest Institute (PFI) in Peshawar. WWF-Pakistan is one of the longest-established National Organisations in the World Wildlife Fund family and I saw this as an opportunity to visit Pakistan, discuss PRI's proposal face-to-face with its staff and meet with WWF-Pakistan, to establish how we could

© Alan Hamilton and Pei Shengji 2024. *History and Future of Plants, Planet and People*
(Alan Hamilton and Pei Shengji)
DOI: 10.1079/9781789248944.0019

collaborate in the future. (In the event, PFI's proposal was not accepted by WWF-International.)

It was while wandering around a garden of medicinal plants in the grounds of PFI that I happened to meet Ashiq Ahmad Khan, a forester who had formerly worked in Pakistan's Forest Department and had been recruited by WWF-Pakistan to help develop its field programme. Ashiq expressed an interest in collaborating with the People and Plants Initiative (PPI), a capacity-building programme in applied ethnobotany that was, at the time, beginning to develop a Hindu Kush-Himalayan component (see Chapter 10). It was from this chance encounter in Peshawar that Pakistan became one of PPI's key countries of field operations.

The PPI project with WWF-Pakistan began with a workshop in 1996 organised by Ashiq at the National Agricultural Research Centre and National Herbarium in Islamabad. The purpose of the workshop, which was attended by Pakistanis with a range of professional interests in ethnobotany, was to discuss the current status of the subject in the country and identify any related capacity-building needs (Aumeerduddy-Thomas et al., 2004). It turned out that, up to then, ethnobotany in Pakistan had been concerned almost exclusively with medicinal plants, with little attention given to plant resources of any other kind. Herbal medicine was widely used and based overwhelmingly on wild-collected, rather than cultivated, plants. It was known that unsustainable harvesting of wild medicinal plants was taking place and there were laws to prevent it, but these had proved largely ineffectual. Hakeems attending the meeting expressed an interest in helping to combat the over-harvesting of wild medicinal plants by issuing guidance to collectors on techniques of sustainable harvesting. (A hakeem is a traditional doctor trained in Unani Medicine – see Section 7.3.3.) No research had been carried out on the socio-economic aspects of medicinal plants' management and use.

It was agreed at the workshop that WWF-Pakistan was in a good position to host a capacity-building project in applied ethnobotany, acting as an intermediary between government agencies, the universities and the communities. It was also agreed that Ayubia National Park in Abbottabad District in (what was then) North-West Frontier Province (NWFP) was a suitable place to train postgraduate students in ethnobotany and investigate how local communities could be more closely involved in the management of natural resources. Ayubia, which had only recently been declared a National Park (in 1984), lies in the foothills of the Himalayas in Northern Pakistan and is conveniently accessible by road from the lowland cities of Lahore, Islamabad and Rawalpindi. For the more affluent social classes, Ayubia is well-known as a resort to which they can retreat during the summer months, when lowland Pakistan can be stiflingly hot. Ayubia offers refreshing mountain air, walks through pinewoods and bazaars catering to the interests of tourists. Every summer, about 100,000 people from the lowlands come to stay in summerhouses and hotels on the periphery of the national park, including in the settlements of Ayubia, Khanspur, Murree and Nathiagali. The climate cools down during the monsoon season (July to mid-August) and is bitterly cold in winter, when a thick cover of snow blankets the park and Ayubia is abandoned to its permanent inhabitants.

Ayubia National Park, with a total area of 3322 ha, ranges in altitude from 2000 to 2800 m. The vegetation of the park is dominated by conifers, the principal one being blue pine (*Pinus wallichiana*), with chir pine (*Pinus roxburghii*) replacing blue pine at lower altitudes and the West Himalayan fir (*Abies pindrow*) growing on summits. Here and there among the pine trees are scattered specimens of the Himalayan yew (*Taxus wallichiana*) and the broadleaved trees, evergreen oak (*Quercus* sp.), Himalayan horse chestnut (*Aesculus indica*) and bird cherry (*Prunus padus*). There are numerous open grassy glades within the park.

19.2   Notes on the National and Local Contexts

19.2.1   The close links between plants, the Indus and the economy

Much of Pakistan is low-lying and climatically too dry to support forest. The only place where tall dense forest is naturally found is on the foothills of the Hindu Kush and Himalayas, where heavy downpours are received during the moist Southwest Monsoon. However, many years of tree-cutting by people and grazing by livestock have resulted in the conversion of much of the original forest cover on these hills into grassland and scrub. Today, only 0.74% of the whole of Pakistan remains forest-covered and this is diminishing at the rate of 1% each year. Illegal tree-felling by a so-called 'timber mafia' with connections to the country's elite has compounded the political difficulty of tackling deforestation (Ashraf, 2017). Figure 19.1 is a photograph taken in November 2005 of the houses and associated agricultural terraces of the village of Mandray, which is situated on the side of a steep hill just outside the boundary of Ayubia National

Fig. 19.1. Mandray Village near Ayubia National Park, Pakistan. Photo: A.H. (November 2005).

Park. It can be seen that grassland covers much of the area, with the cover of trees becoming denser at higher altitude. Before human disturbance, all the area visible in the photo would have been covered by forest. The picture was taken from the road that connects the Park with the district capital of Abbottabad.

The economic future of Pakistan as a whole will depend to a large extent on the management of the waters of the Indus, the great river which dominates the country's hydrology. The Indus begins its life in the mountain ranges of the Hindu Kush and the Himalayas of North Pakistan and neighbouring parts of Afghanistan, India and Tibet in China. Two of its major tributaries are the Kabul River, which flows in from Afghanistan, and the Jhelum River, which originates in Jammu and Kashmir – a territory which de facto lies mostly in India but is claimed by Pakistan. How the land and its plants are managed in all parts of the Indus' catchment influences the benefits that Pakistan receives from the river (Younis and Ammar, 2018; Khan *et al.*, 2021). The extent of forest cover and the way that the forests are managed are major factors determining the seasonality of flows in the Indus and the sediment load in its water.

Fields irrigated by the Indus in lowland Pakistan lie at the heart of the Pakistani economy, producing much of the country's food and also cotton, the raw material which feeds its important textile sector. The Indus has been at the heart of the economy in the western part of the Indian subcontinent for a long time. It was the discovery of ways to harness the water of the Indus to boost agricultural production that gave rise, at about 3000 BCE, to the Indus Valley Civilisation, one of the great ancient civilisations of Eurasia (see Section 3.4). Until recently, Pakistan's irrigation system depended on the unregulated flows of the Indus and its tributaries, but, in the 1960s, two huge dams, the Mangla and Tarbela dams, were constructed in its upper reaches, opening a new era in the relationship between people and the river. The building of the Mangla Dam on the Jhelum River led to around 110,000 people losing their homes. Many were recruited by Britain to work in its cotton mills, which, at the time, were looking for labour (see Section 5.3.3).

The construction of the Mangla and Tarbela dams only became possible after a treaty (the Indus Waters Treaty) was signed in 1960 between India and Pakistan guaranteeing that Pakistan would receive a specified and uninterrupted share of the water originating in the Indus' catchment. Earlier, the partition of India in 1947 had led immediately to disputes breaking out between Pakistan and India on how to share the water of the Indus. The signing of the Indus Waters Treaty was eventually achieved through third-party mediation, with the World Bank playing a major role. However, the functioning of the irrigation system in Pakistan remains precarious. It has been suffering from reduced water storage capacity, caused by the siltation of dams, and soil salinisation, caused by a failure to flush out the salts accumulating in irrigated fields (Qureshi, 2011).

19.2.2   A note on the political and social geography of Pakistan and Ayubia

Pakistan was created in 1947 during the chaos and religiously fuelled violence that marked the end of British rule in India. Muslims fled towards the west and east, where they founded the new state of Pakistan. Hindus and Sikhs fled towards the centre, which is where independent India was born. (East Pakistan detached itself from West Pakistan in 1971 and became Bangladesh.) The legacies of history and geography have left Pakistan largely isolated economically within its region. Its border with India on its east mostly follows the Radcliffe Line, a boundary hastily established by the British as they were leaving India. Few people and goods have crossed this boundary for 50 years. To the west is Afghanistan, a country that has been politically unstable for decades. The natural barriers of the Hindu Kush and Himalayas hinder easy communication with the north, although movement became somewhat eased in 1978 when the Karakoram Highway was opened connecting the city of Gilgit in Pakistan with that of Kashgar in China. This highway passes through a high pass in the mountains that once carried a branch of the Ancient Silk Roads (see Section 3.5.2).

As mentioned in Section 5.1, the British, when they came to rule all India in the 19th century, adopted a 'forward policy' for its northwestern frontier (Churchill, 1898). This was designed to prevent the Russians establishing themselves in Afghanistan and, from there, using the Khyber Pass near Peshawar to mount an invasion of India. One element of this policy consisted of an agreement with Abdur Rahman Khan, who was the emir (ruler) of Afghanistan between 1880 and 1901, that established the boundary between what was then British India and Afghanistan (the Durand Line) at the foot of the Afghan mountains. The other was the periodic mounting of 'punitive expeditions' into Afghanistan to make sure that its foreign policy conformed to British interests.

The Durand Line split the tribal lands of the Pashtun people into a part in Afghanistan and a part in British India, stoking fears among them that this would weaken their cultural identity. In 1901, these fears became intensified when the British split off the NWFP from the rest of the Punjab, which further subdivided their territory (Harrison, 2008). A sense of grievance against the authority of central government smouldered on among the Pashtuns even after 1947, when Pakistan became politically independent. In 2007, this resentment took a violent turn with the founding of the Pakistani Taliban, one of whose goals was (and is) to split off a Pashtun state from the rest of Pakistan. In 2010, the Pakistani Government changed the name of the NWFP to Khyber Pakhtunkhwa Province in an attempt to appease the Pashtuns. However, this was not received well by the other ethnic groups living in the area who feared that their own identities would become subsumed in a Pashtun-dominated province (Abbasi and Kalhoro, 2022; Abbasi et al., 2022).

In 1864, the British created a National Forestry Service in India (see Section 5.5). At the time, Britain was only just beginning to bring the northwestern part of the subcontinent under its control, a milestone in its advance being the military defeat of its Sikh rulers in 1849. During the 1880s, the Indian Forest Service started to establish Forest Reserves in the Galyat Tract, the range of hills that contains Ayubia. This resulted in numerous objections being raised by the local people, who were annoyed because it reduced the easy access to some natural resources that they had previously enjoyed (Ali, 2006). Vociferous objections were raised complaining about the positioning of the boundaries of the Forest Reserves and the restrictions placed on collecting natural resources within them (Azhar, 1989).

At the time of the PPI project, Ayubia was designated as both a Protected Forest and a National Park, which meant that it came under the authority of both the Forest and Wildlife departments. The local communities had no legal rights to use any of the Park's resources. Outside the Park, the forests found within the project's area of operations were broadly classifiable into Reserved Forests, which belonged to the Forest Department, and Guzara forests, which were privately owned, but had restrictions imposed on them by the Forest Department to regulate the cutting of trees. The Forest Department placed no restrictions on the collection of non-timber forest products (NTFPs) in the Guzara forests, which was a matter left to the Guzara-owners to negotiate with the communities. In practice what had happened was that the Guzara forests had become virtually open-access areas for the collection of NTFPs and of grazing by goats. Inevitably, this resulted in serious depletion of those NTFPs that interest the communities, namely dead wood (for use as fuel), fodder (for stall-feeding livestock), wild vegetables, medicinal plants and mushrooms. In comparison, the National Park, which had received at least a modicum of protection, tended to be in a better state of repair as regards the availability of these resources, which made it a tempting target for the communities.

Figure 19.2 is a photograph of a patch of Guzara forest in Malachh Village near the boundary of the Park. Note that there is only a low vegetation cover on the open forest floor, which is because of the intensive collection of grass fodder by the people and grazing by goats. Note that some branches are missing from the pine trees (P. wallichiana). This, according to local information, is due to them being removed by women who have climbed the trees and lopped

Fig. 19.2. Guzara forest on the margin of Ayubia National Park. Photo: A.H. (1996).

them off to be used as fuel. Every year, I was told, two or three women meet their deaths through falling from the trees.

19.3 The Ayubia Project

19.3.1 Project launch, its structure and initial findings

Following the introductory workshop in Islamabad, Ashiq formed a project team to implement the project at Ayubia and rented a house at Nathiagali on the border of the Park to accommodate the team and establish an office (Aumeerduddy-Thomas *et al.*, 2004). The team consisted of Abdullah Ayaz (a forester and the Project Officer), Junaid Khan (social forester), Hasrat Jabeen (sociologist), and Zabida Zaman and Mohammed Sajjad (field assistants from local villages). The field assistants provided direct links to the communities throughout the project and contributed their first-hand knowledge to assist with the fieldwork. Postgraduate students from universities in Pakistan stayed with the field team for varying lengths of time to conduct ethnobotanical research. Supervision for the postgraduates was provided by the staff of their universities, with further guidance from Yildiz Aumeeruddy and Pei Shengji, the successive PPI Regional Coordinators. The project ran from 1996 to 2004.

The project team was guided by an Advisory Committee formed by representatives of the local villages, staff of the local Forest and Wildlife departments, some other interested local parties (such as hotel-owners) and staff from the headquarters of WWF-Pakistan. The findings of the project were fed into a series of National Training Workshops held at Ayubia, each attended by about 25 people drawn from all over Pakistan, mainly postgraduate students, university staff and conservation managers. Among the themes of the workshops were 'Ethnobotany applied to participatory forest management in Pakistan' (Shinwari and Khan, 2001), 'Land tenure and resource ownership in Pakistan' (Shinwari and Khan, 2002) and 'The purposes and teaching of applied ethnobotany' (Hamilton *et al.*, 2003). Through these workshops, it was hoped that the project would help stimulate the development of ethnobotany in the country and its applications to conservation and development.

The focus of the fieldwork of the project, which ran from 1996 to 2004, was on the use of plant resources by the approximately 42,000 people in 6000 households and 12 villages who were living around the perimeter of the Park. Two villages, Pasala and Malachh, were chosen as pilots in which new project activities were first tried out. Most of the households in the project area had small rainfed farms, typically 0.25–0.50 ha in size, on which they grew food for themselves and fodder for their animals. Seven ethnic groups were represented in the project area, the main ones being the Karalls and Abassis, who share a common language called Hindko and are collectively known as Hindkowans.

Many ad hoc committees with flexible membership were convened at the village level to discuss the project, especially before new activities were started in villages. It soon became apparent that some of the most pressing local livelihood issues in relation to the Park revolved around firewood collection, fodder extraction and the use of the open glades in the park to graze cattle and buffalo in the summer months. It also became apparent that, to a considerable extent, the illegal incursions into the park to collect firewood and fodder were being made by women, not men. The project responded to this finding by encouraging the formation of four village-based Women Advisory Committees, so that the women would have a space in which to voice and discuss their concerns. This was considered essential if women were to play a meaningful role in guiding the research programme and devising improved ways to manage the resources.

Fieldwork during the first field season (March–December 1997) was devoted to determining the broad parameters of social organisation at the village level, the use of plant resources by the communities, the formal and informal arrangements that governed access to plant resources, and the impacts of collecting the resources on the ecosystems. It was found that, at village level, the families were grouped into *baradri*, each of which was under the authority of a chief (*badka*). All the *badka* in a village formed a council (*jirga*), the membership of which could overlap with that of the Village Union Council, which was the government administrative unit at village level.

Men were the main decision-makers in the family and were responsible for earning money. Some worked in local bazaars or as seasonal workers in summerhouses and hotels. At home, the men undertook the ploughing of the fields, shared some of the harvesting tasks with the women and were responsible for constructing and maintaining buildings. Women's tasks included collecting firewood and fodder, looking after the livestock, choosing seed for planting, housekeeping and taking care of children. When women entered the Park to collect plant resources, they did so in small groups to provide them with a measure of protection against harassment by the Forest and Wildlife guards, all of whom were men. Research in Pasala and Malachh revealed that the collection groups were formed mainly on the basis of neighbourhood, rather than family relationships, although it was common for the households of relatives to be clustered together. The groups varied in size from three to 40, the larger ones tending to be when there was further to travel, which could be either because their villages were located further from the park or because the groups had decided to go to remoter locations in the park. Women from Pasala visited twice as many collection sites within the park as women from Malachh, which may be because Pasala is a more scattered settlement, spread over a larger area.

While in the Park to collect firewood and fodder, the women sometimes made side-collections of medicinal plants, wild vegetables and mushrooms – the last two, in particular, being generally unavailable outside the park. Herbal medicine was being regularly used by these communities, which had little access to pharmacies and Western medical practitioners. The main wild vegetables collected were the leaves of the ferns, *kandhor* (*Dryopteris blanfordii*) and *kunji* (*Dryopteris stewartii*), and also the leaves of *mirchi* (*Solanum nigrum*), *mushkana* (*Nepeta laevigata*) and *tandi* (*Dipsacus inermis*). The only mushroom collected was *guchi* (morel – *Morchella*), which was sold to travelling merchants for export to the Middle East and Europe (Hamayun et al., 2006). For many women, selling *guchi* represented their only source of cash income.

In one survey of 40 households, it was found that 21 women, 12 men, six boys and one girl collected *guchi* during the mushroom season (March to mid-May), 38% of them on a daily basis.

There was an exceptionally high level of conflict over plant resources between the people and government institutions at Ayubia (Lodhi, 2007). The Park Management Plan at the time of the PPI project mentioned that there had been 1227 offences committed relating to the Park during 1996–2000, the offences listed being lopping branches off trees (977 cases), grazing (62) and theft of timber (188) (Shinwari, 2002). In addition to legally registered offences, the project found that informal punishments were sometimes meted out by the guards, for instance an 'informal fine' of 50–200 Rupees was being paid for carrying home a bundle of firewood (Aumeerduddy-Thomas *et al.*, 2004). For households rich enough to purchase wood (illegally) from the Forest Guards, an 'informal fine' of 2000–5000 Rupees was payable per tree. (At the time of the research, 50 Rupees was equivalent to 1 US dollar.)

19.3.2 Work on fodder and fuelwood

Having discovered in its first year of operation that fodder and fuelwood collection were the two principal plant resources being taken by the villagers from the Park, the project team moved on in the second year to collect more detailed information about these resources. Much of the groundwork with the communities was carried out by Asma Jabeen, a postgraduate student at Quaid-i-Azam University, Islamabad, supervised by Dr Mir Ajab Khan. Asma can be seen in Fig. 19.3 standing on a terraced field in one of the villages near the Park, whose forested hills can be glimpsed in the background. She is standing by the side of a pile of maize stalks, one of the types of fodder used by the communities.

The project team found that the number of livestock owned per household is very low – typically, one buffalo (used for milk production) or one cow and one goat. During winter to early spring (November to March/May), all livestock are stall-fed on feed collected during the previous growing season. The winter feed includes crop residues (e.g. maize stalks), grasses growing on *banna* (field margins) and grasses harvested from three locally recognised land units present in Guzara forests. One of the main grasses found on field margins is *maniara* (*Alopecurus* sp.), a belt of which can be seen to the right of Asma in Fig. 19.3. The three land units in Guzara forests recognised by the communities are *beth* (flat unterraced land, potentially suitable for agriculture), *thaia* (steep unterraced land, potentially suitable for agriculture) and *rakhan* (steep rocky land unsuitable for agriculture, but which has been transformed by people into productive grassland through weeding and periodic burning).

Women had a detailed knowledge of fodder – for instance, about the most suitable types to feed to milking cows and young calves. There was an annual fodder calendar. Milking buffalo and cows were stall-fed throughout the year, fresh feed being given to them on a daily basis from March/May to October.

Fig. 19.3. Asma Jabeen and fodder on a farm near Ayubia National Park. Photo: A.H. (1999).

Only fresh tree fodder was used in early spring (March to April), after which herb fodder began to come on stream. The preferred types of trees used for fodder included *bharmi* (*T. wallichiana*), and the broadleaved trees, evergreen oak, Himalayan horse chestnut and bird cherry. When plant growth starts to become vigorous in spring, the oxen and male buffaloes were taken from their winter stalls and released into the Park to wander about at will to feed in the glades. Goats were also released in spring to roam about outside, but were prevented from entering the park where, the people feared, they might be eaten by leopards. The project team found signs that some of the small ever-green oak trees in the park were being harvested on a rotational basis, which suggested that at least some of the women realised the value of avoiding the over-harvesting of the resources. However, the project team thought that there was little point in trying to spread such traditional conservation practices until the women had greater assurance that they would be able to benefit collectively from their use.

The research on firewood found that most of it is collected in the Park, that deadwood is preferred over fresh wood, and that the bulk of the firewood collected comes from conifers. The average weight of wood placed into storage per household between mid-June and mid-September was about 10 tonnes. Households were burning about twice as much wood per day in winter as sum-mer. Outside the park, only conifer wood was normally available. In addition to the lopping of branches off trees in Guzara forests, another common practice outside the park was to ringbark trees so that they would die, at which stage, according to Forest Regulations, they could be taken as fuel.

It was found that the villagers were not the only users of firewood at Ayubia. Although initially denied by the hotels, it was discovered that, in fact, most of the wood used by the hotels comes from local sources. One survey revealed that the majority of wood stacked in the woodstores of hotels was oak (*Quercus*) and yew (*Taxus*) and that about 13 tonnes of firewood were being used per hotel per year. It was deduced from this that there was active competition between the hotels and the women over the collection of firewood within the Park.

Once the results of the detailed studies of fodder and fuelwood use started to become available, the project moved on to consider what solutions it might be able to offer to meet the conservation challenges. Two potential lines of improvement were pursued with respect to firewood, one the planting of trees outside the Park to provide alternative sources of firewood and the other the promotion of fuel-efficient woodstoves. Discussions were held at Pasala and Malachh to choose species for tree-planting, decide where the nurseries to provide the seedlings were to be located and determine who would be responsible for their management. The project staff were provided with training in tree nursery techniques at the PFI in Peshawar and two women students from the Environmental Sciences Department at the University of Peshawar served internships with the project to assist with the work in the villages. In practice, all the managers of the tree nurseries turned out to be women.

The introduction of the tree nurseries proved to be a learning process for all concerned. Twenty-five nurseries were started in the first year (1999), all of them based on growing black locust (*Robinia pseudoacacia*), which is a small, fast-growing tree native to temperate North America and which is popular with agroforestry projects. Following suggestions from the communities, additional species were added in the second year (2000), including *khor* (*A. indica*), which is a local native, and *darawa* (*Ailanthus altissima*), which is a native of China. This increased the success rate of nursery establishment and was, anyway, a step forward environmentally because it reduced reliance on the black locust tree, which can be a troublesome invasive (Vítková *et al.*, 2017). In 2002, a further adjustment to project procedures was introduced, involving the introduction of a Terms of Partnership Agreement between WWF-Pakistan and each of the nursery owners. This committed WWF-Pakistan to provide targeted support for each of the nursery owners on condition that the owners looked after the nurseries and took tree-planting seriously. This adjustment proved beneficial and, in 2002, 67 new nurseries were established, with a 65% success rate (Aumeerduddy-Thomas *et al.*, 2004).

The first step taken to promote fuel-efficient wood stoves was for Hazrat Jabeen, the project's sociologist, to go to Gilgit in Northern Pakistan to learn from the members of the Aga Khan Development Network about the designs of the fuel-efficient woodstoves that they were promoting. On return to Ayubia, Hazrat requested a local blacksmith to make some copies of each of the designs to try out locally. The selection of households for inclusion in the trial was based on expressed interest and assessments by the project staff of their abilities to keep accurate records of the amounts of firewood that they used. For each household that received an experimental stove, two nearby households were recruited as controls and requested to continue using their existing stoves in their normal ways. At the conclusion of the trial, two of the designs were selected to be manufactured at scale by local blacksmiths, who were provided with training in how to make them. In the first year, the prices of the new stoves were subsidised by a large European Union-funded development project which was working on natural resource issues in Abbottabad District, but thereafter they were sold at market prices. The trials revealed one drawback of the fuel-efficient stoves, which was that they radiated less heat into the rooms compared with the normal stoves, which made the rooms less comfortable during the cold winter months.

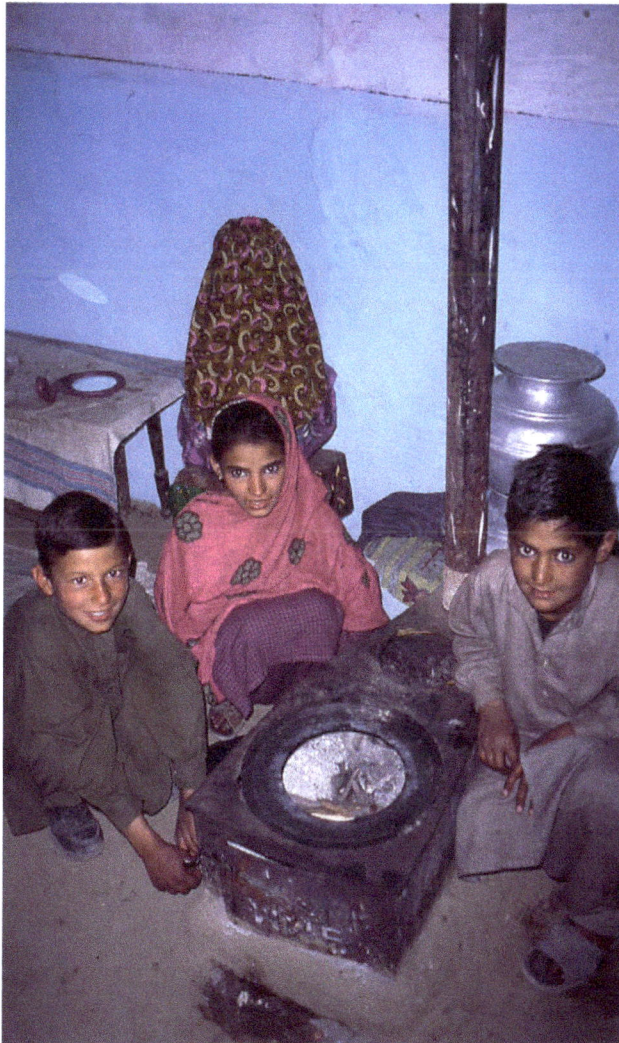

Fig. 19.4. Fuel-efficient stove in a home near Ayubia National Park. Photo: A.H. (2002).

Figure 19.4 is a photograph of one of the fuel-efficient stoves in a home in Toheetabad, which is one of the villages on the margin of Ayubia National Park.

19.3.3   A proposal to advance plant conservation at Ayubia

Before the project ended in 2004, one more step was taken to try and instigate more sustainable use of plant resources at Ayubia. This was in the form of a proposal submitted in 2001 to WWF-Pakistan by Yildiz Aumeeruddy, PPI's Regional Coordinator. The proposal contained three linked elements. One was to regularise the informal system used by the villagers to regulate the collection of fodder in the Park, which was through unwritten understandings between the groups of women as to which glades 'belonged' to which of them. It was suggested that, by formally recognising this reality and incorporating it into the park's management system, more progress would be made in strengthening the ability of the managers of the park to conserve its biodiversity. As an added incentive to encourage the women to obey the rules and stick to their own fodder-collection areas, it was suggested that they should be allowed to collect *guchi* (morel mushrooms), which tend to be found where the forest is dense and the soils are suitably rich in organic matter. This, it was suggested, would give the women an incentive to protect patches of more mature forest.

The second element of the proposal was to formally allocate particular areas of the Guzara and Reserved forests outside the Park to particular households. In that way, it was suggested, they would have an incentive to care for them, create new areas of *rakhan* and plant trees. This was in recognition that the small sizes of the villagers' farms left little room for growing more fodder or for planting trees. The third element of the proposal was to develop a system of mediation to deal with the questions and conflicts that, it was thought, would inevitably arise as the first two elements were implemented. Eventually, it was envisaged, the use of the park to gather firewood and graze livestock would diminish as stocks of firewood and fodder became built up around its perimeter.

WWF-Pakistan's response to this proposition was that it would be unlikely to be acceptable to the government because allowing the legal collection of plant resources within Ayubia would require the law to be changed for all National Parks, which would be such a big change from the status quo that it could have many unforeseen consequences. Having had this proposal for a way forward at policy level rejected, the project devoted more effort in its final years to encouraging universities and forestry schools to include ethnobotany courses in their curricula. No further attempts were made by the project team aimed directly at influencing on-the-ground realities.

19.4   More Turmoil and Conflict in the Ayubia Area

My personal collaboration with Ashiq Ahmad Khan and WWF-Pakistan continued during 2004–2008, which was after I had left WWF-International and was working for Plantlife in its Medicinal Plants Conservation Initiative

(MPCI – see Chapter 11). The field site that Ashiq chose for the work with Plantlife was Swat, a long valley that extends deep into the Hindu Kush about 100 km northwest of Ayubia close to the Afghan border. One of the components of the work at Swat was the promotion of environmental education in primary schools. In 2009, soon after the project ended, Swat was in the world news when the Pakistan Taliban took control of the valley, in the process displacing about 2 million people from their homes. In 2012, a local schoolgirl called Malala, who had courageously been promoting girls' education in Swat, was identified in a school bus and shot by a Taliban gunman. She survived and later went to Britain, where she continued her studies. In 2014 at the age of 17, she became an international celebrity when she became the youngest person ever to receive the Nobel Peace Prize for speaking up for the rights of all children to have an education – girls, as well as boys.

Fig. 19.5. Earthquake-damaged school in Mandray Village. Photo: A.H. (November 2005).

In November 2005, Dr Zabta Khan Shinwari, who had earlier acted as Project Officer for the PPI project at Ayubia, organised a workshop for Plantlife's project at the University of Kohat, of which he was the vice-chancellor. I participated at the workshop at the request of Zabta and afterwards travelled to Ayubia to see the damage that had been inflected there by a massive earthquake that had struck just one month earlier. An estimated 100,000 people had been killed and about 3.5 million more displaced from their homes. I took the photograph of Mandray Village shown in Fig. 19.1 during this visit. When I was at Mandray, I was informed that most of the people who lived in the houses shown on the photograph had gone to Abbottabad, where emergency help was available, but it was added that someone from each of the households had remained behind to guard against theft of their precious stocks of firewood. Figure 19.5 is a photograph of Mandray's primary school, which had been wrecked by the earthquake.

The damage inflicted by the earthquake at Mandray seemed symbolic of the struggles that the people of the Ayubia area have been facing in trying to find a way to live in harmony with nature. The earthquake was not the last event in the area to make the international headlines. In 2011, the US military mounted a lightning raid on a house in Abbottabad and killed Osama bin Laden, the al-Qaeda leader. Bin Laden had retreated to Abbottabad from his former hideout in the mountainous borderlands between Afghanistan and Pakistan after al-Qaeda's suicide attack on the Twin Towers in New York had been answered by a US-led invasion of Afghanistan.

20 Conservation and Healing in Dolpo, Nepal (1997–2004)

WRITTEN IN THE FIRST PERSON BY ALAN HAMILTON

Abstract

An ethnobotanical approach was used in a project in Dolpo, Nepal, in 1997–2004 to research relationships between its people and plants and to determine any related conservation and development issues. With the assistance of WWF, the Nepalese Government was establishing a new protected area, Shey Phoksundo National Park (SPNP), in Dolpo, which is a culturally Tibetan area. The staff of WWF in Nepal were concerned about reports of large-scale commercial harvesting of medicinal and aromatic plants (MAPs) in SPNP and feared that this might be degrading the park's environment and creating a shortage of those medicinal species that the local people relied on for their healthcare. A decision was made at the launch of the project to approach the development issues posed by the commercial MAP collection by placing an emphasis on the links between the plants and the delivery of local healthcare, hoping that this might stimulate the people into paying more attention to conserving the species. Two groups of local people were identified as potentially being keenly interested in the conservation of medicinal plants – women, who assume most of the responsibility for looking after the healthcare of their families, and amchis, who are traditional doctors trained in Tibetan Medicine. Follow-up research revealed deficiencies in local healthcare, to which the project responded by providing training in hygiene and sanitation to the women and assisting the amchis to refresh their knowledge of Tibetan Medicine. It was discovered that traditional conservation measures already existed at Dolpo, for instance to avoid the overgrazing of pastures by yak, and it was decided to build on these to ensure sustainability in the harvesting of MAPs. A system for the overall management of MAPs was devised linked to the improvement of local healthcare services. This included the construction of a traditional medical centre institutionally linked to the National Park. The project was seriously disrupted in 1998 when fighting erupted in the area between the army and Maoist insurgents. This eventually resulted in the destruction of the traditional medical centre, which had been opened for business in 2000. By the early 2010s, when WWF decided to resume its conservation work at SPNP, this time concentrating on 'wildlife', there was no institutional memory among the staff of WWF-Nepal at its head office in Kathmandu about what the ethnobotany project had accomplished.

© Alan Hamilton and Pei Shengji 2024. *History and Future of Plants, Planet and People*
(Alan Hamilton and Pei Shengji)
DOI: 10.1079/9781789248944.0020

20.1   Project Initiation

In 1997, I was requested by Mingma Norbu Sherpa, a Nepalese conservationist working in the office of WWF-US in Washington, to assist with the development of the botanical aspects of a conservation project at Shey Phoksundo National Park (SPNP), which is in the Himalayas of Nepal. Declared a National Park in 1984, SPNP was a place of special interest for animal conservationists, being home to the snow leopard, a famous endangered species, and also the common leopard, the Himalayan wolf, the red panda and the blue sheep. Mingma was concerned about reports of large-scale commercial harvesting of medicinal and aromatic plants (MAPs). This, it was feared, might be degrading the park's environment and creating a shortage of the medicinal species that the resident people relied on for their own healthcare.

Mingma's request to me was made within the context of my involvement with the People and Plants Initiative (PPI), a partnership programme of the World Wildlife Fund (WWF – now World Wide Fund for Nature), the United Nations Educational, Scientific and Cultural Organization (UNESCO) and the Royal Botanic Gardens, Kew, aimed at increasing global capacity in applied ethnobotany (see Chapter 10). Before agreeing to Mingma's request, Yildiz Aumeeruddy, who had been appointed as the Regional Coordinator of PPI for the Hindu Kush-Himalayas, attended a workshop of Nepalese conservationists and ethnobotanists in Kathmandu to identify priorities for capacity-building in ethnobotany in Nepal. This confirmed that the conservation and development issues reported for SPNP were matters of concern in Nepal generally and that SPNP was a good place to base a capacity-building project. The drawback was that SPNP was situated in an extremely remote place, even by Nepalese standards. There are no roads in SPNP. Travel is by foot or, for a few, on horseback.

SPNP, which is the largest national park in Nepal (3555 km^2), is situated in a part of the Himalayas traditionally known as Dolpo. Its inhabitants, the Dolpo-pa, are ethnically Tibetan. Nowadays, the traditional area of Dolpo falls into two governmental administrative districts, Dolpa (capital Dunai, 2000 m) and Mugu. SPNP, which ranges in altitude from 2200 to 6800 m, contains a very wide range of ecosystem types. Temperate forest is found at lower altitudes on its south-facing slopes, where they are fully exposed to the monsoonal rains. There are alpine communities at high altitudes. The most extensive parts of both Dolpo and SPNP, which are in the north, are covered by high-altitude Tibetan desert. Some of the Tibetans at Dolpo live in some of the highest, permanently inhabited, settlements on Earth.

Mingma Sherpa and Dr Chandra Gurung, who headed the WWF office in Nepal (situated in the capital city of Kathmandu), were keen to try a new conservation model at SPNP, different from the one that had been used at the first two National Parks to be established in Nepal, namely Royal Chitwan and Sagarmatha. Royal Chitwan National Park, established in 1973, is in the lowlands and is home to the tiger, the Asian elephant and the Indian rhinoceros. Sagarmatha National Park, established in 1976, includes Mt Everest, the highest mountain in the world. The conservation model that had been used at Chitwan and Sagarmatha was based on the idea of maintaining them as wildernesses. The adoption of this model speaks to the influence of the United States' National Organisation of WWF on the WWF programme in Nepal. As discussed in Section 8.2.2, the idea of preserving areas of wilderness has a strong hold on America's conservation imagination. The reason why WWF-US was so influential over WWF's activities in Nepal was because, for reasons of political history and administrative convenience, all funds received by WWF for its work in Nepal were channelled and coordinated through WWF-US. At the time, China and Latin America were two other parts of the world whose conservation programmes were coordinated by WWF-US, rather than WWF-International.

The conservation model that interested Mingma and Chandra, both of whom had been raised in remote Himalayan villages, was one that was more sensitive to the needs and aspirations of the local people (Bauer, 2004). The model, which was similar to that of the Biosphere Reserve model promoted by UNESCO (see Section 8.3), incorporates the concepts of core areas and buffer zones. Conservation of biodiversity was seen as a high priority for management in core areas and the sustainable use of natural resources to benefit the local people in the buffer zones.

In the way that Mingma, Chandra and others had been developing the concept of the buffer zone in practice in Nepal since 1993, a difference had emerged in the way that it was applied in the lowlands and the Himalayas. For lowland National Parks, like Chitwan, all the designated buffer zones lay around their perimeters, while, in the Himalayas, some buffer zones existed *within* the core areas, as well as others around their perimeters. This Himalayan variant was adopted in recognition of the reality of people's lives in the mountain environment. Relatively flat areas of more fertile land are few and far between, and these tend to be already inhabited by people, with their homes, fields and areas used for the production of fodder. The great bulk of the landscape, away from these pockets of habitation, is composed of steep-sided rocky slopes covered by temperate and alpine habitats that are exploited much less intensively by the people. Removing the inhabitants of the scattered settlements to create total wildernesses was not something that Mingma and Chandra were prepared to do. In 1997, when the PPI project was launched, there were about 10,000 people living in the buffer zones of SPNP, although only about 3500 of them in those that were internal. The park regulations governing the use of resources differed somewhat between the external and internal buffer zones. For instance, the commercial harvesting of MAPs was allowed in the former, but not the latter.

In 1997, when the PPI project got underway, WWF had another project, the Northern Mountains Conservation Project (NMCP), active in SPNP. Funded by the US Agency for International Development (USAID), its purpose was to assist with the establishment of the new National Park. The WWF Field Officer for the NMCP project was Dhana Rai, who had established his base in the village of Ringmo (altitude 3660 m). At the time when the PPI project started, already eight buffer zones had been designated around the periphery of SPNP and another three within it. Ringmo, together with another village,

Pungmo, were situated within one of these, already established, internal buffer zones. It corresponded to the governmental administrative unit, Phoksundo Village Development Committee (VDC). This VDC and its villages of Ringmo and Pungmo were chosen as pilots for the field activities of PPI in the park. (In 2017, the 'Village Development Committee' was re-termed 'Gaunpalika'.)

SPNP is a long way from Kathmandu. To travel there by air involves a flight to Nepalgunj, a steamy low-altitude market town on the border with India, and then a flight in a small plane through steep-sided valleys to land on a short sloping airstrip at Juphal. It is then a short walk to Dunai (altitude 2000 m), which is the capital of Dolpa District. From Dunai, it is a three-day walk up to Ringmo, where travellers from the lowlands are advised to spend time acclimatising to the rarefied air before walking for another day up to the high pasture of Gunasa (altitude, 4000 m and higher). The significance of Gunasa is that it was selected by the project to trial an innovative system for the sustainable harvesting of medicinal plants (see Section 20.3). The people of Ringmo and Pungmo practice transhumance, taking their yak and sheep in spring up to Gunasa and other high pastures, and returning in the autumn to overwinter with their livestock in their permanent villages. There are fields for growing crops, both near the comparatively low altitudes of Ringmo and Pungmo, and also in high pasture areas, such as Gunasa. Two of the major crops are barley and buckwheat.

20.2 An Overview of the Culture and Economy of Dolpo

Research by the PPI project and other information showed that the economy of the Dolpo-pa in 1997 was based on the trio of pastoralism, agriculture and trade. For centuries, many Dolpo-pa, living in villages such as Ringmo and Pungmo on the southern edge of the Himalayas, have taken advantage of their strategic position at the interface between the Indian subcontinent and the Qinghai–Tibet Plateau to be involved in trade (Clarke, 1977; Bauer, 2004). In Section 3.5.2, we mention the tea–horse trading routes that connected lowland China at the eastern end of the Himalayas with Tibet. Towards the western end of the Himalayas, including in Nepal, four commodities that have traditionally featured strongly in trans-Himalayan trade have been rice (from the lowlands to Tibet), salt and wool (from Tibet to the lowlands), and MAPs (mostly from the highlands to the lowlands). The bulk of MAPs for export from Dolpo comes from its southern margin, which is more biologically productive.

Research by the PPI project revealed that a substantial number of the local people living in Southern Dolpo have traditionally been involved in collecting MAPs for sale (Lama *et al.*, 2001; Ghimire *et al.*, 2016). About 20 species have traditionally been collected in large quantities, among them *Nardostachys jatamansi* (**Am** *pang poe*; **Np, Sn** *jatamansi*), *Neopicrorhiza scrophulariiflora* (**Am** *hon glen*; **Np** *kutki*; **Sn** *katuka*), *Rheum australe* (**Am** *chutsa*; **Sn** *pitamulika*) and *Valeriana jatamansi* (**Am** *na pie*; **Np** *samayo*). (Key to vernacular names: **Am** *Amchi*, **Np** *Nepali*, **Sn** *Sanskrit*.) These are all slow-growing perennial herbs, the first three being found in open rocky places in alpine grassland, with *V. jatamansi* tending to occur at somewhat lower altitudes than the others. *R. australe* is a wild relative of the cultivated rhubarb, which was originally imported as a medicinal into Europe from Asia via the Ancient Silk Roads. Among the relatively few species of MAPs that are traded from the lowlands of the Indian subcontinent into the highlands are some that are required in Amchi Medicine, such as the fruits of the trees *Phyllanthus emblica* and *Terminalia chebula*.

There is one highly priced medicinal commodity collected in Dolpo that is traditionally traded northwards into Tibet and then on to China, rather than southwards into the Indian subcontinent. This is *yartsa gumbu* (caterpillar-fungus), which is highly prized in China as a tonic and aphrodisiac. *Yartsa gumbu* consists of the mummified remains of the larvae of certain types of moths, together with the fruiting body of an attached fungal parasite (*Ophiocordyceps sinensis*). The moths are high-altitude species found above the treeline. The presence of *yartsa gumbu* reveals itself to those who are searching for it by the dark brown club-shaped fruiting body of the fungus protruding out of the soil.

About half of the inhabitants of Dolpo are Buddhists and about half follow Bon, the religion that existed in the Himalayas prior to the arrival of Buddhism (see Section 7.3.3). There are numerous *choden* (stupas) and several *gomba* (monasteries) and it is a common practice for some members of families to enter monasteries or convents and become monks or nuns. To the outsider, Buddhism and Bon, as practised at Dolpo, can appear quite similar. One obvious difference is that Buddhists circumambulate stupas in a clockwise direction, while adherents of Bon go the other way round. The religious beliefs of the people, reinforced by communal pilgrimages and other rituals, are a major factor shaping people's relationships with the land in Dolpo. Local lifestyles are nature-friendly, based on the principles of universal compassion and respect being accorded to all forms of life.

Dolpo is a long way away from the centres of political authority in Nepal. The provision of modern (Western) medicine is minimal and the greater part of professional healthcare is delivered by amchis, the practitioners of Tibetan Medicine (otherwise known as Sowa Rigpa or Amchi Medicine – see Table 7.2). An amchi at Dolpo is commonly referred to as a 'lama', an expanded use of the word, which, originally, was reserved for a spiritual master, such as the head of a monastery. Medicine, as practised by the amchis, is connected with their conviction that physical healing and spiritual development are interlinked. At the time of the PPI project, Tibetan Medicine had not yet been fully recognised officially as a medical system in Nepal, but rather was semi-recognised to tolerated (S. Kloos, Austrian Academy of Science, 2023, personal communication).

20.3   Project Implementation

Three of the field team formed for the project were ethnic Tibetans – Yeshi Choden Lama, the WWF Project Officer, and Gyatso Bista and Tshampa Ngawang Gurung, who were expert amchis from Mustang. Situated in the Himalayas about 100 km east of SPNP, the Kingdom of Mustang is a recognised centre of excellence in Amchi Medicine. The botanists in the team were Krishna Kumar Sherpa, a staff member of the Department of Botany of Tribhuvan University in Kathmandu, and Suresh Kumar Ghimire, his research student. Suresh stayed at SPNP for long periods of time during the course of the project and was responsible for carrying out much of day-to-day botanical and ethnobotanical fieldwork. Figure 20.1 is a photograph of Suresh (squatting) identifying medicinal plants collected by Amchi Tangual Lama (foreground) in Dolpo.

A local resource team of eight amchis from various parts of Dolpo was periodically convened to tap into their expertise in botany and local healthcare issues. Training in field techniques in ethnobotany was provided during the course of the project to 11 young local employees of SPNP in an effort to con-solidate links between the park and the communities. Dhana Rai, from his base in Ringmo, supported the project throughout its time of existence and Yildiz Aumeeruddy, the PPI Regional Coordinator, provided guidance on ethnobo-tanical and other scientific methodologies. Liaison was maintained throughout the project with the Kathmandu-based Himalayan Amchi Association (HAA), an organisation formed in 1998 to provide opportunities for amchis in Nepal to network among themselves and improve the healthcare services that they offered.

In June 1997, the project was launched with a planning meeting at Ringmo that was open to anyone who had a special interest in medicinal plants. Discussions at the meeting resulted in a decision being taken to approach the conservation issues posed by the large-scale commercial collecting of medicinal plants by placing an emphasis on the links between conserving the plants and the delivery of local healthcare. The reasoning behind this decision was based

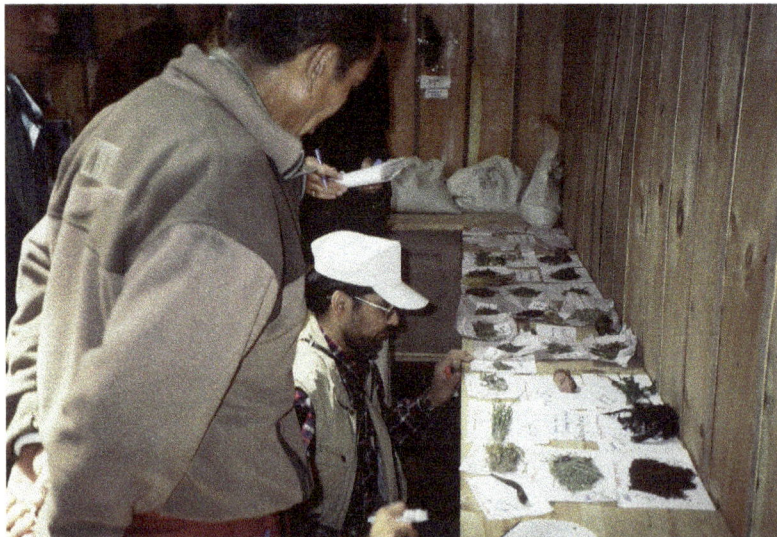

Fig. 20.1. Suresh Ghimire examining plants in Dolpo, Nepal. Photo: Yildiz Aumeeruddy.

on the knowledge that health is a basic human interest and that emphasising the connections between the PPI project and the delivery of local healthcare might stimulate the local people into taking an interest in the project's activities. In Dolpo, locally growing wild medicinal plants are essential ingredients in both the healthcare treatments given in the homes and the professional healthcare services provided by the amchis. Drawing attention to the shortage of local medicinal plants that could be caused by unregulated commercial collection might stimulate people into paying more attention to conserving the species.

Two groups of people living within SPNP were identified as potentially having particularly keen interests in the conservation of medicinal plants. These were the amchis and women. The bulk of the materia medica of the amchis is botanical, much of it collected locally by the amchis themselves. At Dolpo, women are the members of households who assume most of the responsibility for looking after the healthcare of their families. Both amchis and women collect only small amounts of medicinal plants, compared with the large quantities harvested by those who collect the plants to sell in the market.

The planning meeting stimulated the project team into collecting baseline data on several subjects that had been identified as requiring more detailed research. These included the distribution of knowledge of medicinal plants within the communities, the connections between this knowledge and local healthcare concerns, and institutions within the communities relevant to the delivery of local healthcare and to the management of medicinal plants (Ghimire et al., 2000, 2004; Lama et al., 2001). The results of the research confirmed that the amchis are the most knowledgeable members of the communities regarding medicinal plants and identified 373 locally growing species that they used medicinally. It was discovered that the amchis acquire their medical knowledge from a combination of reading Tibetan medical texts, training received through master/pupil relationships based on medical lineages and through their own personal experiences. The lineage is a concept firmly embedded in the traditional worldview of Tibetan Buddhists. It is associated with the idea of reincarnation – the belief that, after death, a person becomes reborn as someone else. According to Buddhist teaching, lineages are traceable back to the Buddha himself and the merit that people accumulate during their lifetimes is passed on to their lineage successors.

The amchis at Dolpo were found to already engage in a number of traditional conservation practices (see Section 8.1). One was to carefully collect only the parts of the plants that they needed, leaving the rest of the plants untouched – a practice which differed from the harvesting techniques of the commercial collectors, who tended to rip up whole plants in their eagerness to fill up their sacks. Another was to take only limited quantities of material from any one place, so as to minimise the effects of collection on the ecosystem. A third, which applied only to those species for which the underground parts of the plants are required, was to make collections only during September and October – a practice prescribed in Tibetan medical texts. An ecological benefit of this is that, by that time of year, the plants have shed their seeds and are beginning to wither, so that the amchis are not interfering with the abilities of the plants to reproduce. To complement the amchis' knowledge of plant biology and sustainable harvesting techniques, the PPI project carried out some supplementary, scientifically organised studies on two of the most important species in local medicine and trade, *jatamansi* (*N. jatamansi*) and *kutki* (*N. scrophulariiflora*) (Ghimire et al., 2005). Included in this work were experiments simulating the collection techniques of the amchis. Susanne Schmitt, a plant conservation officer working with me in WWF-UK, contributed to these experiments.

Some of the results of the project's ethnobotanical work at Dolpo were published in the form of a small book that was available for purchase from WWF-Nepal (Lama et al., 2001). As discussed in Section 6.11, the publication of indigenous knowledge of plants, especially medicinal plants, is a sensitive issue in conservation and sustainable development circles. This sensitivity has arisen from cases when researchers have exploited indigenous knowledge to advantage themselves and businesses, without passing on fair shares of the benefits received to the indigenous communities which provided them with the knowledge. Prior to publication of the book, the amchis of Dolpo were asked about their opinions on this matter. Their response (Lama et al., 2001):

> The transfer of this knowledge to the global community does not pose any ethical problem to the amchis, except regarding specific compounds that have not been fully tested and therefore cannot be used by non-specialists. It is also noted that in the context of Buddhism and Bon, the amchis see this knowledge as an asset to be used for the good of all sentient beings, i.e., human healthcare.

Research in Ringmo and Pungmo revealed the existence of two traditional institutions connected to the management of natural resources. One was the *dratsang*, a lama hierarchical organisation, and the other the *yulgigoje*, a village institution headed by a customary chief designated through his lineage. The customs associated with these institutions include the carrying out of certain rituals prior to the harvesting of crops, the restriction of harvesting to certain days designated by a traditional calendar and rules regulating how pastures are to be managed, with a system of fining in the event of non-compliance. The project team judged that, taken together, this combination of traditional customs is geared towards maintaining the resource base. Research further revealed that the creation of SPNP had had the (presumably unintended) consequence of weakening the authority of the traditional institutions. However, in spite of this, traditional conservation customs were still being followed to some extent and new community rules were still being formulated, for example prohibitions on the grazing of yak in pastures that were being overgrazed.

On the healthcare front, research by the project confirmed that the formal healthcare sector based on Western medicine was unrepresented in SPNP and Dolpo, and that the standards of healthcare, as delivered by families and amchis, had deficiencies. Child mortality was high, which, the project team judged, had much to do with poor standards of hygiene and sanitation. The amchis revealed that they would like to receive more training in Tibetan Medicine, for example to ensure that their identifications of plants were correct; also, that they would appreciate having access to certain authoritative Tibetan medical texts. They expressed concern about the future of their profession. The amchis' ethics stipulate that they do not charge patients fees for their services, but this was becoming increasingly impractical in the modern, more monetarised, world. For instance, the preparation of their herbal medicines sometimes required ingredients produced by plants which grow in the lowlands, and these could only be obtained through purchase.

The project responded to the concerns about deficiencies in home health-care by preparing a healthcare manual for household use, based on advice received from expert amchis augmented by scientific understandings. To back up its use, a number of training sessions for women, facilitated by Gyatso Bista and Tshampa Ngawang Gurung, were held in Phoksundo VDC. The request of the amchis to receive more training in Tibetan Medicine was met by 47 amchis from Dolpo being supported in their attendance in 2001 at the first national workshop of amchis organised by the HAA. Eleven stayed on after the course to participate in a month-long refresher course on the theory and practice of Tibetan Medicine. The amchis' request for medical texts was met by the project purchasing copies of the texts in Kathmandu and dispatching them by donkey train to Dolpo, a journey that takes over a month. The intention was to present the texts to the amchis at a ceremony at Dho Tarap, a village in an external buffer zone of SPNP to the east of Dolpo.

Even at the initial planning meeting at Ringmo in 1997, a way of regulating the harvesting of medicinal plants by the commercial collectors had started to be conceived. Shaped into more concrete form through the project's research, the idea was to develop a co-management system for the medicinal plants in the buffer zones, the system incorporating a Medicinal Plants Management Committee (MPMC) for each village, plus an overarching MPMC to cover all of the villages. Each village committee would monitor the medicinal plants in its area, stipulate places for collection and the amounts of each species to be collected; it would support the park's efforts to exclude unauthorised collec-tion. The overarching MPMC would coordinate the efforts of all the villages, liaise directly with officials of SPNP, and provide advice and guidelines to the village MPMCs regarding the collection, cultivation, management and use of medicinal plants.

This concept incorporated the idea of a physical centre, where records would be kept, consultancy services from expert amchis would be available and apprentices would be trained in Amchi Medicine. A site for building a centre was selected in Phoksundo VDC and Gunasa was chosen as the site to trial the operation of the medicinal plants management system at community level. The project team discussed with the people of Phoksundo VDC who used Gunasa about ways to incorporate the management of medicinal plants into their existing arrangements for managing the pastures. Three amchis were given the opportunity to travel to Mustang to view the traditional hospital and healthcare clinics already existing there and gather more ideas for the design of the new healthcare centre in Phoksundo. Figure 20.2 is a view of Gunasa. It shows a house which is used by the people when they are in residence during the summer months, sheaves of harvested barley and areas of pasture on the slopes behind the buildings.

Fig. 20.2. Gunasa pasture, Shey Phoksundo National Park, Nepal. Photo: A.H. (1997).

20.4   Operational Obstacles and the Project's Demise

This PPI project was originally planned to last for four field seasons (1997–2000), but, through WWF-UK's fund-raising efforts, money was found for a second 4-year phase. The aims set for the second phase were to consolidate the management system for medicinal plants being trialled at Gunasa, explore how it might be adapted for the external buffer zones of the park (in which commercial collection of medicinal plants was, in principle, permitted) and replicate elsewhere the traditional healthcare centre to be built in Phoksundo VDC.

However, despite the money being available for the project to continue and the excellent progress that had been made in the field, obstacles to its future mounted on several fronts. One of them was a decision by WWF-US to focus its support for conservation in the Nepalese Himalayas on Mt Kangchenjunga.

This was because its Conservation Policy Unit had identified Mt Kangchenjunga, but not Dolpo, as lying in a 'Global 200 priority ecoregion'. As explained in Section 8.6, 'Global 200 priority ecoregions' are one of several schemes advanced by international conservation organisations identifying places where, they believe, efforts in biodiversity conservation should be concentrated. More or less simultaneously, the Conservation Policy Unit of WWF-International, in which I was institutionally housed, was dissolved and I lost my job. However, fortunately for me, I was immediately offered a similar position in WWF-UK and so was able to continue my commitment to PPI.

More immediate obstacles to the ability of the PPI project at Dolpo to continue to operate effectively arose from the rapidly deteriorating political climate that was developing within Nepal. At the time, Nepal was still being governed in its traditional way as a feudal state under a Hindu monarchy. However, public discontent was mounting against the existing political system, fuelled by the grinding poverty and the corrupt officials with which the people were faced. Voices were being raised for a new constitution that guaranteed basic human rights, including the right of the people to have some say in the choice of their governments. In 1996, an armed insurgency influenced by Maoist ideology was launched in Nepal with the aim of toppling the monarchy and installing a republic. The political situation became more complicated in 2001, when nine members of the royal family, including the king (Birendra), were assassinated by one of its own members in a royal palace in Kathmandu. The death of King Birendra, who was widely regarded as a reformer and was personally popular, poured petrol on the deteriorating security situation. Thereafter, the armed conflict between the government and the Maoists continued to escalate, leading eventually, in 2006, in the signing of a Comprehensive Peace Accord by the warring parties and, in 2008, the abolition of the monarchy.

The PPI project at Dolpo had been influenced by the deteriorating political climate in Nepal from near its beginning. The first serious incident came in June 1998, when, for the first time ever, the Nepalese army mounted a military offensive against the Maoists, which, coincidentally, happened to be at just the time when the PPI project team had embarked on a several days' walk from Ringmo to Dho Tarap. The latter is a village and important Buddhist monastic centre, situated about halfway between SPNP and Mustang and which, from the park's perspective, lies within one of its external buffer zones. One purpose of the journey was for the project team to present in person to the assembled amchis of Dolpo the Tibetan medical texts that they had requested and had by then been transported by mule train to Dho Tarap. The project planned to use the occasion to have a general discussion with the amchis about how the project should evolve in the future, given its objectives of securing the conservation and sustainable use of Dolpo's medicinal resources. The meeting was intended to be a significant event in the history of Tibetan Medicine in Dolpo, because never before (so it was said) had all the amchis of Dolpo managed to assemble together in one place – an indication of the vastness and difficulty of travel in this remote Himalayan region.

Because of its inherent interest and the significance of the event for the project and for the PPI as a whole, I was with the project team as it headed towards Dho Tarap. Susan, my teenage daughter, who was having a gap year, was accompanying me to see for herself what her dad did for a living. Late in the evening at our first camp, messengers on horseback dispatched by the park arrived bearing the news of the military offensive. They reported that a number of people had been killed in the villages through which we had walked during our journey from Juphal to Ringmo and that it was thought that the army offensive might involve a pincer movement towards Dho Tarap. We were ordered to return to Phoksundo VDC and prepare for evacuation by helicopter. The descent to Phoksundo VDC was necessary because it was at low-enough altitude for helicopters to be able to operate. Whether WWF would be able to find a helicopter at all and, if so, what its size would be, were uncertain, because many of the civilian helicopters in Nepal had been requisitioned by the army for its offensive. WWF had sent a numbered list of those to be evacuated, the number to be taken depending on the size of the helicopter, if any, that they could find, which was scary for some of those ranked lower on the list.

Very early next morning, a planning meeting was held by the PPI team and a few of the large group of amchis and other villagers who were accompanying us on our journey to Dho Tarap. The objective of the meeting was to make some quick decisions about how the project should proceed given the new circumstances. It was decided to dispatch a three-person delegation to proceed to Dho Tarap to inform the amchis about what was happening and advise them to squeeze their meeting into two days, one to discuss the planned traditional healthcare centre in Phoksundo VDC and the other to discuss women's health issues. Some suggestions about the design of the healthcare centre were forwarded for the amchis to consider. These were that the centre should have a large room for meetings, training and to house an herbarium; a small room for medical consultations; quarters for a resident amchi and a trainee; and two stores, one to be used for the bulk storage of medicinal plants.

Three people, all from the village of Pungmo, agreed to be on the delegation. They were Amchi Norbu, Lama Wanggyal (who was involved with a new cultural school in Dolpo, not directly related to the PPI project) and Yanzung, a lady involved with the project's work on women's health issues. After the meeting, we descended hurriedly to Phoksundo and, after an anxious wait, a helicopter came, fortunately large enough to accommodate all those waiting to be evacuated.

This obstacle did not stop the PPI project continuing with its work in SPNP. Suresh Ghimire was not evacuated in the helicopter and remained behind to continue with his field research. The construction of the healthcare centre went ahead, and, in October 2000, an opening ceremony was held attended by amchis from all over Dolpo. Figure 20.3 is a photograph of the opening ceremony. It was taken by Segolene Rocher, a professional French photographer who accompanied Yildiz Aumeeruddy to the meeting to look after her daughter, Hannah-Mahé. The amchis can be distinguished by their saffron robes or yellow jerseys. On 26 September 2000, another incident relating to the turbulent state of Nepalese politics occurred in Dolpo when hundreds of Maoist guerrillas attacked Dunai, the capital of Dolpa District, resulting in some fatalities, the looting of Rs 50 million from the only bank and the freeing of prisoners (Dixit,

Fig. 20.3. Opening ceremony of the traditional healthcare centre in Phoksundo.
Photo: Segolene Rocher (2000).

2017). The bank managed to reopen in December, and, in spite of the difficulties, both the PPI project and the NMPC continued to operate.

All the villages of Dolpo were under Maoist influence from 2000, but, nevertheless, the Maoists allowed the PPI and NMPC projects to continue, being persuaded of their good intentions by the transparent ways in which they operated and the support that they received from the local people (WWF-Nepal Program, 2001). Two new PPI activities went ahead. One was the making of a documentary about the work of the amchis in Dolpo. Filmed by a cameraman under the direction of Yeshi Lama, the two of them had to travel to Dolpo by walking across the mountains from Mustang, the airstrip at Juphal having been put out of action by the Maoist guerillas. The other new activity was an assessment of the potential for extending the work of the project into the external buffer zones of the park. The assessment was made in Pahada VDC, which is situated on the southern border of SPNP. The assessment proved positive, because, by 2002, some farmers had already started cultivation trials of selected species and there seemed to be a good potential for developing community-based cultivation of medicinal plants (Shah, 2005).

From the early 2000s, a serious new problem with the project's efforts to trial a community-based management system for medicinal plants at Gunasa started to arise, due to the large numbers of people from outside the area descending annually on Dolpo between April and June to search for *yartsa gumbu* (Post, 2003). Out of the estimated 40,000 people coming to Dolpo, about 3000 were visiting Gunasa, where the Maoists imposed their own regulations for collecting *yartsa gumbu*, which involved taxing those collectors who were not residents of the park, according to the amounts that they collected. Shops were opened, some selling alcohol, and it was feared by members of the PPI project team that the invasion would cause severe disruption to both the cultural and natural environments. They estimated that it would take many years for the slow-growing, high-altitude, dwarf juniper trees that the collectors were using as fuel to recover. The *yartsa gumbu* Gold Rush, as it has become known, has not been confined to Dolpo. It has become a widespread feature of several parts of the Himalayas. It has been fuelling chaos and violence, as very poor people compete with one another to collect an increasingly scarce resource (Koirala, 2017).

The exceptionally high degree of insecurity that had arisen at Dolpo was related to its geographical proximity to the impoverished mid-hill districts of Rukum and Jajarkot to its south, which were hotbeds of Maoist extremism. If this had been the only reason why WWF-US transferred its attention from Dolpo to Kanchenjunga (rather than its Global 200 argument), it could possibly be seen as sensible, since Kanchenjunga is in a politically less troubled area. Since 1998 at Kanchenjunga, WWF and the Nepalese Department of National Parks and Wildlife Conservation had been developing a 2000 km^2 conservation area, using a similar approach to that being tried at SPNP. On 23 September 2006, a ceremony was held at Kanchenjunga to celebrate the handing over of the responsibility for conserving the biodiversity of the area to the local people. The ceremony was attended by senior government officials, foreign ambassadors, some WWF staff and representatives of USAID (funders of WWF-US's work). After the ceremony, these prominent visitors climbed into a helicopter, together with Mingma Sherpa, Chandra Gurung and Yeshi Lama, to return to Kathmandu, but, soon after take-off, the helicopter crashed and all on board perished.

I happened to be in Kathmandu at the time, attending a workshop on the identification of Important Plant Areas (IPAs) in the Himalayas organised by the Ethnobotanical Society of Nepal and Plantlife (see Chapter 11). Two of those who died were former colleagues of mine in WWF-UK, Dr Jill Bowling Schlaepfer and Jennifer Headley. A memorial service was held for them in the parish church of Godalming and a tree planted in their memory in the local Winkworth Arboretum.

20.5   A Tribute to Three Nepalese Conservationists and a Postscript

Mingma, Chandra and Yeshi did much to advance the causes of conservation and sustainable development in the Himalayas, finding innovative ways to combine conservation of biodiversity, sustainable use of natural resources, and respect for the knowledge and wisdom of the indigenous people. Their legacy lives on materially in the three conservation areas that they, individually or collectively, helped to establish, namely Sagarmatha (Mt Everest) National Park, the Annapurna Conservation Area and the Kanchenjunga Conservation Area – places associated with three of the highest peaks in the world.

Yeshi was born in Sichuan, China, which she left as a child with her mother and father, Bhakha Tulku Rinpoche, in 1959, when they fled on foot to India and Nepal. She progressed well at school and, when she was older, managed to take a course in sociology and anthropology in the United States, which, in turn, helped her to obtain a position on the staff of WWF-Nepal and her role as the Project Officer of the PPI project at Dolpo. Figure 20.4 is a photograph of Yeshi with a Tibetan loom, taken on the roof of a house on the path from Juphal to Ringmo.

Bhakha came from a highly respected Tibetan Buddhist lineage. He was recognised at an early age as the tenth incarnation in the Bharka Tulku lineage, whose ancestral seat is the Bhakha Gomba (monastery) situated in the Powo Valley in Southeast Tibet and just within the Province of Sichuan in China. One

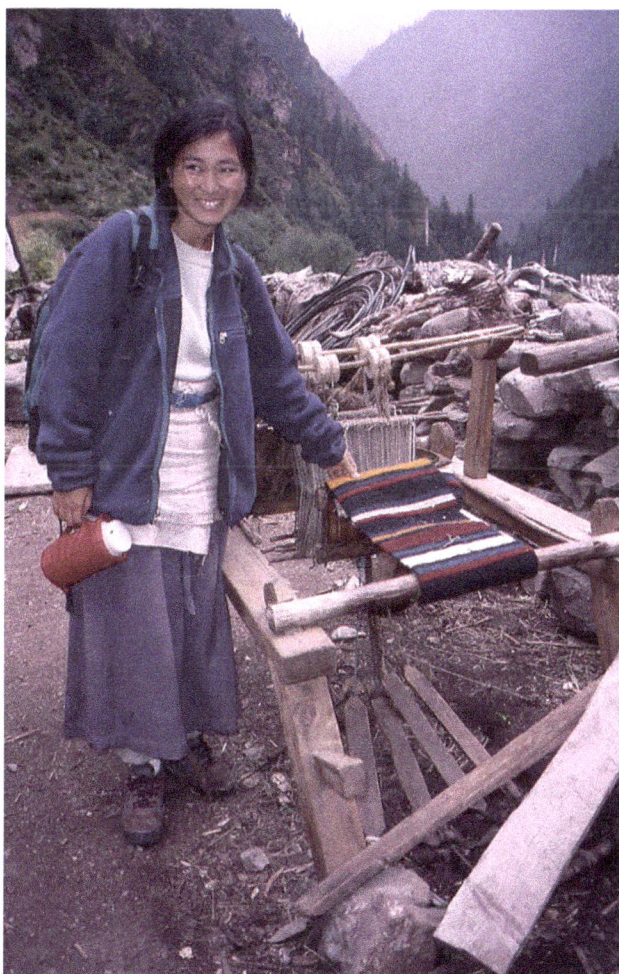

Fig. 20.4. Yeshi Choden Lama with a Tibetan loom near Phoksundo. Photo: Yildiz Aumeeruddy.

of the tracks of the tea–horse exchange route connecting China and Lhasa passes through this valley (see Section 3.5.2). Powo is separated by a high pass of over 4000 m from a naturally well-concealed land called Pemako (Medog County), which is famous both in Tibetan mythology and botany (Cox, 2008). To Tibetans, it is a Shangri-La, an idyllic sacred place where Tibetans can find refuge at times of strife. To botanists, it is a wonderland containing a tremendous diversity of plants. In 1924, the British plant hunter, Frank Kingdon-Ward (1885–1958), penetrated into Pemako, where he discovered a cornucopia of new species to collect and take back home to sell to seed merchants and commercial nurseries. Pemako contains over 154 species of *Rhododendron* and numerous *Primula* and orchids.

In 2002, Yildiz Aumeeruddy helped to organise a regional workshop in Kathmandu entitled 'Himalayan medicinal and aromatic plants, balancing use and conservation' (Thomas *et al.*, 2005). Apart from PPI and WWF, the other organisations involved in its preparation were the Ministry of Forests and Soil Conservation (Nepal), the International Development Research Centre (IDRC, Canada) and the Medicinal and Aromatic Plants Program in Asia (MAPPA). The meeting provided a forum for exchanges of experience between people actively involved in MAP conservation initiatives in the Himalayas, drawing on experiences in Bangladesh, China, India, Nepal and Pakistan. In a conclusion summarising what had been learnt at the workshop, Yildiz Aumeeruddy and Madhav Karki stressed the need for community-based decision-making processes to be combined with the development of national policies and these, in turn, to inform macro-level policies, such as those adopted by aid agencies and international conservation and development organisations. These suggestions helped to inform the suggestions for an ecosystem-based approach to plant conservation (EBPC) described in Chapter 9.

In the late 2010s, more than 10 years after I had left WWF, I was asked to visit the office of WWF-UK, which is situated close to where I live, and tell them about the former ethnobotany project at Dolpo. It turned out that, after a long delay, WWF was restarting activities in SPNP, this time with a focus on 'wildlife' (snow leopard, etc.). Rummaging about in their storerooms, the staff of WWF-Nepal had discovered numerous reports written by those involved in the PPI project, but (so I was told) no one there could remember much about it. During the planning of their new project, a reconnaissance mission to Dolpo had discovered some ruins of a building in Phoksundo and had wondered what the building had once been. It was, in fact, all that remained of the traditional healthcare clinic shown in Fig. 20.3.

21

Conservation and Development Based on Medicinal Plants at Ludian, China, with an Extension to the Meili Snow Mountain Range (from 2006)

Abstract

This chapter is about a project of the Kunming Institute of Botany (KIB) with a Naxi community at Ludian, China. It was started in 2006 under the umbrella of Plantlife's Medicinal Plants Conservation Initiative (MPCI), the overall aim of which was to answer the question 'How can the conservation of biodiversity and sustainable development best be achieved, based on the interests of communities in medicinal plants?' Ludian lies in the upper catchment of the Yangtze River, where serious degradation of the forests began through reckless tree-felling during the Great Leap Forward of 1958–1962. Ludian is a well-known locality for medicinal plants in Yunnan, in which it is known as the 'Home of Medicinal Plants'. With the encouragement of KIB, a Ludian Medicinal Plants Conservation and Development Association was formed, encouragement was given for the establishment of home herbal gardens, and two Medicinal Plants Conservation Areas (MPCAs) were established in forests at village level to conserve medicinal plants and to provide planting stock for cultivation when new species of medicinal plants become popular in the market. Assistance was provided to the communities to bargain for better prices for their farm-grown medicinal produce. The project has received government support at the prefectural level and, in 2020, the Ministry of Agriculture declared Ludian Township to be a model for sustainable rural development in China and a demonstration site for training. An extension of the project to Tibetan communities on the Meili Snow Mountain Range resulted in the creation of a very large MPCA.

21.1 The Locality

This project started as one of the field case studies of Plantlife's Medicinal Plants Conservation Initiative (MCPI) (see Chapter 11) (Pei Shengji *et al.*, 2010). A team from the Kunming Institute of Botany (KIB), working under the guidance of Professor Pei Shengji, has been responsible for developing the project. Much of the day-by-day ethnobotanical field work during the project's opening years was undertaken by Mrs Yang Lixin, who speaks the local language (Naxi) and has worked under Pei's direction for many years. Most project activities have been carried out in Ludian Administrative Village (AV), one of five AVs in Ludian Township, Yulong County, Lijiang City Prefecture, Northwest Yunnan. This part of Yunnan lies within the Hengduan Mountains at the eastern end of the Himalayas. Twelve per cent of the 2700 ha of land in Ludian AV is under agriculture and much of the rest under forest, scrub and pasture. Altitudes range from 2400 to 3800 m.

Figure 21.1 is a photograph of a small part of Ludian AV. The plant growing in the field in the foreground is *Dolomiaea costus*, which is native to the Western Himalayas and is said to have been brought to China by the Buddhist monk and traveller, Xuanzang (602–664 CE). Formerly known as *Saussurea costus* or *Aucklandia lappa*, it is a species listed in the Convention on International Trade in Endangered Species of Wild Fauna and Flora (CITES) and one of the 50 fundamental herbs of Traditional Chinese Medicine (TCM) (Millar *et al.*, 2021). The polythene sheeting in the mid-distance covers a plantation of *Paris yunnanensis* belonging to Mr He Yun. This species is widely used medicinally in the Himalayas but is rarely cultivated.

In 2005, the population of Ludian AV was 5686, distributed over 1374 households in eight villages. Ninety per cent of the people were ethnically Naxi, with Pumi, Yi and Lisu also represented. Surveys by the project in two sample villages revealed that 13% of the people relied exclusively on Naxi Dongba Medicine, 26% used Western medicine and 61% used both (see Section 7.5.1). About 45 species of medicinal plants were being used regularly in local herbal medicine, but many more occasionally (Wang Yuhua, 1999). The Dongba Naxi healthcare practitioners at Ludian collect 60% of the species that they use in the wild, grow 30% in their home gardens and purchase the rest in markets.

Fig. 21.1. Medicinal plant cultivation and forest at Ludian, China. Photo: A.H. (2008).

Ludian Township, especially Ludian AV, is known within Yunnan as the 'Home of Medicinal Plants'. The cultivation and sale of medicinal and aromatic plants (MAPs) to external markets has been a flourishing business here for at least 200 years. It is a Di Dao locality for certain drugs used in TCM, meaning that medicinal materials that are sourced from within it are seen as especially 'authentic'. This adds to their value and makes their sales more reliable. Both cultivated and wild medicinal plants are traded. Farming is the principal economic occupation in Ludian AV. Ninety per cent of the people grow some medicinal plants to sell, providing 10–70% of household income. However, only a few species (about ten) are grown extensively at any one time, those popular in 2005 including *Aconitum stapfianum*, *D. costus*, *Gentiana robusta* and *P. yunnanensis*. The total production of medicinal plants at Ludian in 2008 was 8000 tonnes, which is believed to be among the highest for any community in China. Figure 21.2 is a photograph of a backyard in Ludian showing roots of *D. costus* being processed for the market.

The government's economic policies during the last 70 years have had a big impact on the people and forests of Ludian. As with other parts of China, the Great Leap Forward (1958–1962) saw large numbers of trees felled to fuel small, inefficient, blast furnaces (see Section 6.2.1). The switch over from a command economy to more of a demand economy in 1980 brought little respite for the forests, which continued to be logged unsustainably.

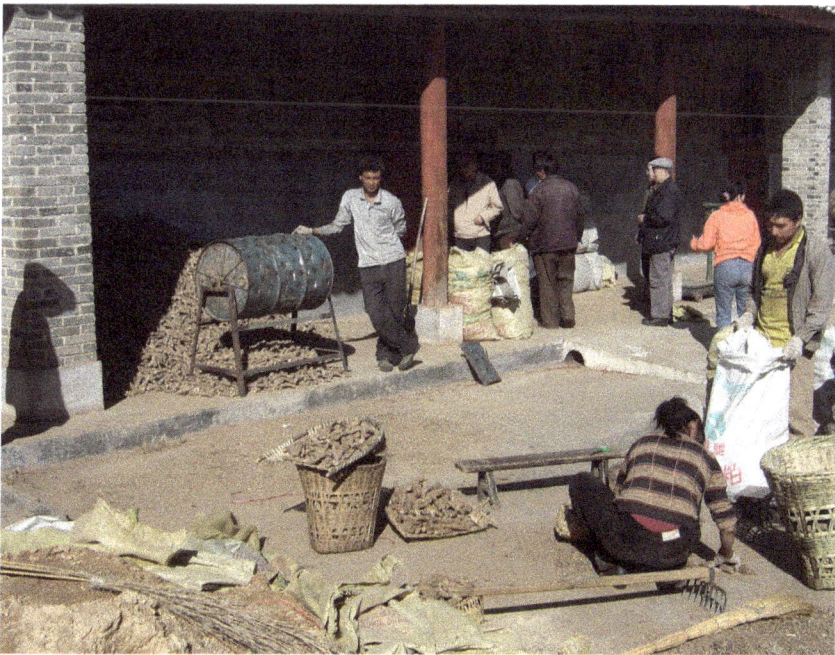

Fig. 21.2. Processing a medicinal plant in a backyard in Ludian. Photo: A.H. (2006).

For China as a whole, unsustainable logging over several decades has resulted in large-scale degradation of catchments. The year 1998 witnessed major flooding in several parts of the country, including in the lower reaches of the Yangtze River (Zong Yongqiang and Chen Xiqing, 2000; Yin Hongfu and Li Changan, 2001). The response of the government to this disaster was to ban logging in natural forests, embark on a huge tree-planting programme and establish a large number of new protected areas (Wang Guangyu *et al.*, 2008, 2012; Viña *et al.*, 2016). The ban on logging caused an upsurge of timber imports into China which, in turn, accelerated deforestation in several parts of the world, including in the Congo Basin (Yang Hongqiang *et al.*, 2010; Fuller *et al.*, 2018). The reliance that China has developed on imported timber is reminiscent of Britain's dependency on imported timber prior to the First World War (see Section 5.5).

Another consequence of the logging ban was to reduce the number of people employed in the timber industry, many of whom turned to the commercial collection of non-timber forest products (NTFPs) instead. This led to a rise in the unsustainable collection of wild medicinal plants (Buntaine *et al.*, 2007; Melick *et al.*, 2007) and there was a collecting boom in Northwest Yunnan in the highly priced matsutake fungus (*Tricholoma matsutake*) (Winkler, 2004; Arora, 2008; Yang Xuefei *et al.*, 2008).

21.2   Project Implementation

In Section 9.5, we mention the value of an ecosystem services framework as a tool for planning, implementing, monitoring and evaluating a plant conservation project. Table 21.1 shows such a framework filled up with data covering the first two to three years of operation of the project at Ludian. The letter **B** (for 'before') indicates the state of affairs at the beginning of the project and the letter **A** (for 'after') shows the project interventions and their results. The table might seem complicated but, starting with an empty framework at the start of the work, as depicted in Table 9.2, or in a somewhat more complex form as in Table 9.3, an ecosystem services framework can serve as a useful aide-memoire for a project team to make sure that nothing important has been missed when implementing a project and when communicating with interested parties.

Plantlife International in Britain partnered with KIB in developing the project during the first two years of its operation (2006–2008). Thereafter, KIB has continued with the work on its own, mainly through Pei Shengji requesting, annually, two ethnobotanists at KIB to go to Ludian to monitor what is happening and provide advice. The project received official recognition in 2020, when the Ministry of Agriculture declared Ludian Township to be a model for sustainable rural development in China and a demonstration site for training. This recognition was based on the relatively high level of local income, the extent of medicinal plant cultivation (4300 ha in 2019) and the existence of two community-managed, forest-covered, conservation areas.

Table 21.1. Ecosystem services framework filled up with data for a project at Ludian, China.
B ('before') = background conditions; **A** ('after') = project interventions and results. Acronyms:
HRS, Household Responsibility System; LCNTC, Lijiang City Nationality Technical College;
LMPCDA, Ludian Medicinal Plants Conservation and Development Association; MAPs,
medicinal and aromatic plants; MPCA, Medicinal Plants Conservation Area; NP, National Park;
NTFP, non-timber forest products; THPs, traditional healthcare practitioners; TNC, The Nature
Conservancy. Adapted from: Pei Shengji *et al.* (2010).

Ecosystem services (categories and subcategories)	Ludian: baseline conditions (**B**) and project initiatives and results (**A**)	
	Conditions at locality (Ludian)	Wider systems influencing locality
Provisioning Products from cultivated plants Products from wild plants	**B**: Farmers and collectors selling MAPs on an individual basis at low prices. **A**: (i) Internet access provided at Ludian to raise local bargaining power. (ii) Idea of establishing a MAPs' marketing cooperative explored	**B**: (i) High market demand for MAPs from Ludian. (ii) HRS applied to farmland from 1979, giving farmers rights of use to particular areas of land (subject to regulations). HRS extended to forests from 2005 (excludes Community Benefit Forests serving as water sources or forests providing ecological protection)
Regulating Delivery of fresh water supplies Regulation of water flows, soil erosion and climate	**B**: (i) Ban on logging imposed in Ludian forests in 1998 (intensive logging reduced forest cover from 80% pre-1960 to 40% in 1990). (ii) Increased soil erosion, drying-up of streams and a less favourable agricultural climate (blamed locally on forest destruction). **A**: Two MPCAs of 300 ha each established in forests at Diannan and Dianbei to safeguard species, serve local medical needs and provide stock for planting. MPCAs retained when forests made Household Responsibility Forests in 2008	**B**: (i) Upper catchment of Yangtze (including the Three Parallel Rivers Area of Yunnan) a priority for natural-area protection and tree-planting since disastrous flooding downriver in 1998. Logging in natural forest prohibited throughout China (but some NTFP collection allowed). (ii) TNC assisting with establishment of Laujunshan NP (Ludian is in buffer zone)

Continued

Table 21.1. Continued.

Ecosystem services (categories and subcategories)		Ludian: baseline conditions (**B**) and project initiatives and results (**A**)	
		Conditions at locality (Ludian)	Wider systems influencing locality
Cultural	Medicinal plants in culture and healthcare Education, training and awareness-raising	**B**: (i) Traditional medicine popular, using many species of MAPs. (ii) Many households grow some MAPs for home treatments and some THPs have species-rich gardens. **A**: (i) Twenty-two new herbal home gardens initiated, to provide herbs for local treatments, serve as education centres on Naxi culture and provide planting materials for farmers. (ii) Workshops: four on Naxi Dongba medical knowledge, five on sustainable harvesting of wild MAPs, one on the use of the Internet for marketing	**B**: (i) Widespread cultural support in China for traditional religions and philosophies (Buddhism, Confucianism, Taoism). (ii) Some healthcare traditions of minorities officially recognised (e.g. Tibetan Medicine), but not others (e.g. Naxi Dongba Medicine). **A**: LCNTC plans to establish a Naxi Dongba herbal garden and a Naxi Dongba hospital
	Social organisation	**B**: Community poorly organised regarding trading in MAPs. **A**: (i) Multidisciplinary project team formed. (ii) Ludian Medicinal Plants Conservation and Development Association formed (later LMPCDA)	**B**: Institution exists at prefectural level supportive of ethnomedicine (LCNTC). **A**: LCNTC instrumental in starting Lijian City Ethnomedicine Association
	Outreach to Diqing Tibetan Autonomous Prefecture	(i) Yongzhi Medicinal Plants Conservation Group formed, stimulated by example of Ludian. (ii) MPCA selected (1000 ha). (iii) Training sessions provided (on cultivation of MAPs and basic Tibetan Medicine). (iv) Six herbal gardens started (none existed previously)	**B**: TNC assisting establishment of proposed Kawagarbo NP (Yongzhi is in buffer zone). **A**: (i) TNC supports project expansion to Yongzhi. (ii) Deqing County Tibetan Medicine Doctor Association formed

21.2.1 Community organisation formed

Concentrated work by the project team at Ludian began in August 2006, when initial discussions between the KIB project team and community members revealed a generally low level of concern about the conservation of medicinal plants. However, some individuals thought differently, especially the Naxi Dongba traditional healthcare practitioners and the (then) headman of Ludian AV, Mr He Yun, a farmer who had developed a 50-ha plantation of the highly priced medicinal plant, *P. yunnanensis* (under the polythene sheeting in Fig. 21.1). This small nucleus decided to form a group, the Ludian Medicinal Plants Conservation Association (LMPCA), and instigate activities within the community. The association was registered officially in February 2007, a constitution agreed (March 2007) and Mr Yang Shengguang (a local herbal doctor and pharmacist) elected as the first chairman. The initial membership was 20.

In May 2009, some adjustments were made to the structure and scope of the project based on ground realities and a growing recognition by the local government of its potential value. Because the LMPCA had become perceived as having an over-representation of traditional doctors, it was decided to adjust its membership to better reflect the whole community. As a signal to show its potential value for development, its name was changed to the Ludian Medicinal Plants Conservation and Development Association (LMPCDA). The project team had come to realise that, in the local socio-economic context, community-based conservation initiatives relating to medicinal plants have to be about maintaining or enhancing the benefits of these resources to the communities to have much chance of success. They also realised that the distinction between wild-collected and cultivated medicinal plants – a division that can exercise the minds of academic plant conservationists – was of little interest to the communities.

At about the time when the project was getting underway, the Chinese Government made a request to communities across the country to choose themes that they would like to prioritise in their own development. In response, Ludian Township opted for 'medicinal plants', so as to capitalise on its reputation as being the 'Home of Medicinal Plants' – a decision that helped to create a more favourable local social climate for the project. The project also started tapping into local concerns about the degraded state of the environment, as was evident in increased soil erosion, the drying-up of streams and a hotter and drier, less favourable, agricultural climate. This deterioration was attributed locally to the overlogging of the forests.

21.2.2 Identification of initial project priorities and sample villages

Since the date of its formation, the LMPCA (later LMPCDA) has taken a leading role in guiding and delivering the project, while remaining receptive to suggestions made by the KIB project team. When it was first formed, two major concerns of its (then) members came quickly to light, which were the growing rarity of some wild-collected species of medicinal plants and the lack of interest among the younger generation in becoming traditional healthcare practitioners. Two villages were selected as pilots for project activities, Diannan and Dianbei. The former, which was somewhat more prosperous, had eight Naxi Dongba doctors in 2007, but there was only one in Dianbei.

21.2.3   Development of herbal home gardens in Ludian Administrative Village

The association decided to encourage the development of herbal home gardens, with three aims in mind, these being to provide: (i) convenient sources of herbs for local treatments; (ii) educational centres for learning about Naxi Dongba culture; and (iii) sources of planting materials in the event of new species of MAPs becoming popular in the market. In 2007, there was only one herbal home garden in each of the pilot villages, both belonging to Naxi Dongba doctors. They contained, *in toto*, 63 species of medicinal plants. Figure 21.3 is a photograph of the herbal garden in Dianbei. Its owner, Mr He Chon Shan, is standing on the left of the peony talking to Dr Yang Lixin on its right.

The encouragement of new herbal gardens had some success. By May 2009, the total number of herbal home gardens in the two villages had expanded to 29 and there were also two 'wild cultivation gardens' (areas of natural vegetation enriched with medicinal plants). The number of species of MAPs cultivated in all the gardens, considered together, was 98. Most of the newly planted species had been transplanted from the wild. The choice of species to plant in the gardens was strongly biased towards those that were currently popular in the market, one of them being *P. yunnanensis*.

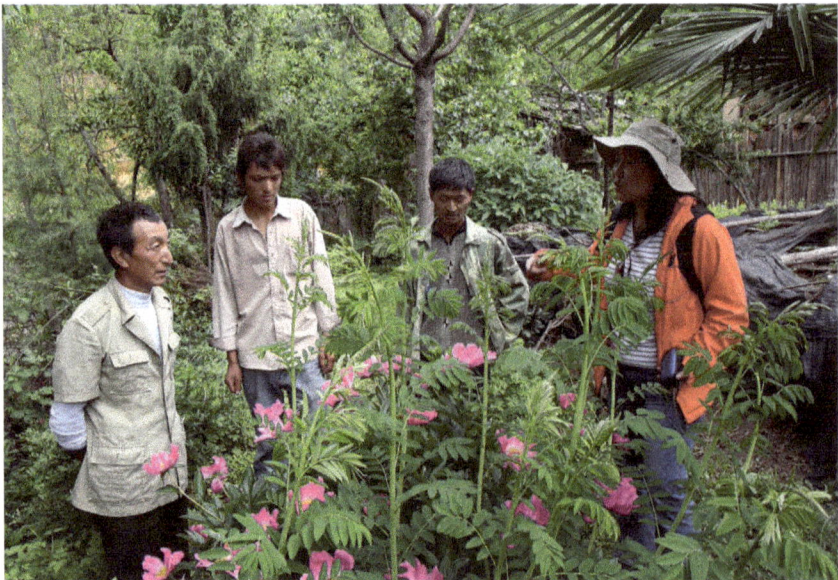

Fig. 21.3. Home herbal garden in Dianbei Village, Ludian. Photo: A.H. (2008).

21.2.4 Building awareness of heritage

One of the wild cultivation gardens developed under the project was close to where the stream that supplies Dianbei with water flows out of a forest. The villagers traditionally gather here to celebrate the Chinese New Year (otherwise known as the Spring Festival), a propitious set of circumstances that made it an ideal place for the project to carry out awareness-raising and educational activities. A growing realisation of the cultural attachment of the communities to certain sites, such as this, led to the project to mount several more ad hoc events at various places of local cultural significance to discuss with the communities the relevance of traditional beliefs and customs to modern times. Among the subjects discussed were the modern relevance of Naxi Dongba medicine and the knowledge of its practitioners about how to harvest medicinal plants sustainably. Perhaps surprisingly, the local youth, among others, were interested to learn more about their Naxi cultural heritage and to share with one another what they knew.

21.2.5 Medicinal Plants Conservation Areas

As mentioned in Section 11.2, in April 2007, a China/India/UK dialogue on conservation of Himalayan medicinal plants was held at KIB, facilitated by Pei Shengji. One of the items discussed was the MPCA, a conservation tool that had been developed in India by the Foundation for Revitalisation of Local Health Traditions (FRLHT). The Chinese team at the meeting wondered whether something similar might be workable in China and later asked the members of the LMPCDA at Ludian for their opinion. The reaction of the LMPCDA was positive and two areas of species-rich forest were proposed to be trialled as MPCAs, one at Diannan (Zhen Gutai MPCA, 330 ha) and the other at Dianbei (Hanjing Ke MPCA, 300 ha). Later, inventories of the two areas revealed that, taken together, they contained a total of 80 species of wild MAPs.

Ludian Township granted legal recognition to the two MPCAs and established a Ludian MPCA Management Committee to provide oversight. The committee consisted of a representative of the LMPCDA, the headmen of the two villages and two elected community members. Its roles included deciding on which species were allowed to be collected within the MPCAs and how the benefits from collection were to be distributed within the communities. Two people were hired as full-time workers for each site, their wages initially paid out of project funds. Regulations governing the management of the MPCAs were agreed through discussions held between the LMPCDA, the Government of Ludian Township and the local Forest Station. One of the regulations was a ban on the harvesting of rare, endangered and endemic species, a list being provided. One of the species was the yew tree, *Taxus yunnanensis* (see Section 6.3.3). Prohibitions were placed on logging, cutting firewood, grazing livestock and lighting fires within the MPCAs, with fines to be imposed in the event of non-compliance. Subsequent to the establishment of the MPCAs, local people have reported that several species of MAPs, not seen for several years, have reappeared, for example *Aristolochia moupinensis*.

The legal status of the two MPCAs changed in 2008, when the forests in which they were situated were declared to be Household Responsibility Forests (see Section 6.2.1). This resulted in 42 households at Diannan and 12 at Dianbei being granted parts of the MPCAs for their exclusive use, although still subject to Forest Department rules. This raised the question as to whether the MPCAs would still be effective for the protection of medicinal plants. The KIB team responded to this development by holding a meeting with 38 of the new rights' holders to find out about their thinking. All of those attending expressed a wish for the MPCAs to continue, their reasoning being that continuing with the MPCAs would reinforce their own authority over their allocated areas, which, in turn, would lead to more abundant harvests of medicinal plants.

21.2.6   Marketing improvements

When the project started, farmers and collectors at Ludian were selling medicinal plants to traders on an individual basis. Their bargaining powers were limited, lacking communal solidarity and ignorant of the prices in the wholesale markets. Typically, they were selling their produce at prices of only about 10% of those prevailing in the major regional market for medicinal plants, which is in Dali City (see Fig. 6.2). Prices received by the farmers could fluctuate greatly. For instance, in US dollar equivalents, the sales price of *D. costus* in 2007 was US$3.4 per kg, but just US$0.7 per kg in 2008. The equivalent sales prices for *G. robusta* were US$2.9 per kg (in 2007) and US$0.6 per kg (in 2008).

The LMPCDA, in discussion with the KIB team, decided that it would be useful to assist sellers of MAPs at Ludian by increasing their access to market information. Accordingly, a workshop on marketing was held (February 2009) that was attended by 28 farmers and at which the idea of using the Internet to gain information about market prices was raised and welcomed. A follow-up workshop on cultivation (also in February 2009), which was attended by 53 farmers, showed that they were confident about their cultivation techniques and confirmed their concerns about marketing. The response of the project was to donate a computer to the association and provide training on the use of the Internet. Within a year, the income received at Ludian through the sale of MAPs had increased by the equivalent of US$4 million.

21.2.7   Project impacts at prefectural and higher administrative levels

At the time when the KIB project started at Ludian, Naxi Dongba Medicine was not an officially recognised traditional medical system in China, in the sense of not receiving government support for the training of traditional doctors or for the development of treatments. However, Lijiang City Nationality Technical College, which is situated within the City of Lijiang, was teaching some aspects of Naxi Dongba Medicine in its courses on TCM. In May 2008, a meeting to discuss the official position of Naxi Dongba Medicine was held at the college, which was attended by college staff, an official from the Lijiang City Sanitary Bureau (responsible for health policy in Lijiang City Prefecture), members of the KIB project team and some traditional doctors from the prefecture. The meeting reached agreement that it would be useful to create a traditional medicine association at prefectural level to cover all of the traditional medical systems that were represented locally. The college stated its intent to establish a Naxi Dongba herbal garden and a Naxi Dongba hospital.

21.3   Project Extension to the Meili Snow Mountain Range

Ludian lies within the part of Northwest Yunnan known as the Three Parallel Rivers, designated by the United Nations Educational, Scientific and Cultural Organization (UNESCO) as a World Heritage Site recognised for its outstanding biodiversity and topographic features. The rivers in question are the Salween, Mekong and Yangtze (locally called Nu, Lancang and Jinsha), which flow southwards from their origins on the Qinghai–Tibet Plateau through steep-sided parallel gorges within the Hengduan Mountains on their journeys to the sea. The Three Parallel Rivers region is a major centre of endemism for Chinese plants (Ying Tsun-Shen et al., 1993; Ma Chang-Le et al., 2007) and is exceptionally culturally diverse, being home to many of the ethnic minorities of West China.

Some of the funds used to support KIB's project at Ludian were provided by The Nature Conservancy (TNC), a US-based conservation organisation that was interested in introducing the concept of the National Park into China (see Section 8.2) (Zhou and Grumbine, 2011). Chinese scientists and officials had become alerted to the inadequacies of the country's existing environmental protection measures in 1998, when, as mentioned in Section 21.1, there was severe flooding in several parts of China that was blamed on a failure to properly safeguard the upper catchments of the major rivers. One of the major causes highlighted was unsustainable logging. There was an openness to using other conservation approaches, such as TNC's offer to introduce the concept of the National Park.

Landscape planners involved in designing more effective environmental protection for the Three Parallel Rivers region were broadly following the lines of UNESCO's Biosphere Reserve model, based on the establishment of core areas for the conservation of biodiversity and surrounding buffer zones where more types of human activities were allowed, especially ones supportive of biodiversity conservation in the core (see Section 8.3) (Li et al., 2023). One function of the buffer zones was to allow connectivity of habitats between the core areas so that species could move between them. According to TNC's landscape planning model, Ludian was conceptualised as lying within the buffer zone of an intended Mt Laojun National Park (Zhou and Grumbine, 2011).

Another place in the Three Parallel Rivers region that TNC had in mind for a National Park was the Meili Snow Mountain Range (MSMR) in Deqin County, Diqing Tibetan Autonomous Prefecture. This range is situated only 350 km east of Pemako, the cultural heartland of the Tibetan people mentioned in

Section 20.5. Appreciating what KIB had managed to achieve at Ludian, TNC enquired of Pei Shengji whether KIB might be interested in extending its ethnobotanical approach to MSMR. KIB agreed and funds were made available for the work by TNC. The bulk of the subsequent ethnobotanical field work at MSMR was conducted by Dr Gao Fu, supplemented by occasional visits by Pei and other scientists from KIB.

The central feature of MSMR is Mt Kawagarbo (6740 m), which is one of the holiest mountains of Tibetan Buddhism and, reportedly, has never been climbed. Every year, tens of thousands of Buddhist pilgrims come to MSMR to circumambulate Kawagarbo. Given this connection, there are some conservationists who believe that it would be unwise to turn Mt Kawagarbo and its neighbourhood into a National Park, because this could attract large numbers of tourists and, thereby, undermine the local culture and destroy the connection between cultural and biological diversity that, to date, had helped to maintain it in a biodiversity-rich state (Wang Guangyu *et al.*, 2012). Another consideration is the threat of further environmental deterioration through climate change, which is already causing exceptionally rapid snowmelt and more intensive rains on the MSMR. The resulting catchment degradation is threatening the livelihoods of large numbers of people living downstream. The Chinese Government recognised the fragility of this situation and, also out of respect for its sacredness, forbade the climbing of Mt Kawagarbo in 2020.

Most of KIB's ethnobotanical work has been undertaken in the village of Yongzhi, which is situated close to a place where pilgrims start their treks around Mt Kawagarbo. The purpose of the first visit of the KIB team to Yongzhi, which was in September 2008, was to discuss the idea of having an ethnobotanical project with the community and to carry out some baseline surveys on the local culture and socio-economy, including in relation to medicinal plants. The KIB team discovered that 600 people were living in Yongzhi, grouped into 106 households in three sub-villages, two ethnically Tibetan and one Lisu. The reaction to the suggestion of having an ethnobotanical project proved to be positive, the people expressing the opinion that it might enhance the conservation of their habitat (which, they said, would make it more attractive to tourists) and that it might increase the populations of some commercially valuable medicinal species, such as *Panax bipinnatifidus* and *Paris yunnanensis*.

There are several ways in which MSMR presents a somewhat different conservation proposition from Ludian. One is that it is inhabited predominantly by Buddhists, with their inherently nature-friendly culture. MSMR contains many sacred sites, which, research at MSMR has shown, are statistically associated with the presence of sacred trees, such as *T. yunnanensis*, and with exceptionally rich floras (in terms of their total numbers of species, their numbers of endemic species and their numbers of useful species) (Anderson *et al.*, 2005; Salick *et al.*, 2007). The same research revealed that the sacred sites tend to be situated relatively close to villages – which is not the pattern expected if the villagers had been taking the easier option of collecting plant resources close to their homes, rather than at a greater distance. This is evidence that the high floristic diversity of MSMR is, at least in part, a product of deliberate human choice. The whole landscape above 4000 m is traditionally regarded as sacred.

In contrast to Ludian, it was found that few people in Yongzhi had much knowledge of herbal medicine and that, in 2008, there was virtually no cultivation of medicinal plants either in fields or gardens. When people needed medical treatment, they either went to a clinic offering modern medicine or consulted the single Tibetan traditional medicine doctor residing in the village. This doctor was the only person in the community who regularly collected local wild medicinal plants to make medicines. However, there was another Tibetan traditional doctor, Dr Ciren Sangzhu, living nearby, but he was only occasionally in residence. A highly respected doctor, Dr Sangzhu was happy to see patients when he was at home. There was no systematic collection of wild medicinal plants to sell, as at Ludian, but the Yongtze community did collect wild medicinal plants to sell when requested to do so by visiting traders. Species in market demand in 2008 included *Cimicifuga foetida*, *Dioscorea panthaica*, *Fritillaria cirrhosa*, *Paeonia delavayi* and the caterpillar-fungus (*Ophiocordyceps sinensis*).

The policy environment for traditional medicine was different in MSMR, compared with Ludian, in that the type of traditional medical system prevalent in the prefecture (Tibetan Medicine) was recognised and supported by the government, while the type prevalent at Ludian (Naxi Dongba Medicine) was not. With the encouragement of TNC and Dr Sangzhu, a group of Tibetan doctors in Deqing County decided to form a new society, the Deqing County Tibetan Medicine Doctor Association, with the dual objectives of sustaining Tibetan Medicine and conserving medicinal plants. The majority of its 36 members worked in government hospitals and clinics that dispensed both Tibetan and Western medical treatments.

The next visit of the KIB team, which was in April–May 2009 (after the winter), resulted in the formation of a new community organisation, the Yongzhi Community Medicinal Plants Conservation Group, with representation from 30 households. Dr Sangzhu was an influential member when available. The group agreed on a series of action points, many similar to the ones that had been undertaken at Ludian. Training sessions on the cultivation of medicinal plants and basic Tibetan Medicine were provided with the help of Dr Sangzhu, a programme to encourage villagers to plant medicinal plants was started and an MPCA was established. The regulations agreed for the management of the MPCA were similar to those at Ludian. By August 2009, six villagers had started to plant herbal gardens, using plants transplanted from the wild, the most developed of them containing 27 species.

The MPCA is huge (1000 ha) and extends over a very wide altitudinal range (2280–3500 m). Several vegetation zones are represented, including (from lower to higher altitudes): abandoned fields, pine forest, broadleaved forest, spruce (*Picea*) forest, bamboo with groves of yew (*Taxus wallichiana*) and dwarf alpine forest. A survey in July 2009 recorded 86 species of medicinal plants within the MPCA, several of them common (e.g. *Dipsacus*, *Gentiana crassicaulis*). The abandoned fields are a product of one of the government's programmes (the Sloping Land Conversion Programme) introduced to improve catchment quality following the disastrous flooding of 1998 (Weyerhaeuser *et al.*, 2005). The loss of these cultivated areas has likely increased the livelihood insecurity of the people, which may be one reason why the Yongzhi community expressed an interest in attracting more tourists.

In 2009, the Chinese office of TNC provided travel funds that allowed Professor Pei, Dr Ciren Sangzhu and Dr Sina Duojie (another Tibetan doctor living in a village in the MSMR) to travel to Thimbu in Bhutan to participate in the Seventh International Congress on Traditional Asian Medicine (IASTAM). This provided an opportunity for this Chinese delegation to share their knowledge of conservation and development of Tibetan Medicine with Tibetan medical doctors from Bhutan, India and Nepal.

References

Aasland, T. (1974) *On the Move-to-the-Left in Uganda 1969–1971*. Research Report No. 26. The Scandinavian Institute of African Studies, Uppsala, Sweden. Available at: https://www.diva-portal.org/smash/get/diva2:276744/FULLTEXT01.pdf (accessed 20 May 2024).

Abbasi, A.M. and Kalhoro, J.A. (2022) Politics of ethnic identities and conflicts: a case study of Hazara and Siraiki ethno-nationalist movements in Pakistan. *Pakistan Social Sciences Review* 6, 15–24.

Abbasi, A.M., Kalhoro, J.A., *et al.* (2022) The role of Hazaras in the creation of Pakistan: the postindependence search for identity. *Pakistan Journal of Social Research* 4, 12–17.

Achigan-Dako, E.G., Sogbohossou, O.E.D., *et al.* (2014) Current knowledge on *Amaranthus* spp.: research avenues for improved nutritional value and yield in leafy amaranths in sub-Saharan Africa. *Euphytica* 197, 303–317. DOI: 10.1007/s10681-014-1081-9.

Acosta Güemes, L.E.A. and Cusumano, A.M. (2022) Brief review of two medical vanguards of the Chinese Tang dynasty (618–907): the Imperial Medical Academy and Sun Simiao. *IASR Journal of Medical and Pharmaceutical Science* 2(3), 20–22.

Adams, J.S. and McShane, T.O. (1992) *The Myth of Wild Africa*. W.W. Norton & Company, New York.

Aerts, R., Van Overtveld, K., *et al.* (2016) Conservation of the Ethiopian church forests: threats, opportunities and implications for their management. *Science of the Total Environment* 551–552, 404–414. DOI: 10.1016/j.scitotenv.2016.02.034.

Agama, A.L. (2002) Dusun communities' efforts to control distribution of a medicinal plant manual around Kinabalu National Park. In: Laird, S.A. (ed.) *Biodiversity and Traditional Knowledge: Equitable Partnerships in Practice*. Earthscan, London, pp. 96–100.

Alcamo, J. and Bennett, E. (2003) *Ecosystems and Human Well-being: A Framework for Assessment*. Island Press, Washington, DC.

Alcorn, J.B. (1995) The scope and aims of ethnobotany in a developing world. In: Schultes, R.E. and von Reis, S. (eds) *Ethnobotany: Evolution of a Discipline*. Chapman and Hall, London, pp. 23–39.

Aldhous, J.R. (1997) British forestry: 70 years of achievement. *Forestry* 4, 283–291.

Al-Hebshi, N.N. and Skaug, N. (2005) Khat (*Catha edulis*) – an updated review. *Addiction Biology* 10, 299–307.

Ali, A. (2006) Management of Guzara forests; policies and their implications in Hazara Division, North-West Frontier Province, Pakistan. In: Gyamtsho, P., Singh, B.K., *et al.* (eds) *Interaction Between Forest Policies and Land Use Patterns in Asia*. International Centre for Integrated Mountain Development (ICIMOD), Kathmandu, pp. 113–124.

Allen, D.E. and Hatfield, G. (2004) *Medicinal Plants in Folk Tradition: An Ethnobotany of Britain and Ireland*. Timber Press, Portland, Oregon.

Amouzou, A., Kozuki, N., *et al.* (2014) Where is the gap?: the contribution of disparities within developing countries to global inequalities in under-five mortality. *BMC Public Health* 14, 216. DOI: 10.1186/1471-2458-14-216.

Anderson, A.M. (1961) Further observations concerning the proposed introduction of the Nile perch into Lake Victoria. *East African Agricultural and Forestry Journal* 26, 195–201.

Anderson, D.M., Salick, J., *et al.* (2005) Conserving the sacred medicine mountains: a vegetation analysis of Tibetan sacred sites in Northwest Yunnan. *Biodiversity and Conservation* 14, 3065–3091.

Anderson, S. (2002) *Identifying Important Plant Areas – A Site Selection Manual for Europe, and A Basis for Developing Guidelines for Other Regions of the World*. Plantlife International, Salisbury, UK.

Andrews, P. and Johnson, R.J. (2019) Evolutionary basis for the human diet: consequences for human health. *Journal of Internal Medicine* 287, 226–237.

Angelo, M.J. (2017) Food security, industrialized agriculture, and a changing global climate: perspectives on the United States and Cuba. *Florida Journal of International Law* 29, 133–155.

Antonelli, A., Fry, C., *et al.* (2020) *State of the World's Plants and Fungi*. Royal Botanic Gardens, Kew, UK.

Apter, D. (1997) *The Political Kingdom in Uganda: A Study of Bureaucratic Nationalism*. Routledge, Abingdon-on-Thames, UK.

Armstrong, K. (2019) *The Lost Art of Scripture: Rescuing the Sacred Texts*. The Bodley Head, London.

Arora, D. (2008) The houses that matsutake built. *Economic Botany* 62(3), 278–290.

Asante, L.A. (2020) Book review: 'Property, Institutions and Social Stratification in Africa' by Franklin Obeng-Odoom. *Africa Spectrum* 55, 351–353. DOI: 10.1177/0002039720974694.

Ashford, R.W., Reid, G.D.F., *et al.* (1990) The intestinal faunas of man and mountain gorilla in a shared habitat. *Annals of Tropical Medicine and Parasitology* 84, 337–340.

Ashley, R., Russell, D., *et al.* (2006) The policy terrain in protected area landscapes: challenges for agroforestry in integrated landscape conservation. *Biodiversity and Conservation* 15, 663–689.

Ashraf, U. (2017) *State, Society and Timber Mafia in Forest Conservation*. International Institute of Social Studies, The Hague, The Netherlands.

Askin, R.A. and Jacobson, S.R. (1996) Palynological change across the Cretaceous–Tertiary boundary on Seymour Island, Antarctica: environmental and depositional factors. In: MacLeod, N. and Keller, G. (eds) *Cretaceous–Tertiary Mass Extinctions: Biotic and Environmental Changes*. W.W. Norton, London, pp. 7–26.

Aubert, M., Lebe, R., *et al.* (2019) Early hunting scene in prehistoric art. *Nature* 576, 442–445.

Aumeerduddy-Thomas, Y., Shinwari, Z.K., *et al.* (2004) *Ethnobotany and The Management of Fodder and Fuelwood at Ayubia National Park*. People and Plants Working Paper No. 13. Division of Ecological Sciences, UNESCO, Paris.

Ausness, R.C. (2022) Corporate misconduct in the pharmaceutical industry. *DePaul Law Review* 71(1), 1–46.

Azhar, R.A. (1989) Communal property rights and depletion of forests in Northern Pakistan. *The Pakistan Development Review* 4, 643–651.

Bacon, K.L. and Swindles, G.T. (2016) Could a potential Anthropocene mass extinction define a new geological period? *The Anthropocene Review* 3, 206–217.

Baggini, J. (2018) *How the World Thinks*. Granta Publications, London.

Bai XueJun and Li Hui (2016) What gave rise to modern human behavior? Perspective from psychology. *Chinese Science Bulletin* 61, 2782–2785.

Baker, W.J. and Couvreur, T.L.P. (2012) Global biogeography and diversification of palms sheds light on the evolution of tropical lineages. II. Diversification history and origin of regional assemblages. *Journal of Biogeography* 40(2), 286–298. DOI:10.1111/j.1365-2699.2012.02794.x.

Ballvé, T. (2012) Everyday state formation: territory, decentralization, and the narco landgrab in Colombia. *Environment and Planning D: Society and Space* 30, 603–622. DOI: 10.1068/d4611.

Banana, A.Y., Vogt, N.D., *et al.* (2007) Decentralized governance and ecological health: why local institutions fail to moderate deforestation in Mpigi district of Uganda. *Scientific Research and Essay* 2, 434–445.

Banerjee, N., Song, L., *et al.* (2015) Exxon's own research confirmed fossil fuels' role in global warming decades ago. *Inside Climate News*, 16 September. Available at: https://insideclimatenews.org/news/16092015/exxons-own-research-confirmed-fossil-fuels-role-in-global-warming/ (accessed 20 May 2024).

Barata, A.M., Rocha, F.R., *et al.* (2016) Conservation and sustainable uses of medicinal and aromatic plants genetic resources on the worldwide for human welfare. *Industrial Crops and Products* 88, 8–11.

Barbaix, S., Kurban, A., *et al.* (2020) The use of historical sources in a multi-layered methodology for karez research in Turpan, China. *Water History* 12, 281–297.

Barber, M., Jackson, S., *et al.* (2014) Working knowledge: characterising collective indigenous, scientific and local knowledge about the ecology, hydrology and geomorphology of Oriners Station, Cape York Peninsula, Australia. *Rangeland Journal* 36, 53–66. DOI: 10.1071/RJ13083.

Bar-On, Y.M., Phillips, R., *et al.* (2018) The biomass distribution on Earth. *Proceedings of the National Academy of Sciences USA* 115, 6506–6511.

Bates, C. and Rowell, A. (2004) *Tobacco Explained: The Truth about the Tobacco Industry in Its Own Words*. Action on Smoking and Health, London. Available at: https://escholarship.org/uc/item/9fp6566b (accessed 23 May 2024).

Bauer, K.M. (2004) *High Frontiers: Dolpo and the Changing World of Himalayan Pastoralists*. Colombia University Press, New York.

Benin, S., Smale, M., *et al.* (2007) The economic determinants of cereal crop diversity on farms in the Ethiopian highlands. *Agricultural Economics* 31, 197–208.

Benson, C.W. (1976) Charles Robert Senhouse Pitman (obituary). *Ibis* 118, 427–428.

Beverley, H., Laitala, K., *et al.* (2019) Microfibres from apparel and home textiles: prospects for including microplastics in environmental sustainability assessment. *Science of the Total Environment* 652, 483–494.

Binggeli, P. (1989) The ecology of *Maesopsis* invasion and dynamics of the evergreen forest of the East Usambaras, and their implications for forest conservation and forestry practices. In: Hamilton, A.C. and Bensted-Smith, R. (eds) *Forest Conservation in the East Usambara Mountains, Tanzania*. IUCN, Gland, Switzerland and Cambridge, pp. 269–300.

BirdLife International (2008) Campaign to save Mabira Forest in Uganda from sugarcane plantation for biofuels. Available at: http://www.birdlife.org/datazone/sowb/casestudy/231 (accessed 28 August 2015).

Bisaso, R. (2017) Makerere University as a flagship institution: sustaining the quest for relevance. In: Teferra, D. (ed.) *Flagship Universities in Africa*. Palgrave Macmillan, Cham, Switzerland, pp. 425–466.

Biswas, A. and Mani, K.R.S. (2008) Relativistic perihelion precession of orbits of Venus and the Earth. *Central European Journal of Physics* 6, 754–758.

Blackie, J.R. and Robinson, M. (2007) Development of catchment research, with particular attention to Plynlimon and its forerunner, the East African catchments. *Hydrology and Earth System Sciences* 11, 26–43.

Blair, T. (2006) *A Global Alliance for Global Values.* The Foreign Policy Centre, London.

Blas, J. and Farchy, J. (2021) *The World for Sale: Money, Power, and the Traders Who Barter the Earth's Resources.* Penguin Books, London.

Blench, R. (2006) *Archaeology, Language, and the African Past.* AltaMira Press, Lanham, Maryland.

Blomme, G., Ploetz, R., *et al.* (2012) A historical overview of the appearance and spread of *Musa* pests and pathogens on the African continent: highlighting the importance of clean *Musa* planting materials and quarantine measures. *Annals of Applied Biology* 162, 314–326.

Boedicker, F. and Boedicker, M. (2009) *The Philosophy of Tai Chi Chuan: Wisdom from Confucius, Lao Tzu, and Other Great Thinkers.* Blue Snake Books, Berkeley, California.

Boerner, L.K. (2019) Industrial ammonia production emits more CO_2 than any other chemical-making reaction. Chemists want to change that. *Chemical & Engineering News* 97(24). Available at: https://cen.acs.org/environment/green-chemistry/Industrial-ammonia-production-emits-CO2/97/i24 (accessed 20 May 2024).

Boroditsky, L. (2018) How does our language shape the way we think? Available at: https://www.edge.org/conversation/lera_boroditsky-how-does-our-language-shape-the-way-we-think (accessed 20 May 2024).

Böttger, A., Vothknecht, U., *et al.* (2018) Plant secondary metabolites and their general function in plants. *Lessons on Caffeine, Cannabis & Co.: Plant-derived Drugs and their Interaction with Human Receptors.* Springer, Cham, Switzerland, pp. 3–17.

Bowles, S. and Choi, J.-K. (2013) Coevolution of farming and private property during the early Holocene. *Proceedings of the National Academy of Sciences USA* 110, 8830–8835.

Bowman, D., Williamson, G., *et al.* (2017) Human exposure and sensitivity to globally extreme wildfire events. *Nature Ecology & Evolution* 1, 0058.

Boyer, P. and Bergstrom, B. (2008) Evolutionary perspectives on religion. *Annual Review of Anthropology* 37, 111–130.

Bozoki, A. and Ádám, Z. (2016) State and faith: right-wing populism and nationalized religion in Hungary. *Intersections: East European Journal of Science and Politics* 2, 89–122.

Brockway, L. (1979) *Science and Colonial Expansion: The Role of the British Royal Botanic Gardens.* Academic Press, London.

Brockway, L. (2002) Science and colonial expansion: the role of the British Royal Botanic Gardens. *American Ethnologist* 6, 449–465.

Brockway, L. (2020) Plant imperialism. Available at: https://www.britishempire.co.uk/science/agriculture/plantimperialism.htm (accessed 20 May 2024).

Brown, M.I. (2013) *Redeeming REDD.* Earthscan, London.

Brown, T.M. and Fee, E. (2008) Spinning for India's independence. *American Journal of Public Health* 98(1), 39.

BSA (1995) *Botany for the Next Millennium.* Botanical Society of America, St. Louis, Missouri.

Buchan, R. (2007) International community and the occupation of Iraq. *Journal of Conflict and Security Law* 12(1), 37–64.

Bullough, O. (2018) *Moneyland.* Profile Books Ltd, London.

Buntaine, M., Mullen, R., *et al.* (2007) Human use and conservation planning in Alpine areas of Northwestern Yunnan, China. *Environment, Development, and Sustainability* 9(3), 305–324.

Bunyan, J. (1678) *The Pilgrim's Progress.* Nathaniel Ponder, London.

Burgarella, C., Cubry, P., *et al.* (2018) A western Sahara centre of domestication inferred from pearl millet genomes. *Nature Ecology & Evolution* 2, 1377–1380.

Burley, J., Mills, R.A., *et al.* (2009) Witness to history: a history of forestry at Oxford University. *British Scholar* 1(2), 236–261.

Bussman, R.W. (2012) I know every tree in the forest: reflections on the life and legacy of Richard Evans Schultes. In: Ponman, B.E. and Bussmann, R.W. (eds) *Medicinal Plants and the Legacy of Richard E. Schultes.* The William L. Brown Center at the Missouri Botanical Garden, St Louis, Missouri, pp. 13–21.

Butynski, T.M. (1984) *Ecological Survey of the Impenetrable (Bwindi) Forest, Uganda, and Recommendations for Its Conservation and Management.* Report submitted to the Ministry of Tourism & Wildlife, Uganda, *et al.* Wildlife Conservation International, New York Zoological Society, New York. Available at: https://www.wildsolutions.nl/wp-content/uploads/Ecological-Survey-of-the-Impenetrable-Bwindi-Forest-Butynski-1984.pdf (accessed 20 May 2024).

Butynski, T.M. (1986) Elephants in the Impenetrable (Bwindi) Forest. *Elephant* 2, 42–43.

Byarugaba, D. (2008) Integrating indigenous knowledge, active conservation and wild food plant use with reference to *Dioscorea* species in Bwindi Forest, southwestern Uganda. *African Journal of Ecology* 48, 539–540.

Caney, S. and Hepburn, C. (2011) *Carbon Trading: Unethical, Unjust and Ineffective?* Centre for Climate Change Economics and Policy, Working Paper No. 59/Grantham Research Institute on Climate Change and the Environment, Working Paper No. 49. University of Leeds, Leeds, UK and London School of Economics and Political Science, London.

Carl, N., Cofnas, N., *et al.* (2016) Scientific literacy, optimism about science and conservatism. *Personality and Individual Differences* 94, 299–302.

Carlson, R. (2016) Estimating the biotech sector's contribution to the US economy. *Nature Biotechnology* 34, 247–255.

Carroll, T. and Jarvis, D.S.L. (2015) The new politics of development: citizens, civil society, and the evolution of neoliberal development policy. *Globalizations* 12(3), 281–304.

Carson, R. (1962) *Silent Spring.* Houghton Mifflin, Boston, Massachusetts.

Carter, M.M., Olm, M.R., *et al.* (2023) Ultra-deep sequencing of Hadza hunter-gatherers recovers vanishing gut microbes. *Cell* 186, 3111–3124.

CBD (1992) *Convention on Biological Diversity.* United Nations, Montreal, Quebec, Canada. Available at: https://www.cbd.int/doc/legal/cbd-en.pdf (accessed 23 May 2024).

CBD (2002) *Global Strategy for Plant Conservation.* Secretariat of the Convention on Biological Diversity, Montreal, Quebec, Canada. Available at: https://www.cbd.int/doc/publications/pc-brochure-en.pdf (accessed 23 May 2024).

CBD (2004) The twelve principles of the ecosystem approach. Available at: https://www.cbd.int/ecosystem/principles.shtml (accessed 22 May 2024).

Ceaser, J.W. (2012) The origin and character of American exceptionalism. *American Political Thought: A Journal of Ideas, Institutions, and Culture* 1, 3–28.

Chase, M.W., Christenhusz, M.J.M., *et al.* (2016) An update of the Angiosperm Phylogeny Group classification for the orders and families of flowering plants: APG IV. *Botanical Journal of the Linnean Society* 181(1), 1–20.

Chaseling, W. (1957) *Yulengor: Nomads of Arnhem Land.* The Epworth Press, London.

Chatterjee, D. (1948) Early history of the Royal Botanic Garden, Calcutta. *Nature* 161, 362–364.

Chen, F.H., Dong, G.H., *et al.* (2015) Agriculture facilitated permanent human occupation of the Tibetan Plateau after 3600 BP. *Science* 347, 248–250.

Chenery, E.M. (1951) *An Introduction to the Soils of the Uganda Protectorate.* Research Division Memoir No. 1. Department of Agriculture, Kampala.

Churchill, W.S. (1898) *Story of the Malakand Field Force.* Longmans, London.

Churchill, W.S. (1908) *My African Journey.* Icon Books, London (republished, 1964).

Clark, P.U., Dyke, A.S., *et al.* (2009) The Last Glacial Maximum. *Science* 325, 710–714.

Clarke, G. (1977) The merchants of Mugu: a village in the Himalaya. *Asian Affairs* 8, 299–305.

Cobley, L.S. (1956) *An Introduction to the Botany of Tropical Crops.* Longmans, London.

Cochrane, L. and O'Regan, D. (2016) Legal harvest and illegal trade: trends, challenges, and options in khat production in Ethiopia. *International Journal of Drug Policy* 30, 27–34.

Coetzee, J.A. (1967) Pollen analytical studies in East and Southern Africa. *Palaeoecology of Africa* 3, 1–146.

Collingham, L. (2017) *The Hungry Empire: How Britain's Quest for Food Shaped the Modern World*. Penguin Random House, London.

Colonial Report (1905) *Uganda, Report for 1904–5. Colonial Reports: Annual*. His Majesty's Stationery Office, London.

Conner, J.K. and Lande, R. (2014) Raissa L. Berg's contributions to the study of phenotypic integration, with a professional biographical sketch. *Philosophical Transactions of the Royal Society of London B: Biological Sciences* 369, 20130250. DOI: 10.1098/rstb.2013.0250.

Constanza, R., d'Arge, R., *et al.* (1997) The value of the world's ecosystem services and natural capital. *Nature* 387, 253–260.

Cooper, D.L.M., Lewis, S.L., *et al.* (2024) Consistent patterns of common species across tropical tree communities. *Nature* 625, 728–734.

Cox, K. (ed.) (2008) *Frank Kingdon Ward's Riddle of the Tsangpo Gorges*. AAC Art Books, Woodbridge, UK.

Cramer, C., Johnston, D., *et al.* (2014) Fairtrade cooperatives in Ethiopia and Uganda uncensored. *Review of African Political Economy* 41, 115–127.

Crone, P. (1989) *Pre-Industrial Societies: Anatomy of the Pre-Modern World*. Blackwell, Oxford.

Cronk, Q. (2016) Plant extinctions take time. *Science* 353, 446–447.

Crowley, R. (2011) *City of Fortune: How Venice Won and Lost a Naval Empire*. Faber and Faber, London.

Crystal, D. (2000) *Language Death*. Cambridge University Press, Cambridge.

Cubry, C., Tranchant-Dubreuil, C., *et al.* (2018) The rise and fall of Africa rice cultivation revealed by analysis of 246 new genomes. *Current Biology* 28, 2274–2282.

Cunningham, A.B. (1993) *African Medicinal Plants: Setting Priorities at the Interface between Conservation and Primary Healthcare*. People and Plants Working Paper No. 1. Division of Ecological Sciences, UNESCO, Paris.

Cunningham, A.B. (1996) *People, Park and Plant Use*. People and Plants Working Paper No. 4. Division of Ecological Sciences, UNESCO, Paris.

Cunningham, A.B., Ayuk, E., *et al.* (2002) *An Economic Evaluation of Medicinal Tree Cultivation: Prunus africana in Cameroon*. People and Plants Working Paper No. 10. Division of Ecological Sciences, UNESCO, Paris.

Cunningham, M., Cunningham, A.B., *et al.* (1997) *Trade in* Prunus africana *and the Implementation of CITES*. German Federal Agency for Nature Conservation, Bonn, Germany.

Dalrymple, W. (2019) *The Anarchy: The Relentless Rise of the East India Company*. Bloomsbury Publishing, London.

D'Andrea, A.C., Kahlheber, S., *et al.* (2007) Early domesticated cowpea (*Vigna unguiculata*) from Central Ghana. *Antiquity* 81, 686–698.

Daniels, J. and Daniels, C. (1993) Sugarcane in prehistory. *Archaeology in Oceania* 28, 1–7.

Darbyshire, I., Lamb, H., *et al.* (2003) Forest clearance and regrowth in Northern Ethiopia during the last 3000 years. *The Holocene* 13, 537–546.

Darwin, C. (1859) *On the Origin of Species by Means of Natural Selection, Or the Preservation of Favoured Races in the Struggle for Life*. John Murray III, London.

Dasandi, N. and Lior, E. (2023) The flag and the stick: aid suspensions, human rights, and the problem of the complicit public. *World Development* 168, 106264.

Davies, P. and Robb, J.G. (2002) The appropriation of the material of places in the landscape: the case of tufa and springs. *Landscape Research* 27, 181–185.

Dawe, M.T. (1906) *A Report on a Botanical Mission through the Forest Districts of Buddu and the Western and Nile Provinces of the Ugandan Protectorate*. His Majesty's Stationery Office, London.

Dawkins, R. (1976) *The Selfish Gene*. Oxford University Press, Oxford.

Degain, C., Meng, B., *et al.* (2017) Recent trends in global trade and global value chains. In: *Measuring and Analyzing the Impact of GVs on Economic Development*. Global Value Chain Development Report 2017. International Bank for Reconstruction and Development/ The World Bank, Washington, DC, pp. 37–68.

deMenocal, P.B. (2014) Climate shocks. *Scientific American* 311(September), 32–37.

Denham, T.P., Haberle, S.G., *et al.* (2003) Origins of agriculture in the Kuk Swamp in the highlands of New Guinea. *Science* 301, 189–193.

Deutsch, S. (2021) Populist authoritarian neoliberalism in Brazil: making sense of Bolsonaro's anti-environment agenda. *Journal of Political Ecology* 28(1), 823–844. DOI: 10.2458/ jpe.2994.

Dev, S. (1999) Ancient–modern concordance in Ayurvedic plants: some examples. *Environmental Health Perspectives* 107, 783–789.

de Vries, R. (2014) *Earning by Degrees: Differences in the Career Outcomes of UK Graduates*. The Sutton Trust, London.

Dhakal, B., Pinard, M.A., *et al.* (2012) Impacts of cardamom cultivation on montane forest eco- systems in Sri Lanka. *Forest Ecology and Management* 274, 151–160.

Di Nicola, V. (2020) The Global South: an emergent epistemology for social psychiatry. *World Social Psychiatry* 2(1), 20–26.

Diamond, A.S. and Hamilton, A.C. (1980) The distribution of forest passerine birds and Quaternary climatic change in tropical Africa. *Journal of Zoology* 191, 379–402.

Diamond, J.M. (2005) *Collapse: How Societies Chose to Fail or Succeed*. Viking Press, New York.

Diamond, P. (2021) Destroying one public service bargain without making another: a comment on Lowe and Pemberton, *The Official History of the British Civil Service, Volume II: 1982–1997. The Political Quarterly* 92(1), 95–100.

Díaz-Forestier, J., León-Lobos, P., *et al.* (2019) Native useful plants of Chile: a review and use patterns. *Economic Botany* 73(1), 112–126.

Dillehay, T. and Ocampo, C. (2015) New archaeological evidence for an early human presence at Monte Verde, Chile. *PLoS ONE* 10(11), e0141923. DOI: 10.1371/journal.pone.0141923.

Dillehay, T.D., Rossen, J., *et al.* (2010) Early Holocene coca chewing in northern Peru. *Antiquity* 84, 939–953.

Dillon, L., Sellers, C., *et al.* (2018) The Environmental Protection Agency in the early Trump administration: prelude to regulatory capture. *American Journal of Public Health* 108(Suppl. 2), 589–394.

Dixit, K. (2017) The turning point. *Nepali Times*, 20–26 October, #880. Available at: https:// archive.nepalitimes.com/article/nation/The-turning-point-war-nepal,3986 (accessed 20 May 2024).

Dobson, A., Lodge, D., *et al.* (2006) Habitat loss, trophic collapse, and the decline of ecosystem services. *Ecology* 87, 1915–1924.

Dompreh, E.B., Asari, R., *et al.* (2020) Do voluntary certification standards improve yields and wellbeing? Evidence from oil palm and cocoa smallholders in Ghana. International *Journal of Agricultural Sustainability* 19, 16–39. DOI: 10.1080/14735903.2020.1807893.

Doornbos, M. (1987) *The Uganda Crisis and the National Question*. Working Papers Series No. 34. International Institute of Social Studies, The Hague, The Netherlands.

Downer, S., Berkowitz, S.A., *et al.* (2020) Food is medicine: actions to integrate food and nu- trition into healthcare. *British Medical Journal* 369, m2482. DOI: 10.1136/bmj.m2482.

Drake, N.A., Blench, R.M., *et al.* (2010) Ancient watercourses and biogeography of the Sahara explain the peopling of the desert. *Proceedings of the National Academy of Sciences USA* 108, 458–462.

Dransfield, J. and Johnson, D.V. (1991) The conservation status of palms in Sabah (Malaysia). In: Johnson, D.V. (ed.) *Palms for Human Needs in Asia*. A.A. Balkema, Rotterdam, The Netherlands, pp. 175–179.

Draper, N. (2010) *The Prince of Emancipation: Slave-Ownership, Compensation and British Society at the End of Slavery*. Cambridge University Press, Cambridge.

Drichi, P. (2002) *National Biomass Study 1996–2002, Technical Report*. Forest Department, Kampala.

Duffie, D. and Stein, J.C. (2015) Reforming LIBOR and other financial market benchmarks. *Journal of Economic Perspectives* 29, 191–212.

Dunbar, A.R. (1965) *A History of Bunyoro-Kitara*. Fountain Publishers, Kampala.

Dunne, J., Mercuri, A.M., *et al.* (2016) Earliest direct evidence of plant processing in prehistoric Saharan pottery. *Nature Plants* 3, 16194.

Durand, J. (1992) Building a national resource: fifty years of Irish forestry. *Irish Forestry* 49, 1–9.

Eaton, S.B. (2006) The ancestral human diet: what was it and should it be a paradigm for contemporary nutrition? *Proceedings of the Nutrition Society* 65, 1–6.

Edmeades, S. and Karamura, D. (2007) Banana taxonomy for Uganda. In: Smale, M. and Tushemereirwe, W.K. (eds) *An Economic Assessment of Banana Genetic Improvement and Innovation in the Lake Victoria Region of Uganda and Tanzania*. International Food Policy Research Institute, Washington, DC, pp. 165–169.

Edmeades, S., Smale, M., *et al.* (2007) Characteristics of banana-growing households and banana cultivars in Uganda and Tanzania. In: Smale, M. and Tushemereirwe, W.K. (eds) *An Economic Assessment of Banana Genetic Improvement and Innovation in the Lake Victoria Region of Uganda and Tanzania*. International Food Policy Research Institute, Washington, DC, pp. 49–71.

Edwards, A.W.F. (2013) Robert Keith Lock and his textbook of genetics. *Genetics* 194, 529–537.

Ehret, C. (1982) Linguistic inferences about early Bantu history. In: Ehret, C. and Posnansky, M. (eds) *The Archaeological and Linguistic Reconstruction of African History*. University of California Press, Berkeley, California, pp. 57–65.

Ehret, C. (1998) *An African Classical Age: Eastern and Southern Africa in World History: 1000 BC to 400 AD*. James Currey, Oxford.

Ehret, C. (2011) *History and the Testimony of Language*. University of California Press, Berkeley, California.

Eltahir, E.A.B. and Bras, R.L. (1994) Precipitation recycling in the Amazon basin. *Quarterly Journal of the Royal Meteorological Society* 120, 861–880.

Elwes, H.J. and Henry, A. (1906–1913) *The Trees of Great Britain and Ireland*. Privately published, Edinburgh.

Entenmann, S.K., Schmitt, C.B., *et al.* (2014) REDD+-related activities in Kenya: actors' views on biodiversity and monitoring in a broader policy context. *Biodiversity and Conservation* 23, 2561–2586.

Erickson, C.L. (2000) The Lake Titicaca Basin: a Pre-Columbian built landscape. In: Lentz, D. (ed.) *Imperfect Balance: Landscape Transformations in the Pre-Columbian Americas*. Columbia University Press, New York, pp. 311–356.

Eshed, V., Gopher, A.G., *et al.* (2004) Musculoskeletal stress markers in Natufian hunter-gatherers and Neolithic farmers in the Levant: the upper limb. *American Journal of Physical Anthropology* 123, 303–315.

Essig, F.B. (2015) *Plant Life: A Brief History*. Oxford University Press, New York.

Ethnologue (2022) Living Languages, 2022. Available at: https://www.ethnologue.com/guides/how-many-languages (accessed 20 May 2024).

Fagan, J.J. and Bhutta, M.F. (2021) General Medical Council report exposes unethical recruitment of doctors in the UK from low-resource countries. *South African Medical Journal* 111(3), 189. DOI: 10.7196/SAMJ.2021.v111i3.15479.

Fantini, E. (2015) Go Pente! The charismatic renewal of the evangelical movement in Ethiopia. In: Prunier, G.P. and Ficquet, É. (eds) *Understanding Contemporary Ethiopia: Monarchy, Revolution and the Legacy of Meles Zenawi*. C. Hurst & Co., London, pp. 123–146.

FAO (2016) *2014 Global Forest Products Facts and Figures.* Food and Agricultural Organization of the United Nations, Rome. Available at: https://openknowledge.fao.org/items/6f08b516-ffd1-4ccc-8f0d-d1ea48576b87 (accessed 20 May 2024).

FAO (2017) *Wood Energy: Basic Knowledge.* Basic learning module. Food and Agricultural Organization of the United Nations, Rome. Available at: http://www.fao.org/sustainable-forest-management/toolbox/modules/wood-energy/basic-knowledge/en/?type=111 (accessed 20 May 2024).

Farnsworth, N.R. and Soejarto, D.D. (1991) Global importance of medicinal plants. In: Akerele, O., Heywood, V., *et al.* (eds) *The Conservation of Medicinal Plants.* Cambridge University Press, Cambridge, pp. 25–51.

Farnsworth, N.R., Akerele, O., *et al.* (1985) Medicinal plants in therapy. *Bulletin of the World Health Organization* 63(6), 965–981.

Fay, M.A. (1978) Did Marx offer to dedicate *Capital* to Darwin?: a reassessment of the evidence. *Journal of the History of Ideas* 39, 133–146.

Fet, V. and Golubovsky, M.D. (2008) Vavilov's vision for genetics was among Stalin's many victims. *Nature* 455, 27.

Finch, J., Wooler, M., *et al.* (2014) Tracing long-term tropical montane ecosystem change in the Eastern Arc Mountains of Tanzania. *Journal of Quaternary Science* 29, 269–278.

Fjær, E.L., Landet, E.R., *et al.* (2020) The use of complementary and alternative medicine (CAM) in Europe. *BMC Complementary Medicine and Therapies* 20, 108.

Fjeldså, J. and Lovett, J.C. (1997) Geographical patterns of old and young species in African forest biota: the significance of specific montane areas as evolutionary centres. *Biodiversity and Conservation* 6, 325–346.

Fjeldså, J., Christidis, L., *et al.* (eds) (2020) *The Largest Avian Radiation: The Evolution of Perching Birds, or the Order Passeriformes.* Lynx Edicions, Barcelona, Spain.

FOE (2012) *Land, Life and Justice: How Land Grabbing in Uganda is Affecting the Environment, Livelihoods and Food Sovereignty of Communities.* Friends of the Earth International, Amsterdam.

Ford, R.I. (1978) *The Nature and Status of Ethnobotany.* Anthropological Papers No. 67. Museum of Anthropology, University of Michigan, Ann Arbor, Michigan.

Fornari, E., Grandi, S., *et al.* (2020) Discounters versus supermarkets and hypermarkets: what drives store-switching? *The International Review of Retail, Distribution and Consumer Research* 30, 555–574.

Frankopan, P. (2015) *The Silk Roads: A New History of the World.* Bloomsbury, London.

Franses, P.H. and van den Heuvel, W. (2019) Aggregate statistics on trafficker–destination relations in the Atlantic slave trade. *International Journal of Maritime History* 31(3), 624–633.

Freeman, D. (2013) Pentecostalism in a rural context: dynamics of religion and development in Southwest Ethiopia. *PentecoStudies* 12, 231–249.

French, H.W. (2021) *Born in Blackness: Africa, Africans and the Making of the Modern World, 1471 to the Second World War.* Liveright, New York.

Frost, M.R. (2004) Asia's maritime networks and the colonial public sphere, 1840–1920. *New Zealand Journal of Asian Studies* 6, 63–94.

Fullagar, R., Field, J., *et al.* (2006) Early and mid Holocene tool-use and processing of taro (*Colocasia esculenta*), yam (*Dioscorea* sp.) and other plants at Kuk Swamp in the highlands of Papua New Guinea. *Journal of Archaeological Science* 33, 595–613.

Fuller, D.Q. (2008) The spread of textile production and textile crops in India beyond the Harappan Zone: an aspect of the emergence of craft specialization and systematic trade. In: Osada, T. and Uesugi, A. (eds) *Linguistics, Archaeology and the Human Past.* Occasional Paper No. 3. Research Institute for Humanity and Nature, Kyoto, Japan, pp. 1–26.

Fuller, T.L., Narins, T.P., *et al.* (2018) Assessing the impact of China's timber industry on Congo Basin land use change. *Area* 51, 340–349. DOI: 10.1111/area.12469.

Furry, T.F. (2008) Bind us together: repentance, Ugandan martyrs, and Christian unity. *New Blackfriars* 89(1019), 39–59. DOI: 10.1111/j.1741-2005.2007.00165.x.

Gaikwad, A.B., Kumari, R., *et al.* (2023) Small cardamom genome: development and utilization of microsatellite markers from a draft genome sequence of *Elettaria cardamomum* Maton. *Frontiers in Plant Science* 14, 1161499. DOI: 10.3389/fpls.2023.1161499.

Galhena, D.H., Freed, R., *et al.* (2013) Home gardens: a promising approach to enhance household food security and wellbeing. *Agriculture & Food Security* 2, 8.

Ganesan, U. (2010) Medicine and modernity: the Ayurvedic Revival Movement in India, 1885–1947. *Studies on Asia* 4(1), 108–131.

Gao Lishi (1998) *On the Dais' Traditional Irrigation System and Environmental Protection in Xishuangbanna.* Yunnan Nationality Press, Kunming, China.

Garba, I., Umar, A.I., *et al.* (2013) Phytochemical and antibacterial properties of garlic extracts. *Bayero Journal of Pure and Applied Sciences* 6(2), 45–48.

García-Jiménez, C.I. and Vargas-Rodriguez, Y.L. (2021) Passive government, organized crime, and massive deforestation: the case of western Mexico. *Conservation Science and Practice* 3(12), e562. DOI: 10.1111/csp2.562.

Garside, R. and Wyn, I. (2021) Tree-planting: why are large investment firms buying Welsh farms? *BBC News*, 6 August. Available at: https://www.bbc.co.uk/news/uk-wales-58103603 (accessed 20 May 2024).

Gassaway, L. (2009) Native American fire patterns in Yosemite Valley: archaeology, dendrochronology, subsistence, and culture change. *Society for California Archaeology Proceedings* 22, 1–19.

Gaynor, K.M., Florella, K.J., *et al.* (2016) War and wildlife: linking armed conflict to conservation. *Frontiers in Ecology and the Environment* 14, 533–542.

Geddes, P. (1915) *Cities in Evolution.* Williams and Norgate, London.

Getachew, G.A., Asfaw, Z., *et al.* (2013) Dietary values of wild and semi-wild edible plants in Southern Ethiopia. *African Journal of Food, Agriculture, Nutrition and Development* 13(2), 7485–7503. DOI: 10.18697/ajfand.57.11125.

Ghimire, S.K., Awasthi, B., *et al.* (2016) Export of medicinal and aromatic plant materials from Nepal. *Journal of Plant Sciences* 10, 24–32.

Ghimire, S.K., Lama, Y.C., *et al.* (2000) *Medicinal Plant Management and Health Care Development in Shey Phoksundo National Park, Dolpa.* WWF Nepal Program, Kathmandu.

Ghimire, S.K., McKey, D., *et al.* (2004) Heterogeneity in ethnoecological knowledge and management of medicinal plants in the Himalayas of Nepal: implications for conservation. *Ecology and Society* 9(3), 6.

Ghimire, S.K., McKey, D., *et al.* (2005) Conservation of Himalayan medicinal plants: harvesting patterns and ecology of two threatened species, *Nardostachys grandiflora* DC. and *Neopicrorhiza scrophulariiflora* (Pennell) Hong. *Biological Conservation* 124, 463–475.

Gibbs, K. (2015) Pottery invention and innovation in East Asia and the Near East. *Cambridge Archaeological Journal* 25, 339–351.

Gibson, E., Futrell, R., *et al.* (2017) Color naming across languages reflects color use. *Proceedings of the National Academy of Sciences USA* 114, 10785–10790.

Giordan, G. and Possamai, A. (eds) (2020) *The Social Scientific Study of Exorcism in Christianity.* Springer, Cham, Switzerland. DOI: 10.1007/978-3-030-43173-0.

Given, D.R. (1994) *Principles and Practice of Plant Conservation.* Timber Press, Portland, Oregon.

Glowka, L., Burhenne-Guilmin, F., *et al.* (1994) *A Guide to the Convention on Biological Diversity.* IUCN, Gland, Switzerland.

Godlaski, T. (2013) Holy smoke: tobacco use among Native American tribes in North America. *Substance Use and Misuse* 48, 1–8.

Gold, C.S., Kiggundu, A., *et al.* (2002) Diversity, distribution and farmer preference of *Musa* cultivars in Uganda. *Experimental Agriculture* 38, 39–50.

Goldsmith, E. and Allen, R. (1972) A blueprint for survival. *The Ecologist* 2 (January Special Issue).

Gong, Z., Chen, H.Z., *et al.* (2007) The temporal and spatial distribution of ancient rice in China and its implications. *Chinese Science Bulletin* 52, 1071–1079.

Gore, A.J.P. (2006) *An Inconvenient Truth*. Rodale Press, Emmaus, Pennsylvania.

Gou Yi, Fan Ruyan, *et al.* (2018) Before it disappeared: ethnobotanical study of fleagrass (*Adenosma buchneroides*), a traditional aromatic plant used by the Akha people. *Journal of Ethnobiology and Ethnomedicine* 14, 79.

Goudswaard, P.C., Witte, F., *et al.* (2007) The invasion of an introduced predator, Nile perch (*Lates niloticus* L.) in Lake Victoria (East Africa): chronology and causes. *Environmental Biology of Fishes* 81, 127–139.

Gowlett, J.A.J. (2016) The discovery of fire by humans: a long and convoluted process. *Philosophical Transactions of the Royal Society of London B: Biological Sciences* 371, 20150164.

Grafen, A. and Ridley, M. (2007) *Richard Dawkins: How a Scientist Changed the Way We Think*. Oxford University Press, Oxford.

Grandmaison, R., Morris, N., *et al.* (2018) The last harvest? From the US fentanyl boom to the Mexican opium crisis. *Journal of Illicit Economies and Development* 1, 312–329.

Greening Godalming (2015) Wood fuel. Available at: https://greengodalming.wordpress.com/woodlands-wood-fuel/ (accessed 20 May 2024).

Greenwood, P.H. (1966) *The Fishes of Uganda*. Uganda Society, Kampala.

Grillo, K., McKeeby, Z., *et al.* (2022) Nderit Ware and the origins of pastoralist pottery in eastern Africa. *Quaternary International* 608–609, 226–242.

Groark, K.P. (2010) The angel in the gourd: ritual, therapeutic, and protective uses of tobacco (*Nicotiana tabacum*) among the Tzeltal and Tzotzil Maya of Chiapas, Mexico. *Journal of Ethnobiology* 30, 5–30.

Grubb, P.J. (1971) Interpretation of the 'Massenerhebung' effect on tropical mountains. *Nature* 229, 44–45.

Guarner, F. and Malagelada, J.-R. (2003) Gut flora in health and disease. *The Lancet* 360, 512–519.

Guedes, J.D.A., Lu Hongliang, *et al.* (2013) Moving agriculture onto the Tibetan plateau: the archaeobotanical evidence. *Archaeological and Anthropological Sciences* 6, 255–269. DOI: 10.1007/s12520-013-0153-4.

Hadley, M. (ed.) (2002) *Biosphere Reserves: Special Places for People and Nature*. UNESCO, Paris.

Haeckel, E. (1866) *Generelle Morphologie der Organismen*. Georg Reimer, Berlin.

Haig, W. (2004) *William Pitt the Younger*. HarperCollins, London.

Hails, C., Humphrey, S., *et al.* (2008) *Living Planet Report*. WWF, Gland, Switzerland.

Hall, C.A.S. and Klitgaard, K.A. (2012) The limits of conventional economics. *Energy and the Wealth of Nations*. Springer, New York, pp. 131–143.

Hall, C.S. and Nordby, V.J. (1973) *A Primer of Jungian Psychology*. Mentor, New York.

Hamayun, M., Khan, S.A., *et al.* (2006) Morel collection and marketing: a case study from the Hindu-Kush mountain region of Swat, Pakistan. *Lyonia* 11, 7–13.

Hamilton, A.C. (1969) The vegetation of southwest Kigezi. *Uganda Journal* 33, 175–199.

Hamilton, A.C. (1972) The interpretation of pollen diagrams from highland Uganda. *Palaeoecology of Africa* 7, 45–149.

Hamilton, A.C. (1974) Distribution patterns of forest trees in Uganda and their historical significance. *Vegetatio* 29, 21–35.

Hamilton, A.C. (1975a) A quantitative analysis of altitudinal zonation in Uganda forests. *Vegetatio* 30, 99–106.

Hamilton, A.C. (1975b) The dispersal of forest tree species in Uganda during the Upper Pleistocene. *Boissiera* 24, 29–32.

Hamilton, A.C. (1976a) Identification of East African Urticales pollen. *Pollen et Spores* 18, 27–66.

Hamilton, A.C. (1976b) The significance of patterns of distribution shown by forest plants and animals in tropical Africa for the reconstruction of upper Pleistocene palaeo-environments: a review. *Palaeoecology of Africa* 9, 63–97.

Hamilton, A.C. (1981a) The Quaternary history of African forests: its relevance to conservation. *African Journal of Ecology* 19, 1–6.

Hamilton, A.C. (1981b) *A Field Guide to Uganda Forest Trees*. Makerere University Printery, Kampala.

Hamilton, A.C. (1982) *Environmental History of East Africa: A Study of the Quaternary*. Academic Press, London.

Hamilton, A.C. (1984) *Deforestation in Uganda*. Oxford University Press, Nairobi.

Hamilton, A.C. (1985) Aspects of the environment of the Mesolithic site. In: Woodman, P.C. (ed.) *Excavations at Mount Sandel 1973–77, County Londonderry*. Department of the Environment for Northern Ireland, Belfast, UK, pp. 105–109.

Hamilton, A.C. (1988a) Conservation of the East Usambara forests, with emphasis on biological conservation. *Acta Universitatis Upsaliensis. Symbolae botanicae Upsalienses* 28, 244–254.

Hamilton, A.C. (1988b) Guenon evolution and forest history. In: Gautier-Hion, A., Bourlière, F., *et al.* (eds) *A Primate Radiation: Evolutionary Biology of the African Guenons*. Cambridge University Press, Cambridge, pp. 13–34.

Hamilton, A.C. (1989) African forests. In: Lieth, H. and Werger, M.J.A. (eds) *Tropical Rain Forest Ecosystems*. Elsevier, Amsterdam, pp. 155–182.

Hamilton, A.C. (1992) History of forests and climate. In: Sayer, J.A., Harcourt, C.S., *et al.* (eds) *The Conservation Atlas of Tropical Forests: Africa*. Macmillan Publishers Ltd, Basingstoke, UK, pp. 17–25.

Hamilton, A.C. (1997) Threats to plants: an analysis of Centres of Plant Diversity. In: Touchell, D.H. and Dixon, K.W. (eds) *Conservation into the 21st Century*. Kings Park and Botanic Garden, Perth, Australia, pp. 309–322.

Hamilton, A.C. (1998) Vegetation, climate and soil: altitudinal relationships on the East Usambara Mountains, Tanzania. *Journal of East African Natural History* 87, 85–89.

Hamilton, A.C. (2001) *Human Nature and the Natural World*. New Millennium, London.

Hamilton, A.C. (2004a) Medicinal plants, conservation and livelihoods. *Biodiversity and Conservation* 13, 1477–1517.

Hamilton, A.C. (2004b) The People and Plants Initiative: the idea, the structure of the programme and the legacy. In: *People and Plants Handbook Issue 9*. WWF, UNESCO and Royal Botanic Gardens, Kew, UK, pp. 1–5.

Hamilton, A.C. (ed.) (2008) *Medicinal Plants in Conservation and Development: Case Studies and Lessons Learnt*. Plantlife International, Salisbury, UK.

Hamilton, A.C. (2011) An evidence-based approach to conservation through medicinal plants. *Medicinal Plant Conservation* 14, 2–7.

Hamilton, A.C. (2012) Evidence-based conservation: best practice in community-based conservation of medicinal plants. Presented at *13th Congress of the International Society of Ethnobiology*, Montpellier, France, May 2012.

Hamilton, A.C. (2013) New developments in plant conservation and the relevance of ethnobotany. *Plant Diversity and Resources* 35, 424–420.

Hamilton, A.C. (2020) *Luganda–English and English–Luganda Dictionary*. Alan Hamilton, Godalming, UK.

Hamilton, A.C. and Aumeeruddy-Thomas, Y. (2013) Maintaining resources for traditional medicine: a global overview and a case study from Buganda (Uganda). *Plant Diversity and Resources* 35, 407–423.

Hamilton, A.C. and Bensted-Smith, R. (1989) *Forest Conservation in the East Usambara Mountains, Tanzania.* IUCN, Gland, Switzerland and Cambridge.

Hamilton, A.C. and Hamilton, P.B. (2006) *Plant Conservation: An Ecosystem Approach.* Earthscan, London.

Hamilton, A.C. and Macfadyen, A. (1989) Climatic change on the East Usambaras: evidence from meteorological records. In: Hamilton, A.C. and Bensted-Smith, R. (eds) *Forest Conservation in the East Usambara Mountains, Tanzania.* IUCN, Gland, Switzerland and Cambridge, pp. 103–108.

Hamilton, A.C. and Perrott, R.A. (1979) Aspects of the glaciation of Mt Elgon, East Africa. *Palaeoecology of Africa* 11, 153–161.

Hamilton, A.C. and Perrott, R.A. (1980) Modern pollen deposition on a tropical African mountain. *Pollen Spores* 22, 437–468.

Hamilton, A.C. and Perrott, R.A. (1981) A study of altitudinal zonation in the Montane Forest Belt of Mt. Elgon, East Africa. *Vegetatio* 45, 107–125.

Hamilton, A.C. and Taylor, D. (1988) *Assessment of the Environmental Impacts of Proposed Forest-Based Developments in Guerrero and Oaxaca, Mexico.* Secretaría de Desarrollo Urbano y Ecología, Mexico City.

Hamilton, A.C. and Taylor, D. (1991) History of climate and forests in tropical Africa during the last 8 million years. *Climatic Change* 19, 65–78.

Hamilton, A.C. and Taylor, D.M. (2018) Palynological evidence for abrupt climatic cooling in equatorial Africa at about 43,000–40,000 cal BP. *Review of Palaeobotany and Palynology* 250, 53–59.

Hamilton, A.C., Baranga, J., *et al.* (1990) *Proposed Bwindi (Impenetrable) National Park: Results of a Public Inquiry and Recommendations for Establishment.* Uganda National Parks, Kampala.

Hamilton, A.C., Cunningham, A., *et al.* (2000) Conservation in a region of political instability: Bwindi Impenetrable Forest, Uganda. *Conservation Biology* 14, 1722–1725.

Hamilton, A.C., Dürbeck, K., *et al.* (2006) Towards a sustainable herbal harvest. *Plant Talk* 43, 32–35.

Hamilton, A.C., Karamura, D., *et al.* (2016) History and conservation of wild and cultivated plant diversity in Uganda: forest species and banana varieties as case studies. *Plant Diversity* 38, 23–44.

Hamilton, A.C., Pei Shengji, *et al.* (2003) *The Purposes and Teaching of Applied Ethnobotany.* People and Plants Working Paper No. 11. WWF, Godalming, UK.

Hamilton, A.C., Pei Shengji, *et al.* (2017) Botanical aspects of eco-civilisation construction. *Plant* Diversity 39, 65–72.

Hamilton, A.C., Taylor, D., *et al.* (2001) Hotspots in African forests as Quaternary refugia. In: Weber, W., White, L.J.T., *et al.* (eds) *African Rain Forest Ecology and Conservation: An Interdisciplinary Perspective.* Yale University Press, New Haven, Connecticut and London, pp. 57–67.

Hamilton, S., Brown, P., *et al.* (1972) Anaesthesia by acupuncture. *British Medical Journal* 3(5822), 352.

Hannah, L. and Hansen, L. (2005) Designing landscapes and seascapes for change. In: Lovejoy, T.E. and Hannah, L. (eds) *Climate Change and Biodiversity.* Yale University Press, New Haven, Connecticut and London, pp. 329–341.

Hansen, H.B. (1984) *Mission, Church and State in a Colonial Setting: Uganda 1890–1925.* Heinemann, London.

Harari, Y.N. (2011) *Sapiens: A Brief History of Humankind.* Penguin Random House, London.

Hardin, G. (1968) The tragedy of the commons. *Science* 162, 1243–1248.

Harper, D.J. (1998) *Mawangdui yi shu yi zhu (The Mawangdui Medical Manuscripts).* Kegan Paul International, London.

Harper, K. (2017) *The Fate of Rome: Climate, Disease, and the End of Empire*. Princeton University Press, Princeton, New Jersey.

Harrison, M., Ssabaganzi, R., *et al.* (2004) Reform of forestry advisory services: learning from practice in Uganda. *Natural Resource Perspectives* (93, August). Available at: https://media.odi.org/documents/2664.pdf (accessed 20 May 2024).

Harrison, S.S. (2008) '*Pashtunistan*': The Challenge to Pakistan and Afghanistan. Area: Security and Defence – ARI 37/2008. Real Instituto Elcano, Madrid.

Harrower, M.J., McCorriston, J., *et al.* (2010) General/specific, local/global: comparing the beginnings of agriculture in the Horn of Africa (Ethiopia/Eritrea) and Southwest Arabia (Yemen). *American Antiquity* 75(3), 452–472.

Hart, T.B., Hart, J.A., *et al.* (1996) Changes in forest composition over the last 4000 years in the Ituri basin, Zaire. In: van der Maesen, L.J.G., van der Burgt, X.M., *et al.* (eds) *The Biodiversity of African Plants*. Kluwer Academic Publishers, Dordrecht, The Netherlands, pp. 541–563.

Harvey, D. (2005) *A Brief History of Neoliberalism*. Oxford University Press, Oxford.

Harvey, J.A., Tougeron, K., *et al.* (2022) Scientists' warning on climate change and insects. *Ecological Monographs* 93, e1553. DOI: 10.1002/ecm.1553.

Haslett, J.R., Berry, P.M., *et al.* (2010) Changing conservation strategies in Europe: a framework integrating ecosystem services and dynamics. *Biodiversity and Conservation* 9, 2963–2977.

Hatfield, G. (1999) *Memory, Wisdom and Healing: The History of Domestic Plant Medicine*. Sutton Publishing Ltd, Stroud, UK.

Hatzisavvidou, S. (2020) Inventing the environmental state: neoliberal common sense and the limits to transformation. *Environmental Politics* 29(1), 96–114.

Hawkes, J.G., Maxted, N., *et al.* (2001) *The Ex Situ Conservation of Plant Genetic Resources*. Kluwer Academic Publishers, London.

He Fangliang (2009) Price of prosperity: economic development and biological conservation in China. *Journal of Applied Ecology* 46, 511–515.

Heckenberger, M.J., Russell, J.C., *et al.* (2008) Pre-Columbian urbanism, anthropogenic landscapes, and the future of the Amazon. *Science* 321, 1214–1217.

Hedberg, O. (1951) Vegetation belts of the East African mountains. *Svensk botanisk tidskrift* 45, 140–202.

Henn, B.M., Gignoux, C., *et al.* (2008) Y-chromosomal evidence of a pastoralist migration through Tanzania to southern Africa. *Proceedings of the National Academy of Sciences USA* 105, 10693–10698.

Henry, A. (1893) *Notes on Economic Botany of China*. Presbyterian Mission Press, Shanghai, China.

Hershey, D.R. (1993) Plant neglect in biology education. *BioScience* 43, 418.

Heugh, K. (2018) Linguistic citizenship: who decides whose languages, ideologies and vocabulary matter? In: Lim, L., Stroud, C., *et al.* (eds) *The Multilingual Citizen: Towards a Politics of Language for Agency and Change*. Multilingual Matters, Bristol, UK, pp. 174–192.

Heuler, H. (2013) British university drops Uganda ties in dispute over gay rights. *Voice of America*, 11 January. Available at: https://www.voanews.com/a/ugana-gay-rights-university/1582067.html (accessed 20 May 2024).

Hewitt, G.M. (2000) The genetic legacy of the Quaternary ice ages. *Nature* 405, 907–913.

Heyerdahl, T. (1961) *The Kon-Tiki Expedition: By Raft Across the South Seas*. Rand McNally, Chicago, Illinois.

Hickman, J. (2018) The political economy of a planetary sunshade. *Astropolitics* 16, 49–58.

Hill, M.E. (1961) *Permanent Way: The Story of the Kenya and Uganda Railway*. East African Literature Bureau, Nairobi.

Hinsley, A., Milner-Gulland, E.J., *et al.* (2020) Building sustainability into the Belt and Road Initiative's Traditional Chinese Medicine trade. *Nature Sustainability* 3, 96–100.

Ho, D. Y.-F. and Chiu, C.-Y. (1994) Component ideas of individualism, collectivism, and social organization: an application in the study of Chinese culture. In: Kim, U., Triandis, H.C., et al. (eds) Individualism and Collectivism: Theory, Method, and Applications. Sage Publications, Thousand Oaks, California, pp. 137–156.

Hoffecker, J.F., Holliday, V.T., et al. (2008) From the Bay of Naples to the River Don: the Campanian Ignimbrite eruption and the Middle to Upper Paleolithic transition in Eastern Europe. Journal of Human Evolution 55, 858–870.

Hofstadter, R. (1944) Social Darwinism in American Thought, 1860–1915. University of Pennsylvania Press, Philadelphia, Pennsylvania.

Hofstede, G. (1980) Culture's Consequences: International Differences in Work-Related Values. Sage Publications, Beverly Hills, California.

Hogan, K.A., Larter, R.D., et al. (2020) Revealing the former bed of Thwaites Glacier using sea-floor bathymetry: implications for warm-water routing and bed controls on ice flow and buttressing. The Cryosphere 14, 2883–2908.

Hollman, A. (1992) Plants in Cardiology. British Medical Journal, London.

Hoszafi, S. (2001) The history of heroin. Acta Pharmaceutica Hungarica 71, 233–242.

Howard, P.C. (1991) Nature Conservation in Uganda's Tropical Forest Reserves. IUCN, Gland, Switzerland.

Hsu, E. (2008) The history of Chinese Medicine in the People's Republic of China and its globalization. East Asian Science, Technology and Society 2(4), 465–484.

Hu Guanjing, Grover, C.E., et al. (2021) Evolution and diversity of the cotton genome. In: Rahman, Mu., Zafar, Y. et al. (eds) Cotton Precision Breeding. Springer, Cham, Switzerland, pp. 25–78.

Hu Huabin, Liu Wenjun, et al. (2008) Impact of land use and land cover changes on ecosystem services in Menglun, Xishuangbanna, Southwest China. Environmental Monitoring and Assessment 146, 147–156.

Huai Huyin and Hamilton, A.C. (2009) Characteristics and functions of traditional homegardens: a review. Frontiers of Biology in China 4, 151–157.

Huai Huyin, Xu Wei, et al. (2011) Comparison of the homegardens of eight cultural groups in Jinping County, southwest China. Economic Botany 65, 345–355. DOI: 10.1007/s12231-011-9172-1.

Hubau, W., Lewis, S.L., et al. (2020) Asynchronous carbon sink saturation in African and Amazonian tropical forests. Nature 579, 80–87.

Hublin, J., Ben-Ncer, A., et al. (2017) New fossils from Jebel Irhoud, Morocco and the pan-African origin of Homo sapiens. Nature 546, 289–292.

Huffman, M.A. and Wrangham, R.W. (1994) Diversity of medicinal plant use by chimpanzees in the wild. In: Wrangham, R.W., McGrew, W.C., et al. (eds) Chimpanzee Cultures. Harvard University Press, Cambridge, Massachusetts, pp. 129–146.

Humphreys, A.M., Govaerts, R., et al. (2019) Global dataset shows geography and life form predict modern plant extinction and rediscovery. Nature Ecology & Evolution 3, 1043–1047.

Humphris, J. and Scheibner, T. (2017) A new radiocarbon chronology for ancient iron production in the Meroe Region of Sudan. African Archaeological Review 34, 377–413.

Hunt, P. (2011) Late Roman silk: espionage and smuggling in the 6th century ce. Philolog, Stanford University, Stanford, California. Available at: https://archive.is/20130626180730/http://traumwerk.stanford.edu/philolog/2011/08/byzantine_silk_smuggling_and_e.html (accessed 20 May 2024).

Iles, L. (2009) Impressions of banana pseudostem in iron slag from eastern Africa. Ethnobotany Research and Applications 7, 283–291.

Iliffe, J. (2017) Africans: The History of a Continent. Cambridge University Press, Cambridge.

Imperato, P.J. (2005) Lords of the fly: sleeping sickness control in British East Africa, 1900–1960 (review). African Studies Review 48, 166–169.

Ioannidis, A.G., Blanco-Portillo, J., *et al.* (2020) Native American gene flow into Polynesia pre-dating Easter Island settlement. *Nature* 583, 572–577.

IPCC (2007) *Climate Change 2007: The Physical Science Basis. Contribution of Working Group I to the Fourth Assessment Report of the Intergovernmental Panel on Climate Change*. Cambridge University Press, Cambridge and New York.

IPCC (2014) *Climate Change 2014: Impacts, Adaptation and Vulnerability*. Cambridge University Press, Cambridge and New York.

IPCC (2021) *Climate Change 2021: The Physical Science Basis*. Cambridge University Press, Cambridge.

ISE (2006) The ISE Code of Ethics. International Society of Ethnobiology, Gainesville, Florida. Available at: https://www.ethnobiology.net/what-we-do/core-programs/ise-ethics-program/code-of-ethics/ (accessed 20 May 2024).

Islam, M.N. (2012) Repacking Ayurveda in post-colonial India: revival or dilution? *Journal of South Asian Studies* 35, 503–519.

IUCN (2012) *IUCN Red List Categories and Criteria, Version 3.1, Second Edition*. IUCN, Gland, Switzerland and Cambridge.

IUCN (2021a) *Forests and Climate Change*. IUCN, Gland, Switzerland.

IUCN (2021b) *Peatlands and Climate Change*. IUCN, Gland, Switzerland.

IUCN/WCMC (1994) *World Heritage Nomination: Bwindi Impenetrable National Park (Uganda)*. IUCN and World Conservation Monitoring Centre, Cambridge.

Iversen, S.T. (1991) *The Usambara Mountains, NE Tanzania: History, Vegetation and Conservation*. Uppsala Universitet, Uppsala, Sweden.

Jablinski, D. (2001) Lessons from the past: evolutionary impacts of mass extinctions. *Proceedings of the National Academy of Sciences* USA 98, 5393–5398.

Jafri, S.H.A. and Margolis, L.S. (1999) The treatment of usury in the holy scriptures. *Thunderbird International Business Review* 41, 371–379.

Jain, L.C. (1929) *Indigenous Banking in India*. Macmillan and Co., Ltd, London.

Jalal, M.A., Bhardhan, K.D.B., *et al.* (2019) Overseas doctors of the NHS: migration, transition, challenges and towards resolution. *Future Healthcare Journal* 6, 76–81.

Jansen, E.G. (1997) *Rich Fisheries – Poor Fisherfolk: Some Preliminary Observations about the Effects of Trade and Aid in the Lake Victoria Fisheries*. Blue Series No. 47. Regional Office for Eastern Africa, IUCN, Nairobi.

Jaspers, K. (1953) *The Origin and Goal of History*. Yale University Press, New Haven, Connecticut.

Jayawardhena, C., Morrell, K., *et al.* (2016) Ethical consumption behaviours in supermarket shoppers: determinants and marketing implications. *Journal of Marketing Management* 32, 777–805. DOI: 10.1080/0267257X.2015.1134627.

Jelen, T.G. and Lockett, L.A. (2014) Religion, partisanship, and attitudes towards science policy. *SAGE Open* 4(1). DOI: 10.1177/2158244013518932.

Jiang Dechun, Klaus, S., *et al.* (2019) Asymmetric biotic interchange across the Bering land bridge between Eurasia and North America. *National Science Review* 6, 739–745.

Jiang Xinyan (2013) Chinese dialectical thinking: the Yin Yang model. *Philosophy Compass* 8, 438–446.

Johnson, D.V. (ed.) (1991) *Palms for Human Needs in Asia*. A.A. Balkema, Rotterdam, The Netherlands.

Johnson, E.M. (2015) Demographics, inequality and entitlements in the Russian Famine of 1891. *Slavonic and East European Review* 93(1), 96–119.

Johnston, L. (1996) *Ideologies*. Broadview Press, Ontario, Canada.

Josupeit, H. (2006) *The Market for Nile Perch*. FAO Globefish Research Programme, Vol. 84. Food and Agriculture Organization of the United Nations, Rome.

Kabogozza, J. (2011) *Forest Plantations and Woodlots in Uganda*. African Forest Forum, Nairobi.

Kachwano, A. (2022) The rationale of clinical medicine and community health professionals in the health sector: Medical Clinical Officers' serve the country. *Social Innovations Journal* 14(4). Available at: https://socialinnovationsjournal.com/index.php/sij/article/view/2026 (accessed 20 May 2024).

Kagoro-Rugunda, G. (2019) Antibacterial activity of plant parts selectively consumed by chimpanzees. *Academia Journal of Medicinal Plants* 7, 243–251.

Kalema, J. and Beentje, H. (2012) *Conservation Checklist of the Trees of Uganda*. Royal Botanic Gardens, Kew, UK.

Kalema, J. and Hamilton, A. (2020) *Field Guide to the Forest Trees of Uganda: For Identification and Conservation*. CAB International, Wallingford, UK.

Kamatenesi, M.M., Hoft, M., *et al.* (2014) Sustainable harvesting of medicinal barks (Rytiginia, Rubiaceae) in multiple-use zones around Bwindi Impenetrable National Park, Uganda. *Advances in Economic Botany* 17, 211–225.

Karamura, D. and Mgenzi, B. (2004) On farm conservation of Musa diversity in the Great Lakes region of East Africa. *African Crop Science Journal* 12, 75–83.

Karamura, D., Mgenzi, B., *et al.* (2004) Exploiting indigenous knowledge for the management and maintenance of Musa biodiversity on farm. *African Crop Science Journal* 12, 67–74.

Karim, S.A. (2010) *The Islamic Moral Economy: A Study of Islamic Money and Financial Instruments*. Brown Walker Press, Boca Raton, Florida.

Kasujja, J.P. (2023) Influence of religion on the political parties' affiliations and elections in Uganda: the case of Iganga District. *International Journal of History and Philosophical Research* 11(2), 28–41.

Kealy, S., Louys, J., *et al.* (2018) Least-cost pathway models indicate northern human dispersal from Sunda to Sahul. *Journal of Human Evolution* 125, 59–70.

Keating, D. (2018) Almost all British train lines are now owned by other EU countries. *Forbes*, 15 August. https://www.forbes.com/sites/davekeating/2019/08/15/almost-all-british-train-lines-are-now-owned-by-other-eu-countries/ (accessed 20 May 2024).

Keefe, P.R. (2017) The family that built an empire of pain: the Sackler dynasty's ruthless marketing of painkillers has generated billions of dollars – and millions of addicts. *The New Yorker*, 23 October. Available at: https://www.newyorker.com/magazine/2017/10/30/the-family-that-built-an-empire-of-pain (accessed 20 May 2024).

Kelbessa, W. (2017) Religious pluralism, tolerance, and public culture in Africa. In: Hogan, J.P. and Akhlaq, S.H. (eds) *The Secular and the Sacred: Complementary and/or Conflictual?* Cultural Heritage and Contemporary Change Series VII, Seminars: Culture and Values, Vol. 35. The Council for Research in Values and Philosophy, Washington, DC, pp. 513–536.

Kendall, R.L. (1969) An ecological history of the Lake Victoria basin. *Ecological Monographs* 39, 121–176.

Keohane, R.O. and Nye, J.S. (1977) *Power and Interdependence*. Little Brown and Company, Boston, Massachusetts.

Kessy, J.F. (1998) Conservation and utilization of natural resources in the East Usambara Forest Reserves: conventional views and local perspectives. PhD thesis, Wageningen Agricultural University, Wageningen, The Netherlands.

Kew (1889) Lists of the staffs of the Royal Gardens, Kew, and of botanical departments and establishments at home, and in India, and in the colonies, in correspondence with Kew. *Bulletin of Miscellaneous Information (Royal Botanic Gardens, Kew)* 1889(29), 122–126.

Khan, M. (2023) Shadow banks should come out into the open or be allowed to go bust. *The Times*, 18 April. Available at: https://www.thetimes.co.uk/article/shadow-banks-should-come-out-into-the-open-or-be-allowed-to-go-bust-bhgvbhfpd (accessed 20 May 2024).

Khan, U., Janjuah, H.T., *et al.* (2021) Natural processes and anthropogenic activity in the Indus River sedimentary environment in Pakistan: a critical review. *Journal of Marine Science and Engineering* 9, 1109.

Killick, D. (2015) Invention and innovation in African iron-smelting technologies. *Cambridge Archaeological Journal* 25, 307–319.

Kim, U., Triandis, H., *et al.* (eds) (1994) *Individualism and Collectivism: Theory, Method, and Applications*. Sage Publications, Thousand Oaks, California.

Kingdon, J. (1990) *Island Africa: The Evolution of Africa's Rare Animals and Plants*. William Collins Sons & Co., London.

Kinloch, B.G. (1957) Tiger fish. *Uganda Wild Life and Sport* 1(2), 16–19.

Knechtges, D.R. (1997) Gradually entering the world of delight: food and drink in early medieval China. *Journal of the American Oriental Society* 117, 229–239.

Kneese, A.V. (1988) The economics of natural resources. *Population and Development Review* 14, 281–309.

Koch, A., Brierley, C., *et al.* (2019) Earth system impacts of the European arrival and great dying in the Americas after 1492. *Quaternary Science Reviews* 207, 13–36.

Kocka, J. (2016) *Capitalism: A Short History*. Princeton University Press, Woodstock, UK.

Koegler, C., Malreddy, P.K., *et al.* (2020) The colonial remains of Brexit: empire nostalgia and narcissistic nationalism. *Journal of Postcolonial Writing* 56(5). 585–592. DOI: 10.1080/17449855.2020.1818440.

Koirala, P. (2017) Yarsagumba fungus: health problems in the Himalayan Gold Rush. *Wildnerness and Environmental Medicine* 28, 267–270.

Komakech, R., Kang, Y., *et al.* (2017) A review of the potential of phytochemicals from *Prunus africana* (Hook f.) Kalkman stem bark for chemoprevention and chemotherapy of prostate cancer. *Evidence-Based Complementary and Alternative Medicine* 2017, 3014019. DOI: 10.1155/2017/3014019.

Koshimizu, K., Ohigashi, H., *et al.* (1994) Use of *Vernonia amygdalina* by wild chimpanzee: possible roles of its bitter and related constituents. *Physiology & Behavior* 56, 1209–1216.

Kosterin, O. (2014) The lost ancestor of the broad bean (*Vicia faba* L.) and the origin of plant cultivation in the Near East. *Vavilov Journal of Genetics and Breeding* 18(4), 831–840.

Kotz, D.M. (2000) Globalization and neoliberalism. *Rethinking Marxism* 12(2), 64–79.

Kourkouta, L., Tsaloglidou, A., *et al.* (2018) History of antibiotics. *Sumerianz Journal of Medical and Healthcare* 1, 51–54.

Kraussmann, F. (2013) Global human appropriation of net primary production doubled in the 20th century. *Proceedings of the National Academy of Sciences* USA 110, 10324–10329.

Kretz, A.J. (2013) From 'kill the gays' to 'kill the gay rights movement': the future of homosexuality legislation in Africa. *Northwestern Journal of Human Rights* 11, 207.

Kuhn, T.N. (1962) *The Structure of Scientific Revolutions*. University of Chicago Press, Chicago, Illinois.

Kupchan, S.M., Komoda, Y., *et al.* (1972a) Maytanprine and maytanbutine, new antileukaernic ansa macrolides from *Maytenus buchananii*. *Journal of the Chemical Society, Chemical Communications* (19), 1065.

Kupchan, S.M., Komofa, Y., *et al.* (1972b) Tumor inhibitors. LXXIII. Maytansine, a novel antileukemic ansa macrolide from *Maytenus ovatus*. *Journal of the American Chemical Society* 94, 1354–1356.

Kyebogola, S., Burras, L.C., *et al.* (2020) Comparing Uganda's indigenous soil classification system with World Reference Base and USDA Soil Taxonomy to predict productivity. *Geoderma Regional* 22, e00296. DOI: 10.1016/j.geodrs.2020.e00296.

Laidlawa, K., Wang, D., *et al.* (2010) Attitudes to ageing and expectations for filial piety across Chinese and British cultures: a pilot exploratory evaluation. *Aging and Mental Health* 14(3), 283–292.

Laird, S.A. (ed.) (2002) *Biodiversity and Traditional Knowledge: Equitable Partnerships in Practice*. Earthscan, London.

Laird, S.A., Wynberg, R., *et al.* (2020) Rethink the expansion of access and benefit sharing. *Science* 367, 1200–1202.

Lama, Y.C., Ghimire, S.K., *et al.* (2001) *Medicinal Plants of Dolpo: Amchis' Knowledge and Conservation*. WWF Nepal Program, Kathmandu.

Langdale-Brown, I., Osmaston, H.A., *et al.* (1964) *The Vegetation of Uganda (Excluding Karamoja) and Its Bearing on Land Use*. Government Printer, Entebbe, Uganda.

Lange, D. (1998) *Europe's Medicinal and Aromatic Plants: Their Use, Trade and Conservation: An Overview*. TRAFFIC International, Cambridge.

Lange, D. (2000) The role of Europe and Germany within the worldwide trade in medicinal and aromatic plants. Presented at *Medicinal Utilization of Wild Species: Challenge for Man and Nature in the New Millennium, EXPO 2000*, Hanover, Germany, 1 June–31 October 2000. WWF-Germany/TRAFFIC Europe-Germany.

Lapper, R. (2021) *Beef, Bible and Bullets*. Manchester University Press, Manchester, UK.

Lauterbach, K. (2020) Fakery and wealth in African charismatic Christianity: moving beyond the prosperity gospel as Script. In: Lauterbach, K. and Vähäkangas, M. (eds) *Faith in African Lived Christianity*. Global Pentecostal and Charismatic Studies No. 35. Brill, Leiden, The Netherlands and Boston, Massachusetts, pp. 111–132.

Laval, G., Peyrégne, S., *et al.* (2019) Recent adaptive acquisition by African rainforest hunter-gatherers of the Late Pleistocene sickle-cell mutation suggests past differences in malaria exposure. *American Journal of Human Genetics* 104, 553–561.

Leacock, L. and Lee, R. (eds) (1982) *Politics and History in Band Societies*. Cambridge University Press, Cambridge.

Leclerc, G.-L. (1749–1804) *Histoire Naturelle*. Imprimerie Royale, Paris.

Lee, C. and Schaaf, T. (eds) (2003) *The Importance of Sacred Natural Sites for Biodiversity Conservation. Proceedings of the International Workshop held in Kunming and Xishuangbanna Biosphere Reserve, People's Republic of China, 17–20 February 2003*. UNESCO, Paris.

Lejju, B.J., Taylor, D., *et al.* (2005) Late-Holocene environmental variability at Munsa archaeological site, Uganda: a multicore, multiproxy approach. *The Holocene* 15, 1044–1061.

Lejju, B.J., Taylor, D., *et al.* (2006) Africa's earliest bananas? *Journal of Archaeological Science* 33, 102–113.

Leonard, M. (1997) *Britain™: Renewing Our Identity*. Demos, London.

Leopold, M. (2021) *Idi Amin: The Story of Africa's Icon of Evil*. Yale University Press, New Haven, Connecticut.

Lester, S., McLeod, K., *et al.* (2010) Science in support of ecosystem-based management for the US West Coast and beyond. *Biological Conservation* 143(3), 576–587.

Lewington, A. (1990) *Plants for People*. The Natural History Museum, London.

Lewis, J.B. (2022) Money finds a way: increasing AML regulation garners diminishing returns and increases demand for dark financing. *Vanderbilt Journal of Transnational Law* 55(2), 529–557.

Lewis, S.L., Sonké, B., *et al.* (2009) Increasing carbon storage in intact African tropical forests. *Nature* 457, 1003–1008.

Li Chunhai, Zheng, Y., *et al.* (2012) Understanding the ecological background of rice agriculture on the Ningshao Plain during the Neolithic Age: pollen evidence from a buried paddy field at the Tianluoshan cultural site. *Quaternary Science News* 35, 131–138.

Li, H., Guo, W., *et al.* (2023) The delineation and ecological connectivity of the Three Parallel Rivers Natural World Heritage Site. *Biology* 12, 3. DOI: 10.3390/biology12010003.

Li Hui-Lin (1979) *Nan-fang ts'ao-mu chuang: A Fourth Century Flora of Southeast Asia*. The Chinese University Press, Hong Kong, China.

Li Rong, Dao Zhiling, *et al.* (2011) Seed plant species diversity and conservation in the northern Gaoligong Mountains in Western Yunnan, China. *Mountain Research and Development* 31, 160–165.

Li Wenhua (ed.) (2016) *Contemporary Ecology Research in China*. Springer, New York.

Liu Li, Field, J., *et al.* (2010) What did grinding stones grind? New light on Early Neolithic subsistence economy in the Middle Yellow River Valley, China. *Antiquity* 84, 816–833.

Liu Li, Wang Jiang, *et al.* (2018) Fermented beverage and food storage in 13,000 y-old stone mortars at Raqefet Cave, Israel: investigating Natufian ritual feasting. *Journal of Archaeological Science: Reports* 21, 783–793.

Liu Li, Wang Jiang, *et al.* (2019) The origins of specialized pottery and diverse alcohol fermentation techniques in Early Neolithic China. *Proceedings of the National Academy of Sciences USA* 116, 12767–12774.

Liu Hongmao, Xu Zaifu, *et al.* (2002) Practice of conservating plant diversity through traditional beliefs: a case study in Xishuangbanna, southwest China. *Biodiversity and Conservation* 11, 705–713.

Liu Liang (2014) Biographical sketch of German forester Fenzel in China. *Journal of Beijing Forestry University (Social Sciences)* 13(3), 12–16.

Liu Weimin, Li Yingying, *et al.* (2010) Origin of the human malaria parasite *Plasmodium falciparum* in gorillas. *Nature* 467, 420–425.

Liu Wenjie, Liu Wenyao, *et al.* (2011) Runoff generation in small catchments under a native rain forest and a rubber plantation in Xishuangbanna, southwestern China. *Water and Environmental Journal* 25, 138–147.

Livingstone, D.A. (1967) Postglacial vegetation of the Ruwenzori Mountains in Equatorial Africa. *Ecological Monographs* 37, 25–52.

Lock, R. (1994) Biology – the study of living things? *Journal of Biological Education* 28(2), 79–80.

Lock, R. (1996) The future of biology beyond the compulsory schooling age or whither post-16 biology? *Journal of Biological Education* 30(1), 3–6.

Lockwood Consultants Ltd (1973) *Forest Resource Development Study, Republic of Uganda.* Canadian International Development Agency, Gatineau, Quebec, Canada.

Lodhi, A. (2007) Conservation of leopards in Ayubia National Park, Pakistan. MSc thesis, University of Montana, Missoula, Montana.

Londono, S.C., Garzon, C., *et al.* (2016) Ethnogeology in Amazonia: surface-water systems in the Colombian Amazon, from perspectives of Uitoto traditional knowledge and mainstream hydrology. In: Wessel, G.R. and Greenberg, J.K. (eds) *Geoscience for the Public Good and Global Development: Toward a Sustainable Future.* Special Paper 520. The Geological Society of America, Boulder, Colorado, pp. 221–232.

Lou Yulie (2017) *The Fundamental Spirit of Chinese Culture.* Zhonghua Book Company, Beijing.

Lovett, J.C. (1996) Elevational and latitudinal changes in tree associations and diversity in the Eastern Arc mountains of Tanzania. *Journal of Tropical Ecology* 12(5), 629–650.

Lovett, J.C. and Wasser, S.K. (eds) (1993) *Biogeography and Ecology of the Rain Forests of Eastern Africa.* Cambridge University Press, Cambridge.

Low, L., Lynam, J., *et al.* (2009) Sweetpotato in sub-Saharan Africa. In: Loebenstein, G. and Thottappilly, G. (eds) *The Sweetpotato.* Springer, Dordrecht, The Netherlands, pp. 359–390.

Lu Houyuan, Zhang Jianping, *et al.* (2016) Earliest tea as evidence for one branch of the Silk Road across the Tibetan Plateau. *Scientific Reports* 6, 18955.

Lucas, G. and Synge, H. (1978) *The IUCN Plant Red Data Book.* IUCN, Morges, Switzerland.

Lui, A. and Lamb, G.W. (2018) Artificial intelligence and augmented intelligence collaboration: regaining trust and confidence in the financial sector. *Information and Communications Technology Law* 27, 267–283.

Lule, J. (2006) *The Hidden Wisdom of the Baganda.* Humbolt and Hartmann, Arlington, Virginia.

Lunyiigo, S.L. (2011a) *Mwanga II: Resistance to Imposition of British Colonial Rule in Buganda 1884–1899.* Wavah Books Ltd, Kampala.

Lunyiigo, S.L. (2011b) *The Struggle for Land in Buganda 1888–2005*. Wavah Books Ltd, Kampala.

Lupo, K.D. (2011) A dog is for hunting. In: Albarella, U. and Trentacoste, A. (eds) *Ethnoarchaeology: The Present and Past of Human–Animal Relationships*. Oxbow Books, Oxford, pp. 4–12.

Lyell, C. (1830–1833) *Principles of Geology*. John Murray, London.

Lyons, K., Richards, C., *et al.* (2014) *The Darker Side of Green: Plantation Forestry and Carbon Violence in Uganda*. The Oakland Institute, Oakland, California.

Ma Chang-Le, Moseley, R.K., *et al.* (2007) Plant diversity and priority conservation areas of Northwestern Yunnan, China. *Biodiversity and Conservation* 16, 757–774.

Ma Xiao-jun, Zhang Li-xia, *et al.* (2017) *Flora of China Dai Medicine*. People's Health Press, Beijing.

MacDonald, K.I. (2010) The devil is in the (bio)diversity: private sector engagement and the restructuring of biodiversity conservation. *Antipode* 42, 513–550.

Mackay, C. (1841) *Memoirs of Extraordinary Popular Delusions*. Richard Bentley, London.

Madsen, R. (2011) Religious renaissance in China today. *Journal of Current Chinese Affairs* 40, 17–42.

Maffi, L. (2007) Biocultural diversity and sustainability. In: Pretty, J. (ed.) *The SAGE Handbook of Environment and Society*. Sage Publications, Thousand Oaks, California, pp. 267–277.

Mair, V.H. and Hoh, E. (2009) *The True Story of Tea*. Thompson, London.

Malthus, T.R. (1798) *An Essay on the Principle of Population*. J. Johnson, London.

Manguin, P.-Y. (2016) Austronesian shipping in the Indian Ocean: from outrigger boats to trading ships. In: Campbell, G. (ed.) *Early Exchange between Africa and the Wider Indian Ocean World*. Palgrave Macmillan, Cham, Switzerland, pp. 51–76.

Mann, N. (2018) A brief history of meat in the human diet and current health implications. *Meat Science* 144, 169–179.

Manning, A.D., Gibbons, P., *et al.* (2009) Scattered trees: a complementary strategy for facilitating adaptive responses to climate change in modified landscapes? *Journal of Applied Ecology* 46, 915–919.

Manschadi, A.M., Oberkircher, L., *et al.* (2010) 'White gold' and Aral Sea disaster: towards more efficient use of water resources in the Khorezm region, Uzbekistan. *Lohmann Information* 45(1), 34–47.

Marchant, R. and Taylor, D. (1998) Dynamics of montane forest in central Africa during the late Holocene: a pollen-based record from western Uganda. *The Holocene* 8, 375–381.

Marchant, R., Mumbi, C., *et al.* (2006) The Indian Ocean dipole – the unsung driver of climatic variability in East Africa. *African Journal of Ecology* 45, 4–16.

Marchant, R., Taylor, D., *et al.* (1997) Late Pleistocene and Holocene history at Mubwindi Swamp, southwest Uganda. *Quaternary Research* 47, 316–328.

Margolis, M.F. (2020) Who wants to make America great again? Understanding evangelical support for Donald Trump. *Religion and Politics* 13, 89–118.

Marino, R. and Gonzales-Portillo, M. (2000) Preconquest Peruvian neurosurgeons: a study of Inca and Pre-Columbian trephination and the art of medicine in ancient Peru. *Neurosurgery* 47, 940–950.

Martin, E.N. (2021) Can public service broadcasting survive Silicon Valley? Synthesizing leadership perspectives at the BBC, PBS, NPR, CPB and local U.S. stations. *Technology in Society* 64, 101451. DOI: 10.1016/j.techsoc.2020.101451.

Martin, G. (1994) *Ethnobotany: A Methods Manual*. Earthscan, London.

Martin, G., Agama, A.L., *et al.* (2002) *Projek Etnobotani Kinabalu: The Making of a Dusun Ethnoflora (Sabah, Malaysia)*. People and Plants Working Paper No. 9. Division of Ecological Sciences, UNESCO, Paris.

Masselos, J. (ed.) (2010) *The Great Empires of Asia*. Thames and Hudson, London.

Matthews, T., Danese, A., *et al.* (2019) Lonely young adults in modern Britain: findings from an epidemiological cohort study. *Psychological Medicine* 49(2), 268–277.

Maundu, P., Johns, T., *et al.* (2002) African leafy vegetables: diversity and use in Sub-Saharan Africa. Presented at *The Sixth Congress of the International Society of Ethnobiology*, Addis Ababa, Ethiopia, 16–20 September 2002.

Mayr, E. (2001) The philosophical foundations of Darwinism. *Proceedings of the American Philosophical Society* 145(4), 488–495.

Mbiti, J. (1969) *African Religions and Philosophy*. Heinemann, London.

Mbiti, J. (1970) *Concepts of God in Africa*. Society for Promoting Christian Knowledge, London.

McAdams, D.P. (1996) Personality, modernity, and the storied self: a contemporary framework for studying persons. *Psychological Inquiry* 7(4), 295–321.

McCoy, M. (1973) A renaissance in Carolinian-Marianas voyaging. *Journal of the Polynesian Society* 82(1), 355–365.

McCulloch, J.S.G. and Robinson, M. (1993) History of forest hydrology. *Journal of Hydrology* 150, 189–216.

McDonald, J. and Veth, P. (2008) Rock-art of the Western Desert and Pilbara: pigment dates provide new perspectives on the role of art in the Australian arid zone. *Australian Aboriginal Studies* 1, 4–21.

McDonald, J. and Veth, P. (2013) Rock art in arid landscapes: Pilbara and Western Desert petroglyphs. *Australian Archaeology* 77, 66–81.

McGlynn, G., Mooney, S., *et al.* (2013) Palaeoecological evidence for Holocene environmental change from the Virunga volcanoes in the Albertine Rift, central Africa. *Quaternary Science Reviews* 61, 32–46.

McGuire, W.P., Rowinsky, E.K., *et al.* (1989) Taxol: a unique antineoplastic agent with significant activity in advanced ovarian epithelial neoplasms. *Annals of Internal Medicine* 111(4), 273–279.

McKenzie, R. and Atkinson, R. (2020) Anchoring capital in place: the grounded impact of international wealth chains on housing markets in London. *Urban Studies* 57(1), 21–38.

McMartin, A. (1961) Sugarcane in Central and East Africa: some observations on its history and present position. *Proceedings of the South African Sugar Technologists' Association* (April), 104–109.

McMichael, C.H., Piperno, D.R., *et al.* (2015) Phytolith assemblages along a gradient of ancient human disturbance in western Amazonia. *Frontiers in Ecology and Evolution* 3, 141.

McNeill, W.H. (1976) *Plagues and People*. Blackwell, Oxford.

McShane, T.O. (1999) *Voyages of Discovery: Four Lessons from the DGIS-WWF Tropical Forest Portfolio*. WWF, Gland, Switzerland.

Medvedev, Z. (1977) Soviet genetics: new controversy. *Nature* 268, 285–287.

Meeks, M.D. (2011) The peril of usury in the Christian tradition. *Journal of Bible and Theology* 65, 128–40.

Melick, D., Yang Xuefei, *et al.* (2007) Seeing the wood for the trees: how conservation policies can place greater pressure on village forests in southwest China. *Biodiversity and Conservation* 16, 1959–1971.

Menzies, N.K. (2021) *Ordering the Myriad Things: From Traditional Knowledge to Scientific Botany in China*. University of Washington Press, Seattle, Washington.

Mergo, L. (2012) The scene does not speak; the demise of the Odaa Bulluq Sacred Forest in Horro Guduru, Northwestern Oromia, Ethiopia. *Journal of Oromo Studies* 19(1&2), 101–137.

Mgaya, E. (2016) Forest and forestry in Tanzania: changes and continuities in policies and practices from colonial times to the present. *Journal of the Geographical Association of Tanzania* 36, 45–58.

Miclotte, L. and Van de Wiele, T. (2019) Food processing, gut microbiota and the globesity problem. *Critical Reviews in Food Science and Nutrition* 60, 1769–1782.

Milillo, P., Rignot, E., *et al.* (2019) Heterogeneous retreat and ice melt of Thwaites Glacier, West Antarctica. *Science Advances* 5(1), eaau3433.

Millar, B.C., Rao, J.R., *et al.* (2021) Fighting antimicrobial resistance (AMR): Chinese herbal medicine as a source of novel antimicrobials – an update. *Letters in Applied Microbiology* 73, 400–407. DOI: 10.1111/lam.13534.

Millar, S. and Woodward, W. (1999) Future for gorillas and tourist industry shrouded in mist. *The Guardian*, 3 March. Available at: https://www.theguardian.com/world/1999/mar/03/uganda.stuartmillar1 (accessed 21 May 2024).

Millard, C. (2005/2006) *sMan* and *Glud*: Standard Tibetan Medicine and ritual medicine in a Bon medical school and clinic in Nepal. *The Tibet Journal* 30/31, 3–30.

Mills, R. (2006) Preserving the past for the future: the importance of archival information in forestry. *Issues in Science and Technology Librarianship*, Spring Supplement. Available at: http://www.istl.org/46-supp/article9.html (accessed 21 May 2024).

Minter, S. (2000) *The Apothecaries' Garden: A New History of the Chelsea Physic Garden.* Sutton Publishing, Stroud, UK.

Mitchell, J.P., Carter, L.M., *et al.* (2016) A history of tillage in California's Central Valley. *Soil and Tillage Research* 157, 52–64.

Monastersky, R. (1995) Iron versus the greenhouse: oceanographers cautiously explore a global warming therapy. *Science News* 148, 220–222.

Moreau, R.E. (1935) A synecological study of Usambara, Tanganyika Territory, with particular reference to birds. *Journal of Ecology* 23, 1–43.

Moreau, R.E. (1966) *The Bird Faunas of Africa and Its Islands.* Academic Press, London.

Morrison, M.E.S. (1968) Vegetation and climate in the uplands of south-western Uganda during the Later Pleistocene Period, 1. Muchoya Swamp, Kigezi District. *Journal of Ecology* 56, 363–384.

Morrison, M.E.S. and Hamilton, A.C. (1974) Vegetation and climate in the uplands of south-western Uganda during the later Pleistocene period, 2: forest clearance and other vegetational changes in the Rukiga Highlands during the past 8000 years. *Journal of Ecology* 62, 1–31.

Mote, F.W. and Twitchett, D. (eds) (1988) *The Ming Dynasty 1368–1644, Part 1.* The Cambridge History of China, Vol. 7. Cambridge University Press, Cambridge.

Mozaffarian, D., Blanck, H.M., *et al.* (2022) A Food is Medicine approach to achieve nutrition security and improve health. *Nature Medicine* 28(11), 2238–2240.

Mrudula, V. (2002) On the work of FRLHT. Paper presented at a *Workshop on Wise Practices and Experiential Learning in the Conservation and Management of Himalayan Medicinal Plants*, Kathmandu, Nepal, 15–20 December 2002, supported by the Ministry of Forest and Soil Conservation, Nepal, the WWF Nepal Program, Medicinal and Aromatic Plants Program in Asia (MAPPA), IDRC, Canada, and the WWF-UNESCO People and Plants Initiative.

Mukembo, S.C. and Edwards, M.C. (2015) Agricultural extension in sub-Saharan Africa during and after its colonial era: the case of Zimbabwe, Uganda, and Kenya. *Journal of International Agricultural and Extension Education* 22, 50–68. DOI: 10.5191/jiaee.2015.22304.

Muller, G.L. and Gutierrez, G. (2015) *On the Side of the Poor: The Theology of Liberation.* Orbis Books, Maryknoll, New York.

Mullins, J.D. (1904) *The Wonderful Story of Uganda.* Church Missionary Society, London.

Mulumba, J.W., Nkwiine, C., *et al.* (2004) Evaluation of farmers' best practices for on-farm conservation of rare banana cultivars in the semi-arid region of Lwengo sub-county, Uganda. *Uganda Journal of Agricultural Sciences* 9, 281–288.

Mumbi, C.T., Marchant, R.A., *et al.* (2008) Late Quaternary vegetation reconstruction from the Eastern Arc Mountains, Tanzania. *Quaternary Research* 69, 320–341.

Muscarella, R., Emilio, T., *et al.* (2020) The global abundance of tree palms. *Global Ecology and Biogeography* 29, 1495–1514.

Mutebi, W.B. (2005) *Towards an Indigenous Understanding and Practice of Baptism amongst the Baganda, Uganda.* Wavah Books Ltd, Kampala.

Muteesa (1967) *Desecration of My Kingdom*. Constable, London.

Muzaffar, S.B., Islam, M.A., *et al.* (2011) The endangered forests of Bangladesh: why the process of implementation of the Convention on Biological Diversity is not working. *Biodiversity Conservation* 20, 1597–1601.

Muzik, T.J. and Cruzado, H.J. (1958) Transmission of juvenile rooting ability from seedlings to adults of *Hevea brasiliensis*. *Nature Communications* 181, 1288.

Myers, N., Mittermeier, R.A., *et al.* (2000) Biodiversity hotspots for conservation priorities. *Nature* 403, 853–858.

Nakkazi, E. (2011) Ugandans mobilize to save Mabira forest from sugarcane plantation. *The Ecologist*, 11 September. Available at: http://www.theecologist.org/campaigning/wildlife/ 1057616/ugandans_mobilise_to_save_mabira_forest_from_sugarcane_plantation.html (accessed 21 May 2024).

Nankindu, P. (2020) The history of educational language policies in Uganda: lessons from the past. *American Journal of Educational Research* 8, 643–652.

Nayyar, G.M.L., Breman, J., *et al.* (2012) Poor-quality antimalarial drugs in southeast Asia and sub-Saharan Africa. *The Lancet Infectious Diseases* 12(6), 488–496.

Negret, P.J., Sonter, L., *et al.* (2019) Emerging evidence that armed conflict and coca cultivation influence deforestation patterns. *Biological Conservation* 239, 108176.

Nelson, E.C. (1986) Introduction. In: *'Notes on Economic Botany of China'* by Augustine Henry. Boethius Press, Kilkenny, Ireland, pp. v–xv.

Neumann, K., Bostoen, K., *et al.* (2012) First farmers in the Central African rainforest: a view from southern Cameroon. *Quaternary International* 249, 53–62.

Neumann, K., Eichhorn, B., *et al.* (2022) Iron Age plant subsistence in the Inner Congo Basin (DR Congo). *Vegetation History and Archaeobotany* 31, 481–509.

Ng, N.Q. (1995) Cowpea. In: Smartt, J. and Simmonds, N.W. (eds) *Evolution of Crops*. Wiley, New York, pp. 326–332.

Nichols, D.J. and Johnson, K.R. (2009) *Plants and the K–T Boundary*. Cambridge University Press, Cambridge.

Nicholson, E., Mace, G.M., *et al.* (2009) Priority research areas for ecosystem services in a changing world. *Journal of Applied Ecology* 46, 1139–1144.

Nicholson, J.W. (1929) *The Future of Forestry in Uganda*. Government Printer, Entebbe, Uganda.

Nic Lughadha, E. (2020) Extinction risk and threats to plants and fungi. *Plants, People, Planet* 2(5), 389–408.

Nielsen, N.H., Philippsen, B., *et al.* (2018) Diet and radiocarbon dating of Tollund Man: new analyses of an Iron Age bog body from Denmark. *Radiocarbon* 60, 1533–1545.

Nogia, P. and Pati, P.K. (2021) Plant secondary metabolite transporters: diversity, functionality, and their modulation. *Fronters in Plant Science* 12, 758202. DOI: 10.3389/ fpls.2021.758202.

Nurse, D. and Philippson, G. (eds) (2003) *The Bantu Languages*. Routledge, London and New York.

Nuwagaba, T.F. (2014) *Totems of Uganda: Buganda Edition*. Taga Nuwagaba and Nathan Kiwere, Kampala.

Oeggl, K. (2009) The significance of the Tyrolean Iceman for the archaeobotany of Central Europe. *Vegetation History and Archaeobotany* 18, 1–11.

Okin, G.S. (2017) Environmental impacts of food consumption by dogs and cats. *PLoS ONE* 12(8), e0181301. DOI: 10.1371/journal.pone.0181301.

Olofinjana, I.O. (2020) Reverse mission: towards an African British theology. *Transformation* 37, 52–65.

Olsen, C.S. (1997) Commercial non-timber forestry in central Nepal: emerging themes and priorities. PhD thesis, Royal Veterinary and Agricultural University, Copenhagen.

Olsen, C.S. and Larsen, H.O. (2003) Alpine medicinal plant trade and Himalayan mountain live-lihood strategies. *The Geographical Journal* 169, 243–254.

Olson, D.M. and Dinerstein, E. (1998) The Global 200: a representation approach to conserving the Earth's most biologically valuable ecoregions. *Conservation Biology* 12(3), 502–515.

Olson, D.M. and Dinerstein, E. (2002) The Global 200: priority ecoregions for global conserva-tion. *Annals of the Missouri Botanical Garden* 89, 199–224.

Olusoga, D. (2016) *Black and British: A Forgotten History*. Macmillan, London.

Onapajo, H. and Isike, C. (2016) The global politics of gay rights: the straining relations between the West and Africa. *Journal of Global Analysis* 6(1), 21–45.

Oosthoek, J. (2010) Worlds apart? The Scottish forestry tradition and the development of forestry in India. *Journal of Irish and Scottish Studies* 3, 69–82.

Ordorika, I. (2021) Student movements and politics in Latin America: a historical reconceptualization. *Higher Education* 83, 297–315.

Oreskes, N. and Conway, E.M. (2023) *The Big Myth: How American Business Taught Us to Loathe Government and Love the Free Market*. Bloomsbury, New York.

Oslisly, R. and White, L. (2007) Human impact and environmental exploitation in Gabon during the Holocene. In: Denham, T.P., Iriate, J., *et al.* (eds) *Rethinking Agriculture: Archaeological and Ethnoarchaeological Perspectives*. Left Coast Press, Inc., Walnut Creek, California, pp. 347–360.

Osmaston, H.A. (1959) *Working Plan for the Bugoma Forest*. Forest Department, Entebbe, Uganda.

Osmaston, H.A. (1968) Uganda. In: Hedberg, I. and Hedberg, O. (eds) *Conservation of Vegetation in Africa South of the Sahara*. Almqvist and Wiksells Boktryckeri AB, Uppsala, Sweden, pp. 148–151.

Ospina, G.A. (2006) War and ecotourism in the national parks of Colombia: some reflections on the public risk and adventure. *International Journal of Tourism Research* 8, 241–246.

Otero, G., Gürcan, E.C., *et al.* (2017) Food security, obesity, and inequality: measuring the risk of exposure in the neoliberal diet. *Journal of Agrarian Change* 18, 536–554.

Otto, T.D., Gilabert, A., *et al.* (2018) Genomes of all known members of a *Plasmodium* sub-genus reveal paths to virulent human malaria. *Nature Microbiology* 3, 687–697.

Pääbo, S. (2003) The mosaic that is our genome. *Nature* 421, 409–412.

Pääbo, S. (2014) *Neanderthal Man: In Search of Lost Genomes*. Basic Books, New York.

Paini, D.R., Sheppard, A.W., *et al.* (2016) Global threat to agriculture from invasive species. *Proceedings of the National Academy of Sciences* USA 113, 7575–7579.

Pakenham, T. (1991) *The Scramble for Africa 1876–1912*. Weidenfeld and Nicolson, London.

Pallaver, K. (2016) Monetary practices and currency transitions in early colonial Uganda. *African Economic History Network*, 18 July. Available at: https://www.aehnetwork.org/blog/monetary-practices-and-currency-transitions-in-early-colonial-uganda (accessed 21 May 2024).

Pascual, M., Ahumada, J.A., *et al.* (2006) Malaria resurgence in the East African highlands: tem-perature trends revisited. *Proceedings of the National Academy of Sciences USA* 103, 5829–5834.

Patterson, J.H. (1928) *The Man-eaters of Tsavo and other East African Adventures*. Macmillan and Co., Ltd, London.

Patwardhan, B. (2014) Bridging Ayurveda with evidence-based scientific approaches in medi-cine. *EPMA Journal* 5, 19. DOI: 10.1186/1878-5085-5-19.

Paul, A.M. (2005) *The Cult of Personality Testing*. Simon and Schuster, New York.

Pearce, F. (2000) Which is more authentic: a game-rich wilderness or cattle pasture? *New Scientist* 167(2251), 30.

Pei Shengji (1982) A preliminary study of the ethnobotany of Xishuangbanna (in Chinese). In: *Collected Research Papers on Tropical Botany*. Yunnan Publishing House, Kunming, China, pp. 16–30.

Pei Shengji (1984) *Botanical Gardens in China*. Harold L. Lyon Arboretum, University of Hawaii Press, Honolulu, Hawaii.

Pei Shengji (1985) Some effects of the Dai people's cultural beliefs and practices upon the plant environment of Xishuangbanna, Yunnan Province, southwest China. In: Hutterer, K.L., Rambo, A.T., *et al.* (eds) *Cultural Values and Human Ecology in Southeast Asia*. Michigan Papers on South and Southeast Asian Studies No. 27. University of Michigan Center for South and Southeast Asian Studies, Ann Arbor, Michigan, pp. 321–339.

Pei Shengji (1988) Plant products and ethnicity in the markets of Xishuangbanna, Yunnan Province, China. In: Rambo, A.T., Gillogly, K., *et al.* (eds) *Ethnic Diversity and The Control of Natural Resources in Southeast Asia*. Michigan Papers on South and Southeast Asia No. 32. University of Michigan Center for South and Southeast Asia, Ann Arbor, Michigan, pp. 119–142.

Pei Shengji (2001) Ethnobotanical approaches of traditional medicine studies: some experiences from Asia. *Pharmaceutical Botany* 39, 74–79.

Pei Shengji (2002) Ethnobotany and modernisation of Traditional Chinese Medicine. Presented at the Regional Workshop on *Wise Practices and Experiential Learning in the Conservation and Management of Himalayan Medicinal Plants*, Kathmandu, Nepal, 15–20 December 2002.

Pei Shengji (2003) Ethnobotany: development dynamics of the discipline and prospects. *Acta Botanica Yunnanica* 14(Suppl.), 1–10.

Pei Shengji (2010) The road to the future? The biocultural values of the Holy Hill Forests of Yunnan Province, China. In: Verschuuren, B., Wild, R., *et al.* (eds) *Sacred Natural Sites: Conserving Nature and Culture*. Earthscan, London, pp. 98–106.

Pei Shengji and Li Yanhui (1981) A taxonomical problem of the genus *Maytenus* Molina and *Gymnosporia* (Wight & Arn.) Benth. & Hook.f. from China. *Acta Botanica Yunnanica* 3(1), 25–31.

Pei Shengji and Youkai, X. (2020) *Plants and Ethnoculture in Xishuangbanna*. Science and Technology Press, Shanghai, China.

Pei Shengji, Chen Sanyang, *et al.* (1994–present) Arecaceae (Palmae). In: Wu Zhengyi, Raven, P.H., *et al.* (eds) *Flora of China, Vol. 23 (Acoraceae through Cyperaceae)*. Science Press, Beijing, pp. 132–157.

Pei Shengji, Hamilton, A.C., *et al.* (2010) Conservation and development through medicinal plants: a case study from Ludian (Northwest Yunnan, China) and presentation of a general model. *Biodiversity and Conservation* 19, 2619–2636.

Pei Shengji, Huai Huyin, *et al.* (2009a) Medicinal plants and their conservation in China with reference to the Chinese Himalaya Region. *Asian Medicine* 5, 275–292.

Pei Shengji, Li Yanhui, *et al.* (1996) Ethnobotanical investigation of plant drugs at local markets in north-west Yunnan of China. In: Pei Shengji, Su Yong-ge, *et al.* (eds) *The Challenges of Ethnobiology in the 21st Century: Proceedings of the Second International Congress of Ethnobiology*. Yunnan Science and Technology Press, Kunming, China, pp. 150–159.

Pei Shengji, Zhang Guoxue, *et al.* (2009b) Application of traditional knowledge in forest management: ethnobotanical indicators of sustainable forest use. *Forest Ecology and Management* 257, 2017–2021.

Peoples, H.C., Duda, P., *et al.* (2016) Hunter-gatherers and the origins of religion. *Human Nature* 27, 261–282.

Pérez-Pardal, L., Royo, L.J., *et al.* (2010) Multiple paternal origins of domestic cattle revealed by Y-specific interspersed multilocus microsatellites. *Heredity* 105, 511–519.

Peripato, V., Levis, C., *et al.* (2023) More than 10,000 pre-Columbian earthworks are still hidden throughout Amazonia. *Science* 382, 103–109.

Perret, C., Powers, S.T., *et al.* (2017) Emergence of hierarchy from the evolution of individual influence in an agent-based model. In: Knibbe, C., Beslon, G., *et al.* (eds) *Proceedings of* the Fourteenth European Conference on Artificial Life, ECAL 2017, Lyon, France, 4–8 September 2017. MIT Press, Cambridge, Massachusetts, pp. 348–355.

Perrier, X., De Langhe, E., et al. (2011) Multidisciplinary perspectives on banana (Musa spp.) domestication. Proceedings of the National Academy of Sciences USA 108, 11311–11318.

Perrings, C. (ed.) (2000) The biodiversity convention and biodiversity loss in Sub-Saharan Africa. In: The Economics of Biodiversity Conservation in sub-Saharan Africa: Mending the Ark. Edward Elgar, Cheltenham, UK, pp. 1–37.

Peterken, G. and Mountford, E. (2017) Woodland Development: A Long-Term Study of Lady Park Wood. CAB International, Wallingford, UK.

Peterson, D.R. (2019) Reading John Mbiti from Uganda. Africa is a Country, 18 October. Available at: https://africasacountry.com/author/derek-r-peterson (accessed 21 May 2024).

Philip, M.S. (1962) The Management of Tropical High Forest. Government Printer, Entebbe, Uganda.

Philippot, L., Raaijmakers, J.M., et al. (2013) Going back to the roots: the microbial ecology of the rhizosphere. Nature Reviews Microbiology 11, 789–799.

Picozzi, K., Fèvre, E.M., et al. (2005) Sleeping sickness in Uganda: a thin line between two fatal diseases. British Medical Journal 331, 1238–1241.

Piquerez, S.J.M., Harvey, S.E., et al. (2014) Improving crop disease resistance: lessons from research on Arabidopsis and tomato. Frontiers in Plant Science 5, 671. DOI: 10.3389/fpls.2014.00671.

Plana, V. (2004) Mechanisms and tempo of evolution in the African Guineo-Congolian rainforest. Philosophical Transactions of the Royal Society B: Biological Sciences 349, 1585–1594.

Plantlife International (2004) Identifying and Protecting the World's Most Important Plant Areas. Plantlife International, Salisbury, UK.

Pócs, T. (1976) Vegetation mapping in the Uluguru Mountains (Tanzania, East Africa). Boissiera 24, 477–498.

Pomeroy, E., Hunt, C.O., et al. (2020) Issues of theory and method in the analysis of Paleolithic mortuary behavior: a view from Shanidar Cave. Evolutionary Anthropology 29, 263–279.

Ponce, L. (2015) Puquios, qanats and springs: water management in ancient Perú. Agricultura Sociedad y Desarrollo 12, 179–296.

Pontzer, H., Wood, B.M., et al. (2018) Hunter-gatherers as models in public health. Obesity Reviews 19(Suppl. 1), 24–35.

Popper, K. (1976) Unended Quest: An Intellectual Autobiography. Fontana/Collins, Glasgow, UK.

Post, M. (2003) Gold rush in Dolpa. Nepali Times, 3 July. Available at: https://archive.nepalitimes.com/news.php?id=2872#:~:text=Gold%20rush%20in%20Dolpa%2D%20Nepali%20Times&text=The%20Himalayan%20viagra%20harvesting%20season, seekers%20to%20this%20harsh%20region.&text=This%20is%20the%20yarchagumba%20 picking,empty%2C%20government%20offices%20are%20deserted (accessed 21 May 2024).

Prescott, W.H. (1843) The History of the Conquest of Mexico. Harper Brothers, New York.

Prescott, W.H. (1847) History of the Conquest of Peru. George Allen and Unwin Ltd, London.

Price, T.D. and Fienman, G.M. (eds) (2010) Pathways to Power. Springer, New York.

Pringle, P. (2008) The Murder of Nikolai Vavilov: The Story of Stalin's Persecution of One of the Great Scientists of the Twentieth Century. Simon and Schuster, New York.

Pringle, R.M. (2005) The origins of the Nile perch in Lake Victoria. BioScience 55, 780–787.

Qiu Qiang, Wang Lizhong, et al. (2015) Yak whole-genome resequencing reveals domestication signatures and prehistoric population expansions. Nature Communications 6, 10283.

Qureshi, A.S. (2011) Water management in the Indus basin in Pakistan: challenges and opportunities. Mountain Research and Development 31, 252–260.

Rackham, O. (1976) *Trees and Woodland in the British Landscape*. J.M. Dent and Sons, London.

Ract, C., Burgess, N., *et al.* (2024) Nature Forest Reserves in Tanzania and their importance for conservation. *PLoS ONE* 19(2), e0281408. DOI: 10.137/journal.pone.0281408.

Radice, H. (2013) How we got here: UK higher education under neoliberalism. *ACME: An International Journal for Critical Geographies* 12(2), 407–418.

Raffer, K. (2011) Neoliberal capitalism: a time warp backwards to capitalism's origins? *Forum for Social Economics* 40(1), 41–62.

Rappaport, R.A. (1984) *Pigs for the Ancestors: Ritual in the Ecology of a New Guinea People*. Waveland Press, Inc., Long Grove, Illinois.

Rawat, J.M., Pandey, S., *et al.* (2021) Preparation of alcoholic beverages by tribal communities in the Indian Himalayan Region: a review on traditional and ethnic consideration. *Frontiers in Sustainable Food Systems* 5, 672411. DOI: 10.3389/fsufs.2021.672411.

Raworth, K. (2017) *Doughnut Economics: Seven Ways to Think Like a 21st-Century Economist*. Chelsea Green Publishing, White River Junction, Vermont.

Rebanks, J. (2020) *English Pastoral: An Inheritance*. Allen Lane, London.

Reich, D. (2018) *Who We Are and How We Got Here: Ancient DNA and the New Science of the Human Past*. Oxford University Press, Oxford.

Rejmánek, M. (2000) Invasive plants: approaches and predictions. *Austral Ecology* 25, 497–506.

Remini, B., Kechad, R., *et al.* (2014) The qanat of Algerian Sahara: an evolutionary hydraulic system. *Applied Water Science* 5, 359–366. DOI: 10.1007/s13201-014-0195-5.

Reynolds, T.W., Stave, K.A., *et al.* (2017) Changes in community perspectives on the roles and rules of church forests in northern Ethiopia: evidence from a panel survey of four Ethiopian Orthodox communities. *International Journal of the Commons* 11, 355–387.

Richards, P.W. (1964) *The Tropical Rain Forest: An Ecological Study*. Cambridge University Press, Cambridge.

Richter, D., Grün, R., *et al.* (2017) The age of the hominin fossils from Jebel Irhoud, Morocco, and the origins of the Middle Stone Age. *Nature* 546, 293–296.

Riggs, T. (ed.) (2015) *Worldmark Encyclopaedia of Religious Practices*. Gale, Farmington Hills, Michigan.

Ritchie, H. and Roser, M. (2024) Land use. Available at: https://ourworldindata.org/land-use#cropland-use (accessed 21 May 2024).

Roberson, E. (2009) *Medicinal Plants at Risk*. Center for Biological Diversity, Tucson, Arizona.

Robertshaw, P., Kamuhangire, E.R., *et al.* (1997) Archaeological research in Bunyoro-Kitara: preliminary results. *Nyame Akuma* 48, 70–77.

Roche, E. (1996) L'influence anthropique sur l'environnement à l'Age du Fer dans la Rwanda ancien. *Geo-Eco-Trop* 20, 73–89.

Rodgers, W.A. and Homewood, K.M. (1982) Species richness and endemism in the Usambara mountain forests, Tanzania. *Biological Journal of the Linnean Society* 18, 197–242.

Rodrigues, A.D. (1993) Linguas indígenas: 500 anos de descobertas e perdas. *DELTA* 9, 83–103.

Rodrigues, P.T., Valdivia, H.O., *et al.* (2018) Human migration and the spread of malaria parasites to the New World. *Scientific Reports* 8, 1993.

Romer, J. (2012) *A History of Ancient Egypt: From the First Farmers to the Great Pyramid*. Penguin Books, London.

Roome, W.J.W. (1927) *Can Africa Be Won?* A & C Black Ltd, London.

Roscoe, J. (1909) Python worship in Uganda. *Man (Journal of the Royal Anthropological Institute)* 57, 88–90.

Roscoe, J. (1911) *The Baganda: An Account of Their Native Customs and Beliefs*. Macmillan, London.

Roscoe, J. (1921) *Twenty-Five Years in East Africa*. Cambridge University Press, Cambridge.

Rucker, R.B. and Rucker, M.R. (2016) Nutrition: ethical issues and challenges. *Nutrition Research* 36, 1183–1192.

Ruggiero, V. (2022) Justificatory narratives: the collapse of Greensill Capital. *Crime, Justice and Social Democracy* 11, 210–221.

Rull, V. (2021) Contributions of paleoecology to Easter Island's prehistory: a thorough review. *Quaternary Science Reviews* 252, 106751. DOI: 10.1016/j.quascirev.2020.106751.

Russell, B. (1979) *History of Western Philosophy*. Unwin Paperbacks, London.

Russell, G. (2014) *Heirs to Forgotten Kingdoms: Journeys into the Disappearing Religions of the Middle East*. Simon and Schuster, London.

Ryves, D.B., Mills, K., *et al.* (2011) Environmental change over the last millennium recorded in two contrasting crater lakes in western Uganda, eastern Africa (Lakes Kasenda and Wandakara). *Quaternary Science Reviews* 30, 555–569.

Sackett, D.L., Rosenberg, W.M.C., *et al.* (1996) Evidence-based medicine: what it is and what it isn't. *British Medical Journal* 312, 71–72.

Sakthivadivel, R., Gomathinayagam, P., *et al.* (2004) Rejuvenating irrigation tanks through local institutions. *Economic and Political Weekly* 39(31), 3521–3526.

Salavert, A., Zazzo, A., *et al.* (2020) Direct dating reveals the early history of opium poppy in western Europe. *Scientific Reports* 10, 20263.

Salick, J., Amend, A., *et al.* (2007) Tibetan sacred sites conserve old growth trees and cover in the eastern Himalayas. *Biodiversity and Conservation* 16, 693–706.

Salick, J., Byg, A., *et al.* (2006) Tibetan medicine plurality. *Economic Botany* 60, 227–253.

Sandom, C., Faurby, S., *et al.* (2014) Global late Quaternary megafauna extinctions linked to humans, not climate change. *Proceedings of the Royal Society B: Biological Sciences* 281, 20133254.

Savage, N. (2022) New yarn from old clothes. *Nature* 611, S20–S21.

Saxena, M., Saxena, J., *et al.* (2013) Phytochemistry of medicinal plants. *Journal of Pharmacognosy and Phytochemistry* 1(6), 168–182.

Scarcelli, N., Cubry, P., *et al.* (2019) Yam genomics supports West Africa as a major cradle of crop domestication. *Science Advances* 5(5), eaaw1947.

Schadeberg, T.C. (2003) Historical linguists. In: Nurse, D. and Philippson, G. (eds) *The Bantu Languages*. Routledge, London and New York, pp. 143–163.

Schama, S. (1995) *Landscape and Memory*. HarperCollins, London.

Schippmann, U. (1998) Summarising remarks and conclusions. In: *First International Symposium on the Conservation of Medicinal Plant Trade in Europe, 22–23 June 1998, Royal Botanic Gardens, Kew, UK*. TRAFFIC Europe, Brussels. Available at: https://portals.iucn.org/library/sites/library/files/documents/Traf-073.pdf (accessed 23 May 2024).

Schippmann, U., Leaman, D.J., *et al.* (2006) A comparison of cultivation and wild collection of medicinal and aromatic plants under sustainability aspects. In: Bogers, R.J., Craker, L.E., *et al.* (eds) *Medicinal and Aromatic Plants*. Springer, Dordrecht, The Netherlands, pp. 75–95.

Schoenbrun, D.L. (1993a) Cattle herds and banana gardens: the historical geography of the western Great Lakes region, ca AD 800–1500. *The African Archaeological Review* 11, 39–72.

Schoenbrun, D.L. (1993b) We are what we eat: ancient agriculture between the Great Lakes. *The Journal of African History* 34, 1–31.

Schoenbrun, D.L. (1998) *A Green Place, A Good Place: Agrarian Change, Gender and Social Identity in the Great Lakes Region to the 15th Century*. Heinemann, Portsmouth, New Hampshire.

Schoenbrun, D.L. (2021) *The Names of the Python: Belonging in East Africa 900 to 1930*. University of Wisconsin Press, Madison, Wisconsin.

Schultes, R.E. and Rauffauf, R.F. (1990) *The Healing Forest: Medicinal and Toxic Plants of the Northwest Amazonia*. Dioscorides Press, Portland, Oregon.

Schultes, R.E. and von Reis, S. (1995) *Ethnobotany: Evolution of a Discipline*. Chapman and Hall, London.

Scoppola, M. (2021) Globalisation in agriculture and food: the role of multinational enterprises. *European Review of Agricultural Economics* 48(2), 1–37.

Sender, R., Fuchs, S., *et al.* (2016) Revised estimates for the number of human and bacteria cells in the body. *PLoS Biology* 14, e1002533. DOI: 10.1371/journal.pbio.1002533.

Sepulchre, P., Ramstein, G., *et al.* (2007) H4 abrupt event and late Neanderthal presence in Iberia. *Earth and Planetary Science Letters* 258, 283–292.

Shah, D.P. (2005) Trade systems and feasibility of cultivation/domestication of MAPs in the Southern Buffer Zone of Shey Phoksundo National Park, Dolpa. In: Thomas, Y., Karki, M., *et al.* (eds) *Himalayan Medicinal and Aromatic Plants, Balancing Use and Conservation. Proceedings of the Regional Workshop on Wise Practices and Experiential Learning in the Conservation and Management of Himalayan Medicinal Plants, December 15–20, 2002, Kathmandu, Nepal*. Ministry of Forests and Soil Conservation, Kathmandu, pp. 506–510.

Shannon, V.P. and J.W. Keller (2007) Leadership style and international norm violation: the case of the Iraq War. *Foreign Policy Analysis* 3(1), 79–104.

Sharma, K., Mahato, N., *et al.* (2018) Systematic study on active compounds as antibacterial and antibiofilm agent in aging onions. *Journal of Food and Drug Analysis* 26(2), 518–528.

Sheil, D., Puri, R., *et al.* (2006) Recognizing local people's priorities for tropical forest biodiversity. *Ambio* 35(1), 17–24.

Sheldrake, M. (2020) *Entangled Life: How Fungi Make Our Worlds, Change Our Minds and Shape Our Futures*. The Bodley Head, London.

Shen Peiqiong, Sun Hangdon, *et al.* (1990) Ethnobotany of fleagrass (*Adenosma buechneroides* Bonati), a traditional cultivated plant of the Hani people, Xishuangbanna, Yunnan, China. In: Posey, D.A. and Overal, W.L. (eds) *Ethnobiology: Implications and Applications. Proceedings of the First International Congress of Ethnobiology, Belém, 1988, Vol. 1*. Museu Paraense Emilio Goeldi, Belém, Brazil, pp. 305–309.

Shinwari, A.K. (2002) *Ethnobotany Project, WWF-Pakistan: A Review*. International Curriculum Development in Applied Ethnobotany, WWF-Pakistan, Nathiagali, Pakistan.

Shinwari, A.K. and Khan, A.A. (2001) *Ethnobotany Applied to Participatory Forest Management in Pakistan*. WWF-Pakistan, Peshawar, Pakistan.

Shinwari, A.K. and Khan, A.A. (2002) *Land Tenure and Resource Ownership in Pakistan*. WWF-Pakistan, Peshawar, Pakistan.

Shiva, V. (ed.) (1996) *Protecting Our Biological and Intellectual Heritage in the Age of Biopiracy*. The Research Foundation for Science, Technology and Natural Resources Policy, New Delhi.

Signer, M. (2009) *Demagogue: The Fight to Save Democracy from Its Worst Enemies*. St Martin's Press, New York.

Simard, S. (2021) *Finding the Mother Tree*. Vintage Books, New York.

Simmons, G.F. and Fenning, C.D. (eds) (2018) *Ethnologue: Languages of Uganda*, 21st edn. SIL International, Dallas, Texas.

Sire, J.W. (2004) *Naming the Elephant: Worldview as a Concept*. IVP Academic, Downers Grove, Illinois.

Smil, V. (1999) China's great famine: 40 years later. *British Medical Journal* 319, 1619–1621.

Smith, A. (1776) *An Inquiry into the Nature and Causes of the Wealth of Nations*. W. Strahan and T. Cadell, London.

Soejarto, D.D. (2012) Medicinal plants and the legacy of Richard Evans Schultes. In: Ponman, B.E. and Bussmann, R.W. (eds) *Medicinal Plants and the Legacy of Richard E. Schultes*. The William L. Brown Center at the Missouri Botanical Garden, St Louis, Missouri, pp. 103–118.

Solis, R.S., Haas, J., *et al.* (2001) Dating Caral, a preceramic site in the Supe Valley on the Central Coast of Peru. *Science* 292, 723–726.

Song Yuanyuan, Wang Ming, *et al.* (2019) Priming and filtering of antiherbivore defences among *Nicotiana attenuata* plants connected by mycorrhizal networks. *Plant, Cell & Environment* 42, 2945–2961.

Soper, R. (1967) Iron Age sites in north-eastern Tanzania. *Azania* 2, 19–36.

Speke, J.H. (1863) *Journal of the Discovery of the Source of the Nile.* J.M. Dent & Co., London.

Spranger, M., Pauw, S., *et al.* (2010) Open-ended semantics co-evolving with spatial language. In: Smith, A.D.M., Schouwstra, M., *et al.* (eds) *Evolution of Language: Proceedings of the 8th International Conference.* World Scientific Press, Singapore, pp. 297–304.

Sprou, M.A., Tukhbatova, R.I., *et al.* (2018) Analysis of 3800-year-old *Yersinia pestis* genomes suggests Bronze Age origin for bubonic plague. *Nature Communications* 9, 2234.

Ssegawa, P. and Kasenene, J.M. (2007) Plants for malaria treatment in southern Uganda: traditional use, preference and ecological viability. *Journal of Ethnobiology* 27, 110–131.

Ssemanda, I., Ryves, D.B., *et al.* (2005) Vegetation history in western Uganda during the last 1200 years: a sediment-based reconstruction from two crater lakes. *The Holocene* 15, 119–132.

Stafford, R., Chamberlain, B., *et al.* (eds) (2021) *Nature-based Solutions for Climate Change in the UK: A Report by the British Ecological Society.* British Ecological Society, London. Available at: https://www.britishecologicalsociety.org/wp-content/uploads/2022/02/NbS-Report-Final-Updated-Feb-2022.pdf (accessed 21 May 2024).

Stark, R. and Liu, E.Y. (2011) The religious awakening in China. *Review of Religious Research* 52, 282–289.

Stewart, K.M. (2003a) The African cherry (*Prunus africana*): can lessons be learned from an over-exploited medicinal tree? *Journal of Ethnopharmacology* 89, 3–13.

Stewart, K.M. (2003b) The African cherry (*Prunus africana*): from hoe-handles to the international herb market. *Economic Botany* 57(4), 559–569.

Stewart, P., Garvey, B., *et al.* (2020) Amazonian destruction, Bolsonaro and COVID-19: neoliberalism unchained. *Capital & Class* 45, 173–181. DOI: 10.1177/0309816820971131.

Stocking, M. and Perkin, S. (1991) Conservation-with-development: an application of the concept in the Usambara Mountains, Tanzania. *Transactions of the Institute of British Geographers, New Series* 17, 337–349.

Storkey, J. and Westbury, D.B. (2007) Managing arable weeds for biodiversity. *Pest Management Science* 63(6), 517–523.

Strahm, W. (1994) Regional overview: Indian Ocean islands. In: Davis, S.D., Heywood, V.H., *et al.* (eds) *Centres of Plant Diversity: A Guide and Strategy for Their Conservation.* IUCN Publications Unit, Cambridge, pp. 265–270.

Stuart, S. (1989) The forest bird fauna of the East Usambara Mountains. In: Hamilton, A.C. and Bensted-Smith, R. (eds) *Forest Conservation in the East Usambara Mountains, Tanzania.* IUCN, Gland, Switzerland and Cambridge, pp. 357–361.

Su Nan-Yao (2023) How to become a successful invader. *Florida Entomologist* 96, 765–769.

Suggs, R.C. (1951) *The Island Civilizations of Polynesia.* Mentor Books, New York.

Sullivan, M.J.P. (2020) Long-term thermal sensitivity of Earth's tropical forests. *Science* 368, 869–874.

Sunstein, C.R. (2008) *Of Montreal and Kyoto: A Tale of Two Protocols.* Environmental Law Institute, Washington, DC.

Supe, A. (2016) Evolution of medical education in India: the impact of colonialism. *Journal of Postgraduate Medicine* 62, 255–259.

Synge, H. (1988) *The Joint IUCN–WWF Plants Conservation Programme: Achievements 1984–1987 and Activities Planned 1988–1990.* Plant Conservation Office, Kew, UK.

Tabuti, J.R.S., Dhillion, S.S., *et al.* (2003) Traditional medicine in Bulamogi County, Uganda: its practitioners, users and viability. *Journal of Ethnopharmacology* 85, 119–129.

Tansley, A.G. (1946) *Our Heritage of Wild Nature: A Plea for Organized Nature Conservation.* Readers Union/Cambridge University Press, London.

Taylor, D.M. (1990) Late Quaternary pollen records from two Ugandan mires: evidence for environmental change in the Rukiga Highlands of southwest Uganda. *Palaeogeography, Palaeoclimatology, Palaeoecology* 80, 283–300.

Taylor, D.M., Hamilton, A.C., *et al.* (2008) Thirty-eight years of change in a tropical forest: plot data from Mpanga Forest Reserve, Uganda. *African Journal of Ecology* 46, 655–667.

Taylor, D.M, Marchant, R., *et al.* (1999) A sediment-based history of medium altitude forest in central Africa: a record from Kabata Swamp, Ndale volcanic field, Uganda. *Journal of Ecology* 87, 303–315.

Tenywa, G. (2013) Oil palm growing threatens Buggala Island forest cover. *New Vision*, 18 April. Available at: https://www.newvision.co.ug/news/1317539/oil-palm-growing-threatens-buggala-island-forest-cover (accessed 21 May 2024).

Teshome, A., Fahrig, L., *et al.* (1999) Maintenance of sorghum (*Sorghum bicolor*, Poaceae) landrace diversity by farmers' selection in Ethiopia. *Economic Botany* 53(1), 79–88.

Thatcher, M. (1989) Speech to United Nations General Assembly. Available at: https://www.mssimonsays.org/uploads/5/8/8/7/5887486/2019_source_1_-_thatcher.pdf (accessed 21 May 2024).

Thomas, E., van Zonneveld, M., *et al.* (2012) Present spatial diversity patterns of *Theobroma cacao* L. in the Neotropics reflect genetic differentiation in Pleistocene refugia followed by human-Influenced dispersal. *PLoS ONE* 7(10), e47676. DOI: 10.1371/journal.pone.0047676.

Thomas, Y., Karki, M., *et al.* (eds) (2005) *Himalayan Medicinal and Aromatic Plants, Balancing Use and Conservation. Proceedings of the Regional Workshop on Wise Practices and Experiential Learning in the Conservation and Management of Himalayan Medicinal Plants, December 15–20, 2002, Kathmandu, Nepal.* Ministry of Forests and Soil Conservation, Kathmandu.

Thompson, J.L. (2010) *Theodore Roosevelt Abroad: Nature, Empire, and the Journey of an American President.* Palgrave Macmillan, New York.

Thompson, K. and Hamilton, A.C. (1983) Peatland and swamps of the African continent. In: Gore, A.J.P. (ed.) *Mires: Swamp, Bog, Fen and Moor, B: Regional Studies.* Elsevier, Amsterdam, pp. 331–373.

Thomson, L.A.J., Englberger, L., *et al.* (2006) Species Profiles for Pacific Island Agroforestry (www.traditionaltree.org): *Pandanus tectorius* (pandanus). Available at: https://raskisimani.com/wp-content/uploads/2013/01/p-tectorius-pandanus.pdf (accessed 21 May 2024).

Thoreau, H.D. (1854) *Walden.* Ticknor and Fields, Boston, Massachusetts.

Tierney, J.E., Lewis, S.C., *et al.* (2011) Model, proxy and isotopic perspectives on the East African Humid Period. *Earth and Planetary Letters* 307, 103–112.

Tierney, J.E., Pausata, F.S.R., *et al.* (2017) Rainfall regimes of the Green Sahara. *Science Advances* 3(1), e1601503.

Titus, A.O., Arogundade, O.O., *et al.* (2023) Micro-morphological study of three members of genus *Plectranthus* L. (Lamiaceae) in Nigeria. *Ife Journal of Science* 25(3), 389–397.

Troup, R.S. (1922) *Forestry in Uganda.* Crown Agents, London.

Tsai Xitao and Pei Shengji (1959) On the integrated utilization of wild plant-resources in Yunnan. *Biology Bulletin* 7, 293–296.

Tsouvalis, J. (2000) *A Critical Geography of Britain's State Forests: An Exploration of Processes of Reality Construction.* Oxford University Press, Oxford.

Turnbull, C. (1961) *The Forest People.* Simon and Schuster, New York.

Tutino, J. (2021) Capitalism, Christianity and slavery: Jesuits in New Spain, 1572–1767. *Journal of Jesuit Studies* 8, 11–36.

Tuxill, J. and Nabhan, G.P. (2001) *People, Plants and Protected Areas*. Earthscan, London.

Twaddle, M. (1993) *Kakungulu and the Creation of Uganda*. James Currey, London.

Twongyirwe, R., Sheil, D., et al. (2015) REDD at the crossroads? The opportunities and challenges of REDD for conservation and human welfare in South West Uganda. *International Journal of Environment and Sustainable Development* 14, 273–298.

Tylor, E.B. (1871) *Primitive Culture: Researches into the Development of Mythology, Philosophy, Religion, Art and Custom*. John Murray, London.

Ullo, S.L. and Sinha, G.R. (2021) Advances in IoT and smart sensors for remote sensing and agriculture applications. *Remote Sensing* 13, 2585. DOI: 10.3390/rs13132585.

UN (2020) *Inequality in a Rapidly Changing World*. Department of Economic and Social Affairs, United Nations, New York.

UNDP (2000) *Millennium Development Goals*. United Nations Development Programme, New York.

UNDP (2015) *Sustainable Development Goals*. United Nations Development Programme, New York.

UNEP (2021) *Religions and Environmental Protection*. United Nations Environment Programme, Nairobi.

Ungar, P.S. (2020) The trouble with teeth. *Scientific American* 322(April), 44.

Uno, G.R. and Bybee, R.W. (1994) Understanding the dimensions of biological literacy. *BioScience* 44, 553–557.

Upadhyay, M.R., Chen, W., et al. (2016) Genetic origin, admixture and population history of aurochs (*Bos primigenius*) and primitive European cattle. *Heredity* 118, 169–176.

Viana, D.S., Gangoso, L., et al. (2016) Overseas seed dispersal by migratory birds. *Proceedings of the Royal Society B: Biological Sciences* 283, 20152406.

Vajda, V. and Bercovici, A. (2014) The global vegetation pattern across the Cretaceous–Paleogene mass extinction interval: a template for other extinction events. *Global and Planetary Change* 122, 29–49.

Van Grunderbeek, M.-C. and Roche, E. (2007) Multidisciplinary evidence of mixed farming during the Early Iron Age in Rwanda and Burundi. In: Denham, T., Iriate, J., et al. (eds) *Rethinking Agriculture: Archaeological and Ethnoarchaeological Perspectives*. Left Coast Press, Inc., Walnut Creek, California, pp. 299–319.

Van Grunderbeek, M.C., Roche, E., et al. (1982) L'age du fer ancien au Rwanda et au Burundi: archeologie et environment. *Journal des Africanistes* 52, 5–58.

Van Noorden, R. (2023) More than 10,000 research papers were retracted in 2023 – a new record. *Nature* 624, 479–481.

Vaz, J. (2006) Seeking spaces for biodiversity by improving tenure security for local communities in Sabah. In: Cooke, F.M. (ed.) *State, Communities and Forests in Contemporary Borneo*. Asia-Pacific Environment Monographs No. 1. ANU Press, Canberra, pp. 133–162.

Verschuuren, B., Ormsby, A., et al. (2022) How might World Heritage status support the protection of Sacred Natural Sites? An analysis of nomination files, management, and governance contexts. *Land* 11, 97. DOI: 10.3390/land11010097.

Verschuuren, B., Wild, R., et al. (eds) (2010) *Sacred Natural Sites: Conserving Nature and Culture*. Earthscan, London.

Victurine, R. and Oryema Lalobo, C. (2001) Building conflict into cooperation: case study of the Mgahinga and Bwindi Impenetrable Forest Conservation Trust. In: *Mobilizing Funding for Biodiversity Conservation: A User-Friendly Training Guide*. Available at: https://www.cbd.int/doc/nbsap/finance/CaseStudy-TrustFunds_UgandaBwindi_Nov2001.pdf (accessed 21 May 2024).

Vidal, J. (2013) Margaret Thatcher: an unlikely green hero? *The Guardian*, 9 April. Available at: https://www.theguardian.com/environment/blog/2013/apr/09/margaret-thatcher-green-hero (accessed 21 May 2024).

Viña, A., McConnell, W.J., *et al.* (2016) Effects of conservation policy on China's forest recovery. *Science Advances* 2(3), e1500965. DOI: 10.1126/sciadv.1500965.

Vítková, M., Müllerová, J., *et al.* (2017) Black locust (*Robinia pseudoacacia*) beloved and despised: a story of an invasive tree in Central Europe. *Forest Ecology and Management* 384, 287–302.

von Hagen, V.W. (1944) *The Aztec and Maya Papermakers*. J.J. Augustin, New York.

Wall, P. (1972) An eye on the needle. *New Scientist* 55(805), 129–131.

Walter, K.S. and Gillett, H.J. (eds) (1997) *1997 IUCN Red List of Threatened Plants*. IUCN, Gland, Switzerland and Cambridge.

Wang Guangyu, Innes, J.L., *et al.* (2008) Towards a new paradigm: the development of China's forestry in the 21st century. *International Forestry Review* 10(4), 619–631.

Wang Guangyu, Innes, J.L., *et al.* (2012) National Park development in China: conservation or commercialization? *Ambio* 41(3), 247–261.

Wang Guo-Dong, Zhai WeiWei, *et al.* (2016) Out of southern East Asia: the natural history of domestic dogs across the world. *Cell Research* 26, 21–33.

Wang Xia (2020) Study on localization of Zitong Wenchang Culture in Japan. In: Mthembu, A. (ed.) *2020 2nd International Conference on Humanities, Cultures, Arts and Design, Sydney, Australia, 18–20 December 2020*. Francis Academic Press, London, pp. 78–84.

Wang Xinyang, Jin Cheng, *et al.* (2020) Plant diversity and species replacement in Chinese Buddhist temples. *Biodiversity Science* 28, 668–677.

Wang Xi-qun (2017) The chronicle of German forester Gottlieb Fenzel in China. *Journal of Beijing Forestry University (Social Sciences)* 16, 21–28.

Wang Yuhua (1999) Sustainable management of medicinal plant resource in northwest of Yunnan: a case study on Ludian Administrative Village in Lijiang County. PhD thesis, Kunming Institute of Botany, Kunming, China.

Warner, J.N. (1962) Sugar cane: an indigenous Papuan cultivar. *Ethnology* 1, 405–411.

Wayland, E.J. (1934) Rifts, rivers, rains and early man in Uganda. *The Journal of the Royal Anthropological Institute of Great Britain and Ireland* 64, 333–352.

Wayland, E.J. (1952) The study of past climates in tropical Africa. In: Leakey, L.S.B. and Cole, S.M. (eds) *Pan African Congress on Pre-History, 1947*. Blackwell, Oxford, pp. 59–66.

WCED (1987) *Report of the World Commission on Environment and Development: Our Common Future*. Oxford University Press, Oxford.

WCS (2016) *Nationally Threatened Species for Uganda*. Wildlife Conservation Society, New York.

Weber, A., Kalema-Zikusoka, G., *et al.* (2020) Lack of rule-adherence during mountain gorilla tourism encounters in Bwindi Impenetrable National Park, Uganda, places gorillas at risk from human disease. *Frontiers in Public Health* 8, 1. DOI: 10.3389/fpubh.2020.00001.

Webster, G. and Osmaston, H.A. (2003) *A History of the Uganda Forest Department 1951–1965*. Commonwealth Secretariat, London.

Welbourn, F.B. (1962) Some aspects of Kiganda religion. *Uganda Journal* 26(2), 171–182.

Wellems, T.E. and Plowe, C.V. (2001) Chloroquine-resistant malaria. *The Journal of Infectious Diseases* 184, 770–776.

Wells, M.P. and McShane, T.O. (2004) Integrating protected area management with local needs and aspirations. *Ambio* 33(8), 513–519.

Wendel, J.F. (2009) Evolution and natural history of the cotton genus. In: Paterson, A.H. (ed.) *Genetics and Genomics of Cotton*. Plant Genetics and Genomics: Crops and Models, Vol. 3. Springer, New York, pp. 3–22.

Wenzlhuemer, R. (2010) *From Coffee to Tea Cultivation in Ceylon, 1880–1900: An Economic and Social History*. Brill, Leiden, The Netherlands.

Werner, D. (1973) *Donde no hay doctor: Una guía para los campesinos que viven lejos de los centros médicos*. Hesperian Foundation, Berkeley, California.

Wesley, J. (1743) *Primitive Physic: Or, An Easy and Natural Method of Curing Most Diseases.* Barr & Co., London.

Wesley, J. (1748) *A Letter to a Friend, Concerning Tea.* A. Macintosh, London (facsimile edition, printed 1825).

West, H.W. (1965) *The Mailo System in Buganda.* Uganda Government, Entebbe, Uganda.

Weyerhaeuser, H., Wilkes, A., *et al.* (2005) Local impacts and responses to regional forest conservation and rehabilitation programs in China's northwest Yunnan province. *Agricultural Systems* 85, 234–253.

White, C. (2022) The rise of OxyContin: how Purdue Pharma and the Sackler Family is responsible for the epidemic behind the pandemic. BA thesis, Dominican University of California, San Rafael, California. DOI: 10.33015/dominican.edu/2022.HIST.ST.04.

White, D.J., Hubacek, K., *et al.* (2018) The water–energy–food nexus in East Asia: a tele-connected value chain analysis using inter-regional input–output analysis. *Applied Energy* 218, 550–567.

White, G. (1789) *The Natural History of Selbourne* (edited by Grant Allen, 1996). Wordsworth Editions Ltd, Ware, UK.

Whitehead, J. (2010) John Locke and the governance of India's landscape: the category of wasteland in colonial revenue and forest legislation. *Economic and Political Weekly* 45(50), 83–93.

WHO, IUCN, *et al.* (1993) *Guidelines on the Conservation of Medicinal Plants.* IUCN, Gland, Switzerland.

Wiggins, S. and Shields, D. (1995) Clarifying the 'logical framework' as a tool for planning and managing development projects. *Project Appraisal* 10, 2–12.

Wild, R. and McLeod, C. (eds) (2008) *Sacred Natural Sites: Guidelines for Protected Area Managers.* IUCN, Gland, Switzerland.

Wild, R. and Mutebi, J. (1996) *Conservation Through Community Use of Plant Resources.* People and Plants Working Paper No. 5. Division of Ecological Sciences, UNESCO, Paris.

Wild, R. and Mutebi, J. (1997) Bwindi Impenetrable Forest, Uganda: conservation through collaborative management. *Nature and Resources* 33, 33–51.

Wild-Wood, E. (2021) Modern African missionaries. A reassessment of their impact in Uganda 1890s–1920s. *Exchange* 50, 270–288.

Williams, C. (2022) Rights over genetic resources and ways of monitoring the value chain. A case study from the Royal Botanic Gardens, Kew. In: Kamau, E.C. (ed.) *Implementation of the Nagoya Protocol: Fulfilling New Obligations among Emerging Issues.* Federal Agency for Nature Conservation, Bonn, Germany, pp. 137–142.

Williams, E. (2011) *Language, Politics and Development in Africa.* British Council, London.

Wilson, P.K. (2010) Centuries of seeking chocolate's medicinal benefits. *The Lancet* 376, 158–159.

Wilson, R.T. (2023) Coping with catastrophe: crop diversity and crop production in Tigray National Regional State in Northern Ethiopia. *African Journal of Agricultural Research* 19, 321–336.

Winchell, F., Stevens, C.J., *et al.* (2017) Evidence for sorghum domestication in fourth millennium BC eastern Sudan: spikelet morphology from ceramic impressions of the Butana group. *Current Anthropology* 58, 673–683.

Winkler, D. (2004) Matsutake mycelium under attack in Southwest China: how the mushrooming trade mines its resources and how to achieve sustainability. Available at: https://mushroaming.com/Matsutake_Conservation_in_sw_China (accessed 21 May 2024).

Withers, P.J.A., Neal, C., *et al.* (2014) Agriculture and eutrophication: where do we go from here? *Sustainability* 6, 5853–5875.

Witkowski, J. (2008) Stalin's war on genetic science. *Nature* 454, 577–579.

Wood, J.J., Beaman, J., *et al.* (1993) *The Plants of Mount Kinabalu (2): Orchids.* Royal Botanic Gardens, Kew, UK.

Wood, M. (2020) *The Story of China: A Portrait of a Civilisation and Its People*. Simon and Schuster, London.

Woodman, P.C. (ed.) (1985) *Excavations at Mount Sandel 1973–77, County Londonderry*. Department of the Environment for Northern Ireland, Belfast, UK.

Wright, L.R. (1966) Notes on the North Borneo dispute. *Journal of Asian Studies* 25, 471–484.

Wrigley, C.C. (1989) Bananas in Buganda. *Azania* 24, 64–70.

Wu Shaobin, Zhang Xiuyue, *et al.* (2010) Phylogenetic position of the takin (*Budorcas taxicolor*) and the yak (*Bos grunniens*) within the family Bovidae. *Zootaxa* 2392, 62–68.

Wulf, A. (2015) *The Invention of Nature: The Adventures of Alexander von Humboldt*. John Murray, London.

WWF and IUCN (1994–1997) *Centres of Plant Diversity: A Guide and Strategy for their Conservation*. IUCN Publications Unit, Cambridge.

WWF Nepal Program (2001) *Northern Mountains Conservation Project: Annual Technical Progress Report (July 2000–June 2001)*. WWF Nepal Program, Kathmandu.

Xiang, J., Hansen, A., *et al.* (2018) Association between malaria incidence and meteorological factors: a multi-location study in China, 2005–2012. *Epidemiology and Infection* 146, 89–99.

Yan Jinshen (2017) *Tibetan Medical Plants*. Yunnan Science and Technology Press, Kunming, China.

Yang Hongqiang, Nie Ying, *et al.* (2010) Study on China's timber resource shortage and import structure: Natural Forest Protection Program outlook, 1998 to 2008. *Forest Products Journal* 60, 408–414.

Yang Lixin, Ahmed, S., *et al.* (2014) Comparative homegarden medical ethnobotany of Naxi healers and farmers in Northwestern Yunnan, China. *Journal of Ethnobiology and Ethnomedicine* 10, 6. DOI: 10.1186/1746-4269-10-6.

Yang Xuefei, He Jun, *et al.* (2008) Matsutake trade in Yunnan Province, China: an overview. *Economic Botany* 62(3), 269–277.

Yarnell, S.L. (1998) *The Southern Appalachians: A History of the Landscape*. Southern Research Station, Forest Service, Asheville, North Carolina.

Yi Zhou, Yang, Q.E., *et al.* (2021) Antibiotic resistome in the livestock and aquaculture industries: status and solutions. *Critical Reviews in Environmental Science and Technology* 51, 2159–2196.

Yin Hongfu and Li Changan (2001) Human impact on floods and flood disasters on the Yangtze River. *Geomorphology* 41, 105–109.

Ying Tsun-Shen, Zhang Yu-long, *et al.* (1993) *The Endemic Genera of Seed Plants of China*. Science Press, Beijing.

Younis, S.M.Z. and Ammar, A. (2018) Quantification of impact of changes in land use–land cover on hydrology in the upper Indus Basin, Pakistan. *The Egyptian Journal of Remote Sensing and Space Science* 21(3), 255–263.

Yufenyuy, M. and Nguetsop, V.-F. (2020) Climate variability and the emergence of malaria: case of Kumbo Central Sub-Division, North West Region, Cameroon. *International Journal of Global Sustainability* 4, 104–127.

Zalasiewiczi, J., Williams, M., *et al.* (2011) The Anthropocene: a new epoch of geological time? *Philosophical Transactions of the Royal Society A: Mathematical, Physical and Engineering Sciences* 369, 835–841.

Zhao, G. and Shao, G. (2002) Logging restrictions in China: a turning point for forest sustainability. *Journal of Forestry* 100, 34–37.

Zhou, D.Q. and Grumbine, R.E. (2011) National parks in China: experiments with protecting nature and human livelihoods in Yunnan province, Peoples's Republic of China (PRC). *Biological Conservation* 144, 1314–1421.

Zhu, H. and Tan, Y. (2022) Flora and vegetation of Yunnan, southwestern China: diversity, origin and evolution. *Diversity* 14, 340.

Zhuang, H., Wang, C., *et al.* (2021) Native useful vascular plants of China: a checklist and use patterns. *Plant Diversity* 43, 134–141.

Zong Yongqiang and Chen Xiqing (2000) The 1998 flood on the Yangtze, China. *Natural Hazards* 22(2), 165–184.

Index of Species

General Index